T0182204

ASTRONOMY AND
ASTROPHYSICS LIBRARY

More information about this series at http://www.springer.com/series/848

George K. Parks

Characterizing Space Plasmas

A Data Driven Approach

 Springer

George K. Parks
Space Sciences Laboratory
University of California, Berkeley
Berkeley, CA, USA

ISSN 0941-7834   ISSN 2196-9698   (electronic)
Astronomy and Astrophysics Library
ISBN 978-3-030-07922-2   ISBN 978-3-319-90041-4   (eBook)
https://doi.org/10.1007/978-3-319-90041-4

This Springer imprint is published by the registered company Springer Nature Switzerland AG
The registered company address is: Gewerbestrasse 11, 6330 Cham, Switzerland

Concepts which have proved useful for ordering things easily assume so great an authority over us, that we forget their terrestrial origin and accept them as unalterable facts... The road of scientific progress is frequently blocked for long periods by such errors.

Albert Einstein, *Phys. Zeitschr.*, **17**, 101, 1916.

If there is something very slightly wrong in our definition of the theories, then the full mathematical rigor may convert these errors into ridiculous conclusions.

Richard Feynman, in Feynman Lectures on Gravitation, *Addison-Wesley*, Reading, MA, 1995.

Preface

Our goal in studying space plasma physics is to quantify and understand the complex dynamics of our solar system plasmas in enough detail to piece together a model that can predict the general behavior of collisionless plasmas in nature. Collisionless space plasmas interact collectively through long-range electromagnetic forces and can exhibit effects arising from both single and groups of particles. It is important to observe the dynamics in Earth's environment because we have access to detailed information on the physics which can be applied to collisionless plasmas elsewhere in the Universe.

Many of the early measurements were interpreted using fluid theory and concepts. For example, solar wind (SW) was interpreted as flowing solar atmosphere that becomes supersonic producing the bow shock in front of Earth. In spite of the fact that questions were raised whether fluid physics correctly describes collisionless plasmas, the fluid picture of the SW became an accepted paradigm and the fluid concepts were adopted as the primary language for describing space plasma phenomena. However, observations have since shown that a complete physical understanding of space plasma phenomena requires kinetic theory and concepts.

The motivation to write this book is to convey a considerable amount of new information about space plasmas to students and researchers. The challenge is to present the material in a logical sense and interest the readers. In this regard, foremost to note is that plasma instruments measure *particles*, and therefore, particle theories and concepts are needed to interpret the data. Second, observations have further established that space plasmas are always moving through interplanetary magnetic fields inducing *electric fields* that drive *currents* from the ionosphere of planets to the distant regions of interplanetary space. Since electrodynamic quantities measured in the moving and stationary frames often look different, the reference frame in which measurements have been made must be clearly stated. Theory of relativity will be used to organize measured quantities in the rest and moving frames.

We use *a data-driven approach* to connect the measurements to the real world. This approach makes full use of an instrument's capability and examines data at the most detailed level an experiment can provide. Students will learn what instruments can measure and the critical role played by observations in developing realistic models. To highlight examples of new results, particle measurements made mostly from the Cluster ion experiment are presented because we are familiar with the experiment, whose data have been extensively studied, analyzed, digested, and published. Unfortunately, this probably has resulted in overlooking important observations made by other missions and we apologize in advance for these omissions.

There is also a problem relying on data obtained during the early days of the space age because it was not known that some measurements were "contaminated." For example, it was not known that the SW in the vicinity of Earth often includes particles leaking out of the magnetosheath and SW reflected from the bow shock. It was also not known that the SW upstream of Earth could have included plasmas from transient coronal mass ejections (CMEs), corotating interaction regions (CIRs), and other solar and interplanetary disturbances. Because the densities, velocities, temperatures, and the amount of He^{++} are different, the bulk quantities computed from the velocity moments have not given "pristine" properties important for developing SW theories and models.

The SW measurements improved considerably using a microprocessor-controlled experiment flown on Cluster. This experiment swept the high voltage only near the peak of the SW distribution thereby minimizing the contributions of the contaminating population to SW measurements, providing more accurate information about the SW distributions. This Cluster instrument has measured ion distributions immediately upstream and downstream of the bow shock allowing us to see that the SW can sometimes penetrate the bow shock with little change in temperature, defying all shock theories.

Identical instruments flown on *multi-spacecraft* missions have further given us new insights into how space plasma dynamics depend on space and time. Four spacecraft measurements on Cluster have allowed application of the "curlometer" technique to compute currents across boundaries using Ampere's law. Numerous spacecraft separation strategies have included a *string of pearl* configuration, and they have shown how the upstream structures steepen nonlinearly as they are convected with the SW toward the bow shock.

Moreover, polar orbiting satellites have established at the same time existence of *parallel electric field* and *currents* in the auroral acceleration region and the relationship of the auroral dynamics to the *global current system*. Cluster has now measured auroral electron and ion beams even in the distant plasma sheet telling us how the magnetosphere and ionosphere are coupled. This discovery has allowed us to begin study of the problem Hannes Alfvén suggested nearly 50 years ago: How do parallel electric fields drive field-aligned currents that connect the Earth's ionosphere to the distant magnetosphere? Studying the magnetosphere and ionosphere as a unit is necessary for revealing information on how the dynamics of both regions affect one another.

In selecting what topics to include, a choice was made that this book is meant to discuss some of the basic problems fundamental for space plasma physics. Some topics are of "discovery" nature and appear for the first time in a space physics textbook. Others include different perspectives and interpretations of old problems and models that were previously considered *incontestable*. To discuss the topics as comprehensively as possible, we have included some relevant material that appeared previously in Parks (2004).[1]

This book focuses on space plasma measurements made mainly in the vicinity of Earth. However, the physical concepts used to interpret data are general and applicable to other solar systems and astrophysical plasmas. To engage the readers, follow up questions are included to stimulate additional discussions. Some of these questions will not always have definite answers, and furthermore, the readers may not agree with our suggested solutions. An extensive list of references is presented including the original research papers so the readers can track how the field has progressed.

This book is an outgrowth of numerous discussions with students and colleagues in the course of many years of teaching and research. It is self-contained but assumes that the readers have basic knowledge of space plasmas at an introductory level. It will supplement the material presented in many introductory space plasma physics textbooks.

Chapters

Chapter 1 reviews the Lorentz equation of motion and Maxwell's equations to set up a framework for interpreting space plasma observations. These equations are presented emphasizing the importance of self-consistency. Because this book considers data fundamental, a short review is also included at the end of the chapter to familiarize the readers with the basic principles of how the two most commonly used instruments work for measuring particles in the energy range from a few electron volts (eV) to a few tens of keV: electrostatic analyzers (ESAs) and Faraday cups (mainly SW ions). We show what parameters are measured and how the macroscopic quantities are computed from the measured quantities. SW measured by ESAs and Faraday cups does not distinguish charge, mass, or velocity. We discuss an important *caveat* that allows H^+ and He^{++} ions to be distinguished in velocity space.

Chapter 2 examines the Lorentz equation and particle acceleration mechanisms. Special theory of relativity shows that when $\mathbf{E}$ is perpendicular to $\mathbf{B}$, one can always find a frame in which the fields are either electric or magnetic. A complete solution of the relativistic Lorentz equation is discussed when $\mathbf{E} \cdot \mathbf{B} = 0$ for arbitrary magnitudes of $|\mathbf{E}|$ and $|\mathbf{B}|$. The solutions here show that when $|\mathbf{E}| \geq |\mathbf{B}|$, a particle can be accelerated in the $\mathbf{E} \times \mathbf{B}$ direction rather than just drift in that direction as

---

[1]Parks, George K. Physics of Space Plasmas, An Introduction, 2nd edn. Westview Press, A Member of Perseus Books, Boulder (2004).

when $|\mathbf{B}| > |\mathbf{E}|$. This acceleration mechanism is important in the neighborhood of magnetic neutral points where $|\mathbf{B}| \sim 0$. Other well-known acceleration mechanisms are also discussed including Fermi and inverse Compton mechanisms of interest in astrophysical plasmas. We conclude with a brief discussion of waves and wave–particle interaction including cyclotron resonance theory important for accelerating and precipitating particles in Earth's magnetosphere.

Chapter 3 discusses how particles escape stellar atmospheres, which for our Sun is called solar wind (SW). We show observational properties and discuss the advantages and disadvantages of MHD fluid and particle interpretations and examine how much of the SW observations could be explained by existing models. The SW electron distributions show the core, halo, and super halo electrons that behave very differently from SW ions. Existing particle models do not explain the mechanisms for the different features. We speculate and suggest possible mechanisms for developing new SW models. We conclude this chapter with discussion of Debye-scale structures, *double layers* in the SW, and discuss their possible importance to the SW problem.

Chapter 4 deals with shocks in collisionless plasmas, and we focus on Earth's bow shock. It has been studied for more than 50 years and is the best known collisionless shock in nature. Particle and field instruments have measured in situ behavior, and data are readily available. We discuss how collisionless shocks differ from ordinary fluid shocks noting that the foreshock region produced by the interaction of the SW with the bow shock is an integral part of the shock dynamics. Examples are given from recent observations, and we discuss controversial and unresolved issues. We attempt to answer the question whether the bow shock can generate entropy using Cluster data to compute entropy. We conclude this chapter by discussing a guiding center model that shows how the reflected SW particles can be accelerated at the bow shock.

Chapter 5 focuses on the transport of particles across boundaries produced by discontinuities in currents and magnetic fields. A short history is given about the importance of neutral sheets existing across boundaries produced by oppositely directed magnetic fields. We discuss how the MHD concept of "merging" or "connecting" oppositely directed magnetic fields creates "open" boundaries so that particles from one side can be transported freely into the other side. However, we remind the readers that Dungey has warned us that field lines are *not real*, and connecting field lines is a *visualization tool* meant to convey the more complicated physics of how large-scale currents are induced to modify the existing magnetic topology. Examples of how particles are accelerated in the neighborhood of neutral points are discussed using the Lorentz equation. The chapter concludes with a discussion of how a self-consistent physics can be applied to study the steady-state Harris current sheet boundary problem.

Chapter 6 discusses observations of electric field and currents with emphasis on the importance of the components parallel to the direction of the magnetic field. We discuss how the different plasma regions are coupled by parallel currents ($\mathbf{J}_\parallel$) and electric field ($\mathbf{E}_\parallel$) and include recent observations of $\mathbf{J}_\parallel$ in the plasma sheet during current sheet instabilities relevant for understanding auroras and coupling of the ionosphere and magnetosphere. We also present observations of electrostatic solitary waves of Debye scales in the same region where $\mathbf{E}_\parallel$ and $\mathbf{J}_\parallel$ exist. This chapter concludes with a discussion of auroral kilometric radiation (AKR) theory intimately associated with the aurora.

Chapter 7 (the last chapter) introduces physics and concepts needed to explain many unsolved problems in space plasmas. We remind the readers that one of our ultimate goals is to understand collective interaction of space plasmas that can produce large-scale structures, such as global auroras, CMEs, and astrophysical structures as those observed in Crab Nebula. We speculate how synergistic effects might be important in such collective interactions. We then remind the readers that some basic concepts such as Landau damping are often misunderstood and applied incorrectly. Another topic is to understand how space plasmas are heated, observed from the SW to the distant geomagnetic tail, which has not been adequately explained. Heating represents diffusion of particles in velocity space, and the question still unanswered is how that can be achieved in plasmas that are nearly collisionless. We briefly discuss how collisions modify the distribution function and then show how electrons can be accelerated to run-away energies important for accelerating electrons to relativistic energies. To move space plasma physics forward, we recommend that future research address some of the problems discussed in this chapter.

Berkeley, CA, USA   George K. Parks

Acknowledgements

The author has benefited greatly from numerous discussions with Suiyan Fu, Ensang Lee, Michael McCarthy, Zhongwei Yang, Ying Liu, Jae Jin Lee, Naiguo Lin, Jinbin Cao, Yong Liu, Q.Q. Shi, Ling Hua Wang, Iannis Dandours, Henri Reme, Patrick Canu and Peter Yoon. Special thanks go to Ensang Lee for making results available before they were published, Michael McCarthy for the solution to the differential equation describing the Harris current sheet, and Zhongwei Yang for unpublished PIC results of the bow shock. We thank Sam Harrison, the editor of this book, for his interest, patience, and guidance.

Contents

Chapter 1
Basic Equations and Concepts

1.1 Introduction

Space plasma physics took a big leap soon after the second world war when declassified reports from plasma physics research were made available to astronomy, physics, and astrophysics communities. At about the same time, advanced radar technologies developed during the war years also became available enabling astronomers to explore regions of space not accessible previously. Several scientific meetings were held devoted specifically to discussing how space observations could be interpreted using the laws of plasma physics. This was in the early 1950s before the launch of the first spacecraft.

Space observations subsequently showed us that we live inside a magnetic bubble called magnetosphere formed by the solar wind interacting with the geomagnetic field. Earth's magnetosphere includes a tail in the anti-solar direction which forms a current sheet that extends to distances of more than $300\,R_E$ (earth radii). The current sheet is connected to the polar ionosphere which has clarified many questions about the aurora. The last 10 years have also been interesting for understanding particle motion described by the Lorentz equation for cases when $|\mathbf{E}| \geq |\mathbf{B}|$. Solutions show that the particles can be *accelerated* in the $\mathbf{E} \times \mathbf{B}$ direction rather than only drift when $|\mathbf{E}| \leq \mathbf{B}|$. This acceleration mechanism could be important in current sheets where $|\mathbf{E}| \neq 0$ but $|\mathbf{B}| \sim 0$. Current sheets are important because large-scale disturbances in space, solar, and astrophysical plasmas often involve the dynamics in the current sheet.

This chapter will present a short review of the well-known equations and concepts to set up a framework for interpreting space plasma observations. Our goal is to put the observations into the context of the fundamental laws. In this regard, an important fact about plasma observations is that all plasma instruments measure *particles* (not fluids). Thus, interpretation of space plasma data requires particle theories and concepts. To help readers understand the data, a short review is given at the end of the chapter on the basic principles of how the two most commonly used

© Springer Nature Switzerland AG 2018

G. K. Parks, *Characterizing Space Plasmas*, Astronomy and Astrophysics Library,
https://doi.org/10.1007/978-3-319-90041-4_1

instruments work for measuring particles in the energy range a few electron volt (eV) to a few tens of keV: electrostatic analyzers (ESA) and Faraday cups (mainly in the SW). We show what parameters are measured and how the measured parameters are used to compute macroscopic quantities.

Until recently, particle description of space plasma was considered *"impossible"* because characterizing a particle using the Lorentz equation requires information on the position **r** and velocity **v**. For a plasma consisting of N particles, there will be $6N$ number of coupled equations. Even for tenuous space plasmas, the number of equations is so large that the equations cannot be solved analytically. This problem however has been alleviated to some extent by supercomputers that can track the particles. Particle simulation models can emulate a plasma system and although the models at present are constrained by limited computer capabilities and full 3D models cannot be simulated yet, they can still help us to interpret complex space plasma features. Computer technology is continually advancing and simulation models will become more and more capable. With the quality of data improving and *in situ* measurements extended to electron inertial scales, it is imperative that we analyze space plasma data using the Lorentz-Maxwell equations.

1.2 Fundamental Equations

This book will treat space plasmas as collisionless. A unique feature about *collisionless space plasmas* is that they can exhibit effects arising from both individual particles and many particles interacting collectively. Charged particles interact through long-range electromagnetic fields and obey the well-known Lorentz and Maxwell equations.

Space plasmas are nearly always moving relative to the observer frame of reference (spacecraft) and because the measured charged particle and electromagnetic quantities can appear different in different frames, it is important to clearly state what coordinate frame is being used. For example, consider an electromagnetic field that consists of both electric **E** and magnetic **B** fields. What one sees, however, depends on the reference frame where measurements are made. An electric field measured in one reference frame may appear as both electric and magnetic fields in another reference frame. This is even true in nonrelativistic cases. Theory of relativity (Einstein 1920) will be used as a framework to interpret space plasma observations.

We will use the standard notation and let S' be an inertial frame moving with velocity V relative to S-frame which is the rest frame of the observer (Møller 1952). Quantities measured in the moving frame will be denoted with a prime ($'$). By inertial frame, we mean that a particle in that frame under no force travels in a straight line with a constant speed. Various quantities measured in the different frames are related to each other by the theory of relativity.

1.2.1 Lorentz Equation

The momentum of a charged particle is governed by the Lorentz equation which in tensor form is

$$\frac{dp_\mu}{d\tau} = \frac{q}{m} F_{\mu\nu} p_\nu \tag{1.1}$$

where τ is proper time, $dt = \gamma d\tau$, and p_ν is the four-momentum vector given by

$$p_\nu = \begin{pmatrix} p_x \\ p_y \\ p_z \\ i\mathcal{E}/c \end{pmatrix} \tag{1.2}$$

where p_x, p_y, p_z are components of three-dimensional momentum $\mathbf{p} = \gamma m_o \mathbf{v}$ in an inertial coordinate system defined by the Cartesian coordinates (x, y, z), $\mathcal{E} = \gamma m_o c^2$ is the total energy of the particle, $\gamma = (1 - V^2/c^2)^{1/2}$, m_o is the rest mass, c is the speed of light, and $i = \sqrt{-1}$. Note that the numerical values of the charge q and speed c are the same in all inertial frames. The covariant electromagnetic field tensor given by $F_{\mu\nu}$ in (1.1) is

$$F_{\mu\nu} = \begin{pmatrix} 0 & cB_z & -cB_y & iE_x/c \\ -cB_z & 0 & cB_x & iE_y/c \\ cB_y & -cB_x & 0 & iE_z/c \\ -iE_x/c & -iE_y/c & -iE_z/c & 0 \end{pmatrix} \tag{1.3}$$

where the electric $\mathbf{E}$ and magnetic $\mathbf{B}$ fields are given in the Cartesian coordinate system (x, y, z).

Consider the kth particle of a plasma system that has many particles. This particle has a charge q_k, mass m_k, and is moving with velocity $\mathbf{v}_k$ at a point in space $\mathbf{r}_k$ at time t where it senses $\mathbf{E}$ and $\mathbf{B}$ fields. The familiar space-time equations for this particle obtained from (1.1) are

$$\frac{d\mathbf{p}_k}{dt} = q_k(\mathbf{E} + \mathbf{v}_k \times \mathbf{B})$$

$$\frac{d\mathcal{E}_k}{dt} = q_k \mathbf{v}_k \cdot \mathbf{E} \tag{1.4}$$

where $\mathbf{p}_k$ is the momentum of the kth particle. (The relative velocity of the reference frames is represented by $\mathbf{V}$ and should not be confused with the particle velocity which is represented by the lower case $\mathbf{v}$.) If a plasma has N particles, we have N of these equations coupled through $\mathbf{E}$ and $\mathbf{B}$ fields.

The covariant formulation includes in addition to the momentum equation, an equation describing the time variation of the total particle energy. The second equation in (1.4) tells us that the total energy of a charged particle can be changed *only* by moving along the electric field **E**. Magnetic field **B** does *no* work and cannot change the energy of the particle. This result is fundamental. All physical mechanisms that accelerate particles in space must involve the electric field **E**. The readers are encouraged to discuss the challenges as to how to identify from where this electric field comes and how the dynamics produce this electric field.

1.2.2 Maxwell Equations

Electric and magnetic fields obey Maxwell's equations, which are best written using tensors because tensor equations have the same form in all inertial frames. Maxwell equations in the contravariant four tensor form are

$$\frac{\partial F_{v\sigma}}{\partial x^\sigma} + \frac{\partial F_{\mu v}}{\partial x^\mu} + \frac{\partial F_{\sigma\mu}}{\partial x^v} = 0$$

$$\frac{\partial F^{\mu v}}{\partial x^v} = J^\mu \tag{1.5}$$

where the contravariant electromagnetic field tensor $F^{\mu v}$ is given by

$$F^{\mu v} = \begin{pmatrix} 0 & B_z & -B_y & -iE_x/c \\ -B_z & 0 & -cB_x & -iE_y/c \\ B_y & -B_x & 0 & -iE_z/c \\ iE_x/c & iE_y/c & iE_z/c & 0 \end{pmatrix} \tag{1.6}$$

The four vector J^μ is

$$J^\mu = \begin{pmatrix} J_x \\ J_y \\ J_z \\ ic\rho_c \end{pmatrix} \tag{1.7}$$

where J_x, J_y, J_z are the current densities in the (x, y, z) directions and ρ_c is the charge density.

The familiar Maxwell equations obtained from the above equations are (in MKS units)

$$\nabla \cdot \mathbf{B} = 0 \tag{1.8}$$

$$\nabla \times \mathbf{H} = \mathbf{J} + \partial \mathbf{D}/\partial t \tag{1.9}$$

$$\nabla \cdot \mathbf{D} = \rho_c \tag{1.10}$$

$$\nabla \times \mathbf{E} = -\partial \mathbf{B}/\partial t \tag{1.11}$$

where $\mathbf{D}$ and magnetic $\mathbf{H}$ fields are $\mathbf{D} = \epsilon \mathbf{E}$ and $\mathbf{B} = \mu \mathbf{H}$ where ϵ and μ are electric permittivity and magnetic permeability of plasma. Particles inside collisionless plasmas move in vacuum, and the permittivity and permeability are constants equal to $\epsilon_o = 8.85 \times 10^{-12}$ F/m and $\mu_o = 4\pi \times 10^{-7}$ H/m. Note that $1/\epsilon_o\mu_o = c^2$, where $c = 3 \times 10^8$ m/s is the speed of light. Maxwell equations are differential equations that describe the electric $\mathbf{E}$ and magnetic $\mathbf{B}$ fields at point $\mathbf{r}$ in space at the instant of time t.

1.3 Statistical Equations

Space plasmas have many particles and characterizing them requires as many equations as there are particles. However, it is not required that we know the trajectories of every particle. Rather than to follow the fate of every particle, a practical way to study space plasmas is to examine the statistical properties using the laws of statistical physics. Statistical properties of the particles are contained in the distribution function which can be constructed from individual particle motions. It is fortunate that space plasma experiments can measure directly the distribution function of the particles.

1.3.1 Boltzmann Equation

An ensemble of particles described by the distribution function $f(\mathbf{r}, \mathbf{p})$ obeys the Boltzmann equation

$$\frac{\partial f}{\partial t} + \mathbf{v} \cdot \frac{\partial f}{\partial \mathbf{r}} + \frac{d\mathbf{p}}{dt} \cdot \frac{\partial f}{\partial \mathbf{p}} = \left(\frac{\partial f}{\partial t}\right)_c \tag{1.12}$$

where $\mathbf{v}$ is the particle velocity and $\mathbf{p} = \gamma m\mathbf{v}$ is the relativistic momentum obeying the first equation in (1.4) which is $d\mathbf{p}/dt = (q/m)(\mathbf{E} + \mathbf{v} \times \mathbf{B})$ and the right-hand term is due to collisions. Here the distribution function depends on the momentum and coordinate, $f = f(\mathbf{p}, \mathbf{r})$ and the equation is relativistically correct. Equation (1.12) together with Maxwell equations $\nabla \times \mathbf{E} = -\partial \mathbf{B}/\partial t, \nabla \times \mathbf{H} = \partial \mathbf{D}/\partial t + \sum qn \int f\mathbf{v}d\mathbf{p}, \nabla \cdot \mathbf{D} = \rho_c$, and $\nabla \cdot \mathbf{B} = 0$ can be used to describe self-consistently an ensemble of relativistic particles.

1.3.2   Vlasov Equation

For collisionless plasmas in space, the right-hand term vanishes. We will now focus on the motion of *nonrelativistic particles* which constitute most of space plasma particles in the vicinity of Earth. Moreover, we assume one particle distribution function consisting of only one particle species (ions and electrons require separate distribution function, each described by the collisionless Boltzmann equation). Note however, the Vlasov equation can also be used for solving general relativity problems (Rasio et al. 1989; Widrow and Kaiser 1993). For the nonrelativistic particles, the motion of the particles can be defined by its position $\mathbf{r}$ and velocity $\mathbf{v}$. Each particle can then be represented by a point in $(\mathbf{r}, \mathbf{v})$ space. This space is six-dimensional with coordinates (x, y, z, v_x, v_y, v_z) and the probability density of points in this $(\mathbf{r}, \mathbf{v})$ space at time t is given by the distribution function $f(\mathbf{r}, \mathbf{v})$. The expected number of particles at time t in $(\mathbf{r}, \mathbf{v})$ space with coordinates $\mathbf{r}$ and $\mathbf{r} + d\mathbf{r}$ and velocity $\mathbf{v}$ and $\mathbf{v} + d\mathbf{v}$ is given by $f(\mathbf{r}, \mathbf{v}) \, d\mathbf{r} \, d\mathbf{v}$. The distribution function is a useful quantity because it can give us information on how many particles there are in the neighborhood of the point $\mathbf{r}$ and how many particles lie in a given range of velocity $\mathbf{v}$.

The collisionless Boltzmann transport equation is called Vlasov equation which is

$$\frac{\partial f}{\partial t} + \mathbf{v} \cdot \frac{\partial f}{\partial \mathbf{r}} + \mathbf{a} \cdot \frac{\partial f}{\partial \mathbf{v}} = 0 \tag{1.13}$$

If the particles are subject only to electromagnetic force, $\mathbf{a} = (q/m)(\mathbf{E} + \mathbf{v} \times \mathbf{B})$. We will generally consider a plasma system consisting of electrons and one ion species (protons), hence there will be two equations, one describing the electron distribution function $f^-(\mathbf{r}, \mathbf{v}, t)$ and the other ion distribution function $f^+(\mathbf{r}, \mathbf{v}, t)$. The differential equation (1.13) has seven variables $(x, y, z, v_x, v_y, v_z, t)$ and it must be solved together with the Maxwell equations.

Question 1.1 Treating space plasmas as collisionless is an approximation which is generally correct. However, we see plasma heating in space. Discuss why the Vlasov equation is not valid if heating is involved. How should the Vlasov equation be modified?

1.3.3   Equivalence of Vlasov and Lorentz Equation

For nonrelativistic particles in the absence of collisions, the Vlasov description of the particles can be shown to be equivalent to describing the particles using the Lorentz equation. To see this, consider the equations

$$\mathbf{F} = m(d\mathbf{v}/dt)$$

$$\mathbf{v} = d\mathbf{r}/dt \tag{1.14}$$

Let the solutions of these equations be

$$\mathbf{v} = \mathbf{v}(c_1, \ldots .c_6, t)$$

$$\mathbf{r} = \mathbf{r}(c_1, \ldots .c_6, t) \tag{1.15}$$

where the c's are integration constants. Let us suppose that these equations can be solved for the c's given by $c_i = c_i(\mathbf{r}, \mathbf{v}, t)$, where $i = 1, 2, \ldots 6$. Any function of the c's, $f(c_1, c_2, \ldots c_6)$, is then a solution to the Boltzmann equation. Substitute $f(c_i)$ into the Boltzmann equation and obtain

$$\sum_{i=1}^{6} \frac{\partial f}{\partial c_i} \left(\frac{\partial c_i}{\partial t} + \mathbf{v} \cdot \nabla c_i + \mathbf{a} \cdot \nabla_v c_i \right) = \sum_{i=1}^{6} \frac{\partial f}{\partial c_i} \frac{dc_i}{dt}$$

$$= 0 \tag{1.16}$$

where the right side vanishes since c_i's are constants.

  If the c's are not true constants but they are adiabatic invariants, f is still a solution but accurate to within the approximation of the adiabatic theory. The constants and invariants are functions of $(\mathbf{r}, \mathbf{v}, \mathbf{t})$ and the distribution function includes information on the particle trajectory. The particles follow a path in the $(\mathbf{r}, \mathbf{v})$ space so as to conserve the constants. Note that if the constants are independent of time, the solutions apply to the stationary Vlasov equation. The equivalence of particle orbit theory and the Boltzmann equation without collisions allows us to describe space plasma dynamics using the Lorentz particle concepts.

1.3.4 Self-consistency

The coupled Lorentz-Maxwell equations are complete and self-consistent in the sense that if the charge $\rho_c(\mathbf{r}, t)$ and current $\mathbf{J}(\mathbf{r}, t)$ densities are given, Maxwell equations can be solved for $\mathbf{E}$ and $\mathbf{B}$ fields that are uniquely defined in space and time, given the initial conditions. However, in plasmas, $\rho_c(\mathbf{r}, t)$ and $\mathbf{J}(\mathbf{r}, t)$ are not always known. On the other hand, if $\mathbf{E}$ and $\mathbf{B}$ were given, the equations of motion of each particle can be solved and $\rho_c(\mathbf{r}, t)$ and $\mathbf{J}(\mathbf{r}, t)$ computed and using them in Maxwell's equations yields solutions of electromagnetic fields. The solutions obtained are self-consistent because the particle motions produce the required electromagnetic fields that in turn are necessary to create the particle motions. We realize that use of self-consistent theory to analyze and interpret data is not always possible, but the ultimate aim is to achieve this goal (Fig. 1.1).

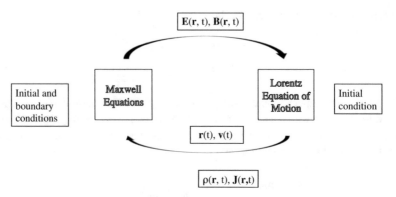

Fig. 1.1 Self-consistent coupled Lorentz and Maxwell's equations (from Schmidt, *Academic Press*, New York, NY, 1966)

1.4   Electric and Magnetic Field in Space

Maxwell equations show that a magnetic field **B** is generated when the electric field **E** is changing in time and electric field **E** is generated when the magnetic field **B** is changing in time. Moreover, special theory of relativity shows that electric field in one frame can appear as magnetic field in another frame and vice versa (see below). Electric and magnetic fields are vector point functions. A vector point function is a function that assigns a vector to each point of some region of space. If to each point (**r**) of a region R in space there is assigned a vector $\mathbf{F} = \mathbf{F}(\mathbf{r})$, then **F** is a vector point function.

The general definition of the electric field is obtained from Ampere's equation $\partial \mathbf{B}/\partial t = -\nabla \times \mathbf{E}$. The divergence free property of magnetic field allows us to write $\mathbf{B} = \nabla \times \mathbf{A}$, where the vector potential **A** is given by

$$\mathbf{A}(\mathbf{r}, t) = \frac{\mu_o}{4\pi} \int \frac{\mathbf{J}(\mathbf{r}', t) d^3 r'}{|\mathbf{r} - \mathbf{r}'|} \tag{1.17}$$

We can then rewrite Ampere's equation as

$$\nabla \times \left(\mathbf{E} + \frac{\partial \mathbf{A}}{\partial t} \right) = 0 \tag{1.18}$$

The vector $(\mathbf{E} + \partial \mathbf{A}/\partial t)$ has a zero curl and therefore it can be written as a gradient of a scalar,

$$-\nabla \phi = \left(\mathbf{E} + \frac{\mathbf{A}}{\partial t} \right) \tag{1.19}$$

where ϕ is the electromagnetic scalar potential function given by

$$\phi(\mathbf{r}, t) = \frac{1}{4\pi \epsilon_o} \int \frac{(\rho', t) d^3 r'}{|\mathbf{r} - \mathbf{r}'|} \tag{1.20}$$

Fig. 1.2 A sketch showing
how the electric force in
conductors vanishes. The
charges move so that the
internal electric field becomes
equal in magnitude and
directed opposite to the
applied field and the force on
every particle vanishes

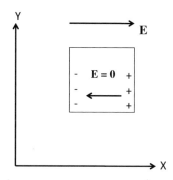

Rearranging (1.19), we find

$$\mathbf{E} = -\nabla\phi - \partial\mathbf{A}/\partial t \tag{1.21}$$

This equation shows that while $\mathbf{B}$ is derived from the vector potential only ($\mathbf{B} = \nabla \times \mathbf{A}$), $\mathbf{E}$ requires both a vector and scalar potential.

1.4.1 Static Electric Field

To understand the physical meaning of Eq. (1.21), we first discuss how particles in plasmas respond to the Coulomb force (neglect collective effects). Consider an isolated plasma blob at rest that is uniform in space and charge neutral with equal number of protons (p^+) and electrons (e^-). Apply an electric field $\mathbf{E} = (E, 0, 0)$ across the blob (Fig. 1.2). There is no magnetic field, $\mathbf{B} = 0$.

Question 1.2 An *i*th particle inside the blob at any instant is identified by the position $\mathbf{r}_i$ and velocity $\mathbf{v}_i$ obeying the Lorentz equation, $\mathbf{F} = md\mathbf{v}/dt = q\mathbf{E}$ where q is the charge, m is mass, and $\mathbf{E}$ is electric field acting on the particle at the position where the particle is. An observer in the S-frame sees the electrons move toward $-X$ and ions $+X$ directions. What does this observer in the S-frame learn about the fundamental dynamics?

Electrostatic Field

A quantitative formulation of the above discussion is to consider a *static point charge* $q_k = q_k\delta(\mathbf{r} - \mathbf{r}_k)$ located at $\mathbf{r}_k$. Here $\delta(\mathbf{r} - \mathbf{r}') = \delta(x - x')\delta(y - y')\delta(z - z')$ is a product of three Dirac delta functions. This point charge produces an electric field given by Coulomb's law and another particle located at point $\mathbf{r}$ will experience a force $\mathbf{F}_E = q\mathbf{E}$, where $\mathbf{E}(\mathbf{r}) = q_k(\mathbf{r} - \mathbf{r}_k)/|\mathbf{r} - \mathbf{r}_k|^3$. The electric field at that point can be defined in terms of the force per unit charge,

$$\mathbf{E} = \lim_{q \to 0} \frac{\mathbf{F}_E}{q} \tag{1.22}$$

Question 1.3 What does this simple equation show about $\mathbf{E}$? How should this answer be modified if there is more than one particle? What if there is a continuous distribution of point charges in a small volume?

Magnetostatic Field

In the presence of a magnetic field $\mathbf{B}$, a charge executes a circular motion due to the magnetic force $\mathbf{F}_B = q\mathbf{v} \times \mathbf{B}$. The magnitude of $\mathbf{F}_B$ depends on the magnitude and direction of $\mathbf{v}$ and can be defined as force per unit current. The force is maximum when $\mathbf{v}$ is perpendicular to $\mathbf{B}$ and minimum when it is parallel to $\mathbf{B}$. The intensity of the magnetic field in terms of the maximum force $|\mathbf{F}_B|_{max}$ is,

$$|\mathbf{B}| = \lim_{q \to 0} |\mathbf{F}_B|_{max}/qv \tag{1.23}$$

Question 1.4 What is the definition of the direction of $\mathbf{B}$? Direction for a continuous distribution of current? Is there a current in the frame moving with the charge? Is there a magnetic field in that frame? What about the charge q? What can you conclude about the importance of magnetic field relative to the electric field?

1.4.2 Induced Electric Field

Let us now return to Eq. (1.21). This equation raises a fundamental question about the source of electric fields in space. As shown by the above example, a plasma does not normally support free charges inside it. Hence, the scalar potential ϕ which depends on the free charge must vanish inside equilibrium plasmas. The electric field $\mathbf{E}$ in space plasmas comes from the second term $\partial \mathbf{A}/\partial t$ indicating that *inductive electric field* is fundamental.

Michael Faraday in 1820 discovered that an electromotive force $\mathcal{EMF}$ can be generated in conductors. Let us discuss briefly what his experiment consisted of and what he learned from it. His experiment included a piece of conducting wire whose ends were hooked to a galvanometer and placed next to a magnet. When the wire was moved across the magnet with a velocity $\mathbf{V}$, the galvanometer needle deflected indicating that a current was flowing in the wire. We can understand this effect knowing that the force that moved the electrons in the wire is $\mathbf{F} = q(\mathbf{E} + \mathbf{V} \times \mathbf{B})$ and in this case, it is the $q\mathbf{V} \times \mathbf{B}$ force that pushed the electrons along the wire.

A similar effect was observed when the magnet was moved over the wire with a velocity $\mathbf{V}$. The galvanometer again detected a current flow in the wire. But in this case one cannot use the same argument because there is no $\mathbf{V} \times \mathbf{B}$ force acting

Fig. 1.3 A sketch showing a plasma blob with an arbitrary contour C traveling with a velocity V. $\mathcal{EMF}$ is induced along C. The total magnetic flux enclosed by C is Φ

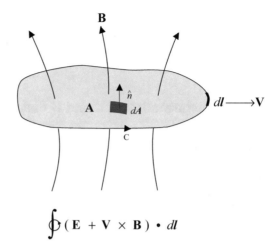

on the particles. (Note that moving a magnet does not mean field lines are moving.) According to the Lorentz equation, the only other force that can push the electrons is $q\mathbf{E}$. Faraday thus discovered that when a particle senses a changing magnetic field, an electric field $\mathbf{E}$ is generated.

In Faraday's experiments, an $\mathcal{EMF}$ is generated and it drives a current. The relationship between the electric field and the changing magnetic field is

$$\nabla \times \mathbf{E} = -\frac{\partial \mathbf{B}}{\partial t} \tag{1.24}$$

To understand the meaning of this equation, consider a contour C of an arbitrary shape that is bounded by a surface A with $\mathbf{n}$ as the unit normal vector (Fig. 1.3). Let this contour move with a velocity $\mathbf{V}$ or let the contour be stationary and let the source of the magnetic field move. The $\mathcal{EMF}$ around the contour is defined as

$$\mathcal{EMF} = \oint_C (\mathbf{E} + \mathbf{V} \times \mathbf{B}) \cdot d\mathbf{l} \tag{1.25}$$

where the electric field is measured in the frame of the contour element $d\mathbf{l}$ moving with the velocity $\mathbf{V}$ (S' frame). The quantity on the right side of (1.25) is force per unit charge multiplied by the distance along which a charge moves. Hence, the $\mathcal{EMF}$ is the work done by the force per unit charge integrated around the contour C. Stoke's theorem allows us to write $\oint_C (\mathbf{E}+\mathbf{V} \times \mathbf{B}) \cdot d\mathbf{l} = \int \nabla \times (\mathbf{E}+\mathbf{V} \times \mathbf{B}) \cdot d\mathbf{S} = -\int (\partial \mathbf{B}/\partial t) \cdot d\mathbf{S}$ in the nonrelativistic limit. The magnetic flux Φ enclosed by the contour C is $\int \mathbf{B} \cdot \mathbf{n} \, dS$. Hence, $\int (\partial \mathbf{B}/\partial t) \cdot \mathbf{n} \, dS = (\partial/\partial t) \int \mathbf{B} \cdot \mathbf{n} \, dS$, and we arrive at

$$\mathcal{EMF} = -d\Phi/dt \tag{1.26}$$

which is Faraday's law. The negative sign comes from Lenz's law, which states that the $\mathcal{EMF}$ is in such a direction as to oppose the change of the magnetic flux.

Question 1.5 Equation (1.26) is telling us that the $\mathcal{EMF}$ is related to the total rate of change of the magnetic flux. Discuss the meaning of the right side. Can there be $\mathcal{EMF}$ in steady state?

1.5   Transformation of E and B Fields

Plasmas in the neighborhood of a planet or star are moving. For example, the solar wind is streaming away from the Sun and the plasma that surrounds a planet or star may also be corotating. These motions affect electromagnetic field measurements and the theory of relativity quantifies the relationship of measurements made in coordinate frames in relative motion. Experiments are made on platforms that are also moving and the platforms are often rotating (non-inertial frame). We begin with a review of some of the important results of this theory to help us interpret the space measurements correctly.

   If we are given two equivalent reference systems that are inertial, the principle of relativity asserts that the laws of electromagnetism are identical in these systems. H.A. Lorentz in 1904 derived the theory of transformation for space and time that extended to electromagnetic fields and obtained the relationship of electromagnetic fields in the different inertial systems. Application of the Lorentz theory shows how currents produce magnetic fields, how changing magnetic fields induce electromotive forces, and how moving charges are deflected perpendicular to the magnetic field direction.

1.5.1   Special Theory of Relativity

Consider **E** and **B** measured in S frame and **E**′ and **B**′ of the same fields measured in S' moving with **V** in the X-direction relative to the S frame. The components of the electric **E**′ and **B**′ measured in S' frame in terms of **E** and **B** measured in S frame are obtained from

$$F'_{\mu\nu} = a_{\mu\lambda} a_{\nu\sigma} F_{\lambda\sigma} \tag{1.27}$$

where

$$a_{\mu\nu} = \begin{pmatrix} \gamma & 0 & 0 & i\beta\gamma \\ 0 & 1 & 0 & 0 \\ 0 & 0 & 1 & 0 \\ -i\beta\gamma & 0 & 0 & \gamma \end{pmatrix} \tag{1.28}$$

Here $\beta = V/c$. Use of this equation and $F_{\mu\nu}$ given in (1.3) shows

$$E'_x = E_x$$
$$E'_y = \gamma(E_y - V B_z)$$
$$E'_z = \gamma(E_z + V B_y)$$
$$B'_x = B_x$$
$$B'_y = \gamma\left[B_y + (V/c^2)E_z\right]$$
$$B'_z = \gamma\left[B_z - (V/c^2)E_y\right] \tag{1.29}$$

where V is along the x-axis.

The vectorial form of the transformation equations is

$$\mathbf{E}'_\parallel = \mathbf{E}_\parallel$$
$$\mathbf{E}'_\perp = \gamma(\mathbf{E} + \mathbf{V} \times \mathbf{B})_\perp$$
$$\mathbf{B}'_\parallel = \mathbf{B}_\parallel$$
$$\mathbf{B}'_\perp = \gamma\left(\mathbf{B} - \frac{\mathbf{V}}{c^2} \times \mathbf{E}\right)_\perp \tag{1.30}$$

The subindices ($\parallel$) and ($\perp$) here refer to directions relative to $\mathbf{V} = V\hat{x}$. For nonrelativistic situation ($\gamma = 1$) and to order V/c^2, the magnetic field is the same in the two frames but the electric field has a different expression, $\mathbf{E}' = (\mathbf{E} + \mathbf{V} \times \mathbf{B})$. Hence, the reference frame must be stated when discussing electric fields.

The transformation equations for current and charge densities are

$$J'_x = \gamma(J_x - V\rho), \quad J'_y = J_y, \quad J'_z = J_z$$
$$\rho'_x = \gamma(\rho_x - V J_x/c^2), \quad \rho'_y = \rho_y, \quad \rho'_z = \rho_z \tag{1.31}$$

Lorentz Invariants

Invariant quantities of electromagnetic fields can be obtained from the four vectors, noting that $F_{i,k}F^{i,k}$ is invariant. It can be shown that the following quantities are invariant:

$$B^2 - E^2/c^2 = B'^2 - E'^2/c^2$$
$$\mathbf{E} \cdot \mathbf{B} = \mathbf{E}' \cdot \mathbf{B}'$$
$$J^2 - \rho^2 c^2 = J'^2 - \rho'^2 c^2 \tag{1.32}$$

These quantities remain unchanged in all inertial reference coordinates. These invariants indicate that an electromagnetic field that is purely magnetic in S may be in part magnetic and in part electric in S'. Also, if the electric field is more intense in S than in S', the magnetic field is also more intense in S than in S'. Finally, if $\mathbf{E}$ and $\mathbf{B}$ are orthogonal in S, they are orthogonal in S'.

1.5.2 Lorentz Transformation When $\mathbf{E} \cdot \mathbf{B} = 0$

Lorentz transformation theory shows that when the fields are perpendicular to each other, $\mathbf{E} \cdot \mathbf{B} = 0$, it is always possible to find a coordinate system in which either $\mathbf{E}$ or $\mathbf{B}$ vanishes. We now examine $\mathbf{E} \cdot \mathbf{B} = 0$ for $|\mathbf{E}| < |\mathbf{B}|$, and $|\mathbf{E}| \geq |\mathbf{B}|$. For the first case, it is well known that the particles simply drift in the $\mathbf{E} \times \mathbf{B}$ direction. For the second case, particles can be accelerated in the $\mathbf{E} \times \mathbf{B}$ direction. This acceleration could be important when particles are moving near magnetic neutral regions where $|\mathbf{B}| \sim 0$.

To see how the solutions of the relativistic Lorentz equation, $d\mathbf{p}/dt = q(\mathbf{E} + \mathbf{v} \times \mathbf{B})$, with uniform and stationary electric $\mathbf{E}$ and magnetic $\mathbf{B}$ fields depend on whether $|\mathbf{E}| < |\mathbf{B}|$ or $|\mathbf{E}| \geq |\mathbf{B}|$, let us examine the transformation equations given in Eq. (1.30). We remind the readers that the unprimed quantities are measured in S frame and primed in S' frame moving with a constant velocity V relative to S frame. The subindices $\parallel$ and $\perp$ are directions relative to $\mathbf{V}$, *not* $\mathbf{B}$.

1.5.3 S'-Frame Electric Field $\mathbf{E}'_\perp = 0$

Suppose we require that $\mathbf{E}'_\perp = 0$ in the S' frame. The second equation in Eq. (1.30) shows that the required transformation velocity for $\mathbf{E}'_\perp = 0$ is $\mathbf{V} = (\mathbf{E} \times \mathbf{B})/B^2$. (Take the cross product with $\mathbf{B}$ and expand the triple cross product and let $\mathbf{E} \cdot \mathbf{B} = 0$.) There are no electric fields parallel to $\mathbf{B}$. All vectors, $\mathbf{E}$, $\mathbf{B}$, and $\mathbf{V}$, are perpendicular to each other. For this case Eq. (1.30) reduces to

$$\mathbf{E}'_\parallel = 0$$

$$\mathbf{B}'_\parallel = 0$$

$$\mathbf{E}'_\perp = \gamma (\mathbf{E} + \mathbf{V} \times \mathbf{B})_\perp = 0$$

$$\mathbf{B}'_\perp = \frac{1}{\gamma} \mathbf{B}_\perp$$

$$= (1 - E^2/B^2)^{1/2} \mathbf{B}_\perp \qquad (1.33)$$

where the last equation of Eq. (1.33) used the relation $V = |\mathbf{V}| = |\mathbf{E}|/|\mathbf{B}|$.

Question 1.6 What do these equations show about the field present in S' frame? Are there any restrictions about $|\mathbf{E}|$ and $|\mathbf{B}|$? What kind of motions are present in S and S' frames?

1.5.4 S'-Frame Magnetic Field $\mathbf{B}'_{\perp} = 0$

We now seek a reference frame in which the magnetic field $\mathbf{B}'_{\perp} = 0$. The fourth equation of Eq. (1.30) shows the required transformation velocity in this case for $\mathbf{B}'_{\perp} = 0$ is $\mathbf{V} = \mathbf{E} \times \mathbf{B}/E^2$. As in Eq. (1.33), we find $\mathbf{E}'_{\parallel} = 0$ and $\mathbf{B}'_{\parallel} = 0$, but the perpendicular components become

$$\mathbf{E}'_{\perp} = \frac{1}{\gamma}\mathbf{E}_{\perp}$$
$$= (1 - B^2/E^2)^{1/2}\mathbf{E}_{\perp}$$
$$\mathbf{B}'_{\perp} = \gamma(\mathbf{B} - \mathbf{V} \times \mathbf{E}/c^2)$$
$$= 0 \qquad\qquad (1.34)$$

Thus, in the S' frame, the magnetic field is zero and the particle is acted on only by an electrostatic field $\mathbf{E}'_{\perp}$ which is stronger than in S frame by the factor γ. The first equation shows that in addition to $\mathbf{E} \cdot \mathbf{B} = 0$, we also require that for γ to be real, $|\mathbf{E}| > |\mathbf{B}|$.

Question 1.7 The case $|\mathbf{E}| > |\mathbf{B}|$ is *not* generally discussed in introductory space plasma physics text books. Why is this problem important? Describe the motion of the particles in the S'-frame.

1.5.5   Inductive Electric Field via Lorentz Transformation

To physically relate the induced electric field to the theory of relativity, consider the example of a plasma blob in the presence of a magnetic field $\mathbf{B}$ (Fig. 1.4). Let the dimension of the blob be much larger than the cyclotron radius of the ions so the plasma can be treated as if all the particles are located at the centers of gyration. For simplicity, we ignore the magnetic field produced by the gyrating particles (no diamagnetic effect). Let the blob move with a velocity $\mathbf{V} = V\hat{\mathbf{x}}$ (S-frame) where $V \ll c$. There is no applied electric field, $\mathbf{E} = 0$ in this frame (Fig. 1.4). An observer in the laboratory frame (S-frame) sees the charges inside the blob respond to the Lorentz force $\mathbf{F} = q\mathbf{V} \times \mathbf{B}$ giving rise to a force in the Y-direction $(-qv_x B_z)$ so the ions move in the $-Y$ direction and the electrons in the $+Y$ direction. These charges moving in opposite directions set up an electric field $\mathbf{E}$ at the edges and the charges will stop moving only when the magnetic force exactly *balances* the electric force and the total force $\mathbf{F}$ on the particles inside the plasma vanishes.

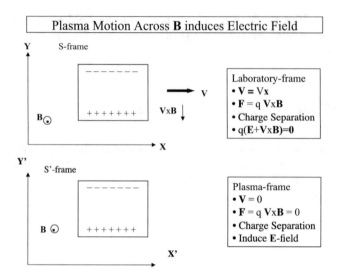

Fig. 1.4 A sketch showing a plasma blob moving uniformly in magnetic field B. The top plots show the observations in the *S*-frame and the bottom *S'*-frame

Question 1.8 What does an observer see in the coordinate frame moving with the plasma blob (*S'*-frame)? If there is no $\mathbf{V} \times \mathbf{B}$ force, what moves the particles?

1.5.6   *Frames Where* **E** *and* **B** *Are Parallel*

An interesting question concerns the coordinate frames in which **E** is parallel to **B**. This problem was discussed by Landau and Lifshitz (1962) who showed that a system of reference can be found in which **E** is parallel to **B**. To find such reference frames, note that the scalar product $\mathbf{E} \cdot \mathbf{B} = \mathbf{E}' \cdot \mathbf{B}'$ is a Lorentz invariant. Electric and magnetic fields will exist simultaneously in the rest and moving frames and the motion of particles must be considered in combined electric and magnetic fields.

To discuss the frames in which the electric and magnetic fields are parallel, let **E** and **B** fields be in the rest *S*-frame and **E'** be parallel to **B'** in the moving *S'*-frame. The velocity of *S'*-frame relative to *S*-frame is found by requiring that $\mathbf{E}' \times \mathbf{B}' = 0$. We choose *V* to be along the *x*-direction, $\mathbf{V} = V_x$, and let **E** and **B** in the *S*-frame be perpendicular to each other, $\mathbf{E} \cdot \mathbf{B} = 0$. In the *S'*-frame, $E'_x = B'_x = 0$ and also the x-component of $\mathbf{E}' \times \mathbf{B}' = 0$. Thus,

$$\mathbf{E}' \times \mathbf{B}' = (E'_y B'_z - E'_z B'_y)\hat{\mathbf{x}} - (E'_x B'_z - E'_z B'_x)\hat{\mathbf{y}} + (E'_x B'_y - E'_y B'_x)\hat{\mathbf{z}}$$
$$= (E'_y B'_z - E'_z B'_y)\hat{\mathbf{x}}$$
$$= 0 \qquad\qquad\qquad\qquad\qquad (1.35)$$

Now use the relations of the Lorentz transformation equations and obtain

$$0 = (E_y - V B_z)(B_z - (V/c^2)E_y) - (E_z + V B_y)(B_y + V/c^2 E_z) \qquad (1.36)$$

The above equation can be rewritten as

$$\frac{V^2}{c^2}(E_y B_z - E_z B_y) - \frac{V}{c}(E_y^2 + E_z^2 + B_y^2 + B_z^2) + (E_y B_z - E_z B_y) = 0 \qquad (1.37)$$

Question 1.9   Show that Eq. (1.37) above can be reduced to

$$\left(\frac{V}{c}\right)^2 - \left(\frac{V}{c}\right)\left[\frac{|E|^2/c^2 + |B|^2}{|E \times B|}\right] + 1 = 0 \qquad (1.38)$$

and discuss the meaning of this solution.

The above results have important implications about the existence of electromagnetic waves with electric and magnetic fields parallel to each other. Essentially all textbooks on electricity and magnetism state that plane wave solutions of Maxwell equations are those with electric field perpendicular to the magnetic field. However, as shown above, there are inertial systems in which parallel electric and magnetic fields could exist. The first paper that showed a general solution of the wave that included **E** ∥ **B** is Chu and Ohkawa (1982). A number of papers refuted this paper in the beginning but it has now been resolved. The solutions with **E** ∥ **B** are consistent with Maxwell equations and the main physical differences are that in the case of parallel fields, there is no Poynting flux, hence while the wave energy can be transported, there is no radiation (see Evangelidis 2004; Gray 1992).

1.5.7   General Theory of Relativity

Rotating frames are *noninertial* and the results of the special theory formulated for inertial frames do not carry over. Effects of rotation on electrodynamics were first studied by Schiff (1939). It is important to understand how rotation affects electromagnetism since most planetary, solar, and astrophysical bodies rotate and electromagnetic field measurements are often made on platforms that also rotate. Physics performed in noninertial frames must take account of general relativity effects and the prescription for the transformation is different from the one given by the special theory of relativity. The topic of electromagnetism involving noninertial frames has been discussed in Parks (2004) and interested readers are recommended to consult the book.

The two principles important for the general relativity are the principle of covariance and the principle of equivalence. The covariance principle states that the laws of physics can be expressed so that they are the same in any coordinate frame of reference which also applies to the special theory. The equivalence principle essentially states that acceleration and gravitation are equivalent. An observer who makes measurements in a stationary frame with a gravitational field is expected to obtain the same result as an observer who makes measurements in an accelerated frame of reference in the absence of gravitational fields. This principle further hypothesizes that it is always possible to transform a local space-time point in an accelerated frame to another coordinate system so that the effect of gravity will not appear in this frame (see Landau and Lifschitz 1962; Møller 1952 and Tolman 1934 cited at the end of this chapter).

For convenience, we repeat here Maxwell equations written in tensor form,

$$\frac{\partial F_{\mu\nu}}{\partial x^{\sigma}} + \frac{\partial F_{\nu\sigma}}{\partial x^{\mu}} + \frac{\partial F_{\sigma\mu}}{\partial x^{\nu}} = 0 \tag{1.39}$$

and

$$\frac{\partial F^{\mu\nu}}{\partial x^{\nu}} = J^{\mu} \tag{1.40}$$

where $F_{\mu\nu}$ and $F^{\mu\nu}$ are covariant and contravariant forms of the electromagnetic field tensor and J^{μ} is the four current-density given by $(\mathbf{J}, \rho)$ where $\mathbf{J}$ and ρ are the usual current and charge densities.

A recipe for understanding how Maxwell equations are structured in a rotating frame uses the fact that the square of the distance is an invariant. Then the metric is transformed and one obtains an expression in the rotating frame. This allows us to calculate the electromagnetic field tensor in the rotating frame. Use of (1.39) will then yield two of the Maxwell equations in the rotating frame. The other two equations with the source terms require that we calculate the contravariant form of the field tensor. Subsequently, we can compare the field tensors between the rotating and nonrotating frames to obtain information on the relationship of the field quantities.

We now state the final results about how electromagnetic fields transform, omitting tedious algebra of derivations found in Parks (2004). As before, we denote quantities measured in the rotating frame with a prime ($'$). The relationships of electromagnetic field quantities in the rest and rotating frames of reference are given by

$$\mathbf{E}' = \mathbf{E} + (\mathbf{V} \times \mathbf{B}')$$

$$\mathbf{B}' = \mathbf{B} \tag{1.41}$$

Now note that $\mathbf{V} = (\boldsymbol{\omega} \times \mathbf{r}')$ at each point in space where $\boldsymbol{\omega}'$ is the rotation vector. Here $\mathbf{V}$ is therefore not a constant and depends on the distance $\mathbf{r}$ even for constant rotations. These transformation equations are identical in form to the ones derived for the special theory of relativity for $V/c \ll 1$. Whereas in the special theory of relativity the equations are approximately equal for $V/c \ll 1$, the two equations in

Eq. (1.41) are exact to all orders of V/c to distances $r\omega < c$. Another important result is that the magnetic field is invariant under rotation. The relationship of electric and magnetic fields in rotating and rest frames given by (1.41) are applicable to many classes of space, solar, and astrophysical problems.

Historically, the electrodynamics of rotation received much attention at the turn of the twentieth century when it was shown that rotating conducting magnets induce electric fields (Chapter 2 of Parks 2004).

Question 1.10 Consider a unipolar inductor consisting of a conducting magnet spinning around its axis. A wire loop is connected between the equator and the pole through a galvanometer by means of brushes. Numerous experiments of this type conducted in the laboratories showed that the instant the magnet is set into rotation, a current will flow in the stationary wire detected by the galvanometer. This current flows because of the electromotive force ($\mathcal{EMF}$). Where was the $\mathcal{EMF}$ induced? Inside the rotating magnet or in the stationary wire? Is the $\mathcal{EMF}$ produced because a conductor moved through a magnetic field or can one look at field line motion inducing the $\mathcal{EMF}$ in the stationary wire loop?

1.6 Macroscopic Equations

A complete description of plasmas requires information on the velocity distribution function for each species of particles. However, in practice, one can begin use of macroscopic or "bulk" parameters to study the average behavior of the plasmas. The physics is simplified considerably since the number of variables has been reduced from $6N$ to 3 or 4. Moreover, if the velocity distribution is Maxwellian, we can obtain information on the temperature T from the thermal energy.

The macroscopic variables are derived parameters, *calculated* from the measured distribution function and they show up in macroscopic momentum and energy conservation equations. The density n is obtained from

$$n = \int f(\mathbf{v})d^3v \qquad (1.42)$$

In practice the integral is replaced by the sum of the number of energy channels contained in the instrument (see Sect. 1.7 where details are given about measurements). The mean velocity is obtained from

$$n\mathbf{V} = \int \mathbf{v}f(\mathbf{v})d^3v \qquad (1.43)$$

pressure and heat flux tensors from

$$\mathbf{P} = m\int (\mathbf{v} - \mathbf{V}) \cdot (\mathbf{v} - \mathbf{V})f(\mathbf{v})d^3v \qquad (1.44)$$

and

$$Q = (m/2) \int (\mathbf{v} - \mathbf{V})|\mathbf{v} - \mathbf{V}|^2 f(\mathbf{v}) d^3 v \tag{1.45}$$

The temperature tensor is obtained from the pressure tensor $\mathbf{P}$ by assuming *a priori* the equation of state, $\mathbf{P} = n m k_B \mathbf{T}$ and that $f(\mathbf{v})$ is a Maxwellian distribution. In space data analysis, use is made of both fundamental $f(\mathbf{v})$ and macroscopic bulk parameters derived from the distribution function to interpret observations.

We now derive the macroscopic energy and momentum conservation equations for a system of charged particles in electromagnetic fields. By macroscopic, we mean that the microscopic valuables have been averaged over the distribution function of the particles. Macroscopic variable such as pressure, bulk velocity, and density are used with conservation "fluid" equations. The physics is simple and intuitive but the applicability of these quantities to collisionless plasmas may not always give a correct picture of the dynamics. The macroscopic equations are useful when the plasma is simple consisting of one ion species and the distributions can be approximated by a Maxwellian. In this case, they give an average picture about the macroscopic behavior of space plasmas.

1.6.1 Conservation of Momentum

The conservation of the momentum equation is obtained by averaging the kth component of the Lorentz equation, $d(m\mathbf{v})_k/dt = q(\mathbf{E} + \mathbf{v} \times \mathbf{B})_k$ over the distribution function $f(\mathbf{r}, \mathbf{v}, t)$ to describe the motion in the average sense. After a lot of algebra (see Jackson 1975), this leads to the conservation of the momentum density equation for particle species α as

$$\frac{\partial}{\partial t} \left(\sum_\alpha \Pi_k^\alpha + G_k \right) + \frac{\partial}{\partial x_i} \left(\sum_\alpha \Pi_{ik}^\alpha + T_{ik} \right) = 0 \tag{1.46}$$

Integration of this equation over a fixed volume V bounded by a surface A gives

$$\frac{\partial}{\partial t} \int_V \left(\sum_\alpha \Pi_k^\alpha + G_k \right) dV + \frac{\partial}{\partial x_i} \int_A \left(\sum_\alpha \Pi_{ik}^\alpha + T_{ik} \right) dA = 0 \tag{1.47}$$

where we have defined

$$\Pi_k = \int m v_k f(\mathbf{r}, \mathbf{v}, t) d^3 v \tag{1.48}$$

as the kth component of the *average momentum vector* and

$$\Pi_{ik} = \int m v_i v_k f(\mathbf{r}, \mathbf{v}, t) d^3 v \tag{1.49}$$

as the (ik)th component of the *momentum transfer tensor*. We emphasize that this equation is generally *not* pressure although they have the same units. (It is pressure only if the distribution function is a Maxwellian.) We can also write these equations as $\Pi_k = nm \langle v_k \rangle$ and $\Pi_{ik} = nm \langle v_i v_k \rangle$. We have also defined T_{ik} as the electromagnetic stress tensor given by

$$T_{ik} = \left(\frac{\epsilon_0 E^2}{2} + \frac{B^2}{2\mu_0} \right) \delta_{ik} - \left(\epsilon_0 E_i E_k + \frac{B_i B_k}{\mu_0} \right) \tag{1.50}$$

and

$$G_k = \epsilon_0 \mu_0 (\mathbf{E} \times \mathbf{H})_k = \frac{(\mathbf{E} \times \mathbf{H})_k}{c^2} \tag{1.51}$$

is the momentum density of the electromagnetic field (Poynting vector). In these equations, the summation convention of repeated indices is used. The first term represents the local change of momentum and the second term, change of momentum in the reference frame of a volume of particles in motion. Equation (1.49) is reduced to ordinary pressure tensor only when the distribution function is Maxwellian. The momentum flux and momentum transfer tensor can be directly computed using space plasma data. Physically, this integral equation states that the rate of change of the total momentum (mechanical plus electromagnetic) in the volume is equal to the rate at which the momentum flows out or into the surface bounding this volume. The quantities in Eq. (1.47) can all be computed from the primary quantities measured by experiments.

1.6.2 Conservation of Energy

The total energy of a particle is $\mathcal{E} = T + m_o c^2$, where $m_o c^2$ is the rest energy and T is the kinetic energy. This equation shows $d\mathcal{E}/dt = dT/dt$. The total work done on the particle is $W = \int \mathbf{F} \cdot \mathbf{v} dt = q \int (\mathbf{E} + \mathbf{v} \times \mathbf{B}) \cdot \mathbf{v} dt$. Assume that all of the work done on the particle goes into increasing the kinetic energy of the particle. Then the kinetic energy of the particle increases at the rate

$$\frac{d\mathcal{E}}{dt} = q\mathbf{E} \cdot \mathbf{v} \tag{1.52}$$

since $q(\mathbf{v} \times \mathbf{B}) \cdot \mathbf{v} = 0$. The magnetic field $\mathbf{B}$ does *no* work and the kinetic energy of charged particles can only change from work done by the *electric field* $\mathbf{E}$. This result is quite general and indicates that for charged particles to be accelerated in solar and astrophysical phenomena, the processes must involve electric fields. One of the main challenges in space plasma phenomena is to identify the source of such electric fields and how they are produced.

Let us multiply the above equation with the distribution function $f(\mathbf{r}, \mathbf{v}, t)$ and integrate over the velocity space. This yields again with a lot of algebra (see Parks 2004), where we have included different particle species.

$$\frac{\partial}{\partial t} \sum_\alpha \mathcal{E}_p + \frac{\partial}{\partial x_j} \sum_\alpha \mathcal{Q}_j = \mathbf{J} \cdot \mathbf{E} \qquad (1.53)$$

This energy conservation equation is also called *Poynting's* theorem. Here we have defined the average kinetic energy of the particles as

$$\mathcal{E}_p = \int \frac{1}{2}mv^2 f(\mathbf{r}, \mathbf{v}, t) d^3v \qquad (1.54)$$

and the heat flow vector as

$$\mathbf{Q} = \int \frac{1}{2}mv^2 \mathbf{v} f(\mathbf{r}, \mathbf{v}, t) d^3v \qquad (1.55)$$

This equation can also be rewritten as

$$\frac{\partial}{\partial t} \left(\sum_\alpha \mathcal{E}_p + \mathcal{E}_f \right) + \frac{\partial}{\partial x_j} \left(\sum_\alpha \mathcal{Q}_j + \mathcal{S}_j \right) = 0 \qquad (1.56)$$

where we have defined $\mathbf{S}$ as the Poynting flow vector and $\mathcal{E}_f$ as the energy density of the electromagnetic field. Integrating this equation over a fixed volume V which is bounded by the area A yields

$$\frac{\partial}{\partial t} \int_V \left(\sum_\alpha \mathcal{E}_p + \mathcal{E}_f \right) dV + \frac{\partial}{\partial x_j} \int_S \left(\sum_\alpha \mathcal{Q}_j + \mathcal{S}_j \right) dA = 0 \qquad (1.57)$$

This conservation of energy equation states that the rate of change of energy of the particles and electromagnetic fields in a given volume V is equal to the rate at which the energy flows out or into the surface by heat transfer and electromagnetic field radiation.

All of the quantities in Eq. (1.57) can be determined by spacecraft experiments that measure $\mathbf{E}$ and $\mathbf{B}$ fields and the distribution function $f(\mathbf{r}, \mathbf{v}, t)$ of ions and electrons. Moreover, the Cluster mission and the MMS mission consisting of four identical spacecraft can determine gradients (∇) and divergences ($\nabla\cdot$). This requires careful cross calibration of the experiments on the four spacecraft which is not an easy task but doable.

1.6.3   Issues with Moment Equations

The moment equations are very useful, but it is important to recognize that the description of space plasmas using the moments equation is not self-consistent. For example, observations in the plasma sheet have shown (Cao et al. 2013) that just looking at bulk parameters can give incorrect physics about the transport processes. Large scale space plasma dynamics have been interpreted using MHD theory and concepts. MHD equations come from conservation equations derived from moments of the collisionless Boltzmann equation. While MHD theory has the advantage that the dynamics can be described with only a few macroscopic parameters and equations, the theory is incomplete.

One must be careful about the velocity moments computed because they do not always reflect the true state of the velocity distribution. This is especially true when the counts are low giving a low signal-to-noise ratio. In such cases, the moments, while useful, can give the illusion of a high temperature and can affect bulk values due to the inability to define the core/peak of the distribution independent of the noise.

Question 1.11 The procedure of taking moments always produces more unknowns than the number of equations and the equations do not form a closed set. Discuss the meaning of this sentence and what other assumptions must be made when space plasmas are studied using the moments equations.

1.7   Plasma Measurements

The two instruments that are most commonly used to measure plasmas in space are Electrostatic Analyzers (ESAs) and Faraday cups (mainly for the solar wind). We first discuss the basics of ESAs and then Faraday cups, what quantities measured are primary, and how the bulk parameters are calculated from the primary measurements including the assumptions made in these measurements.

1.7.1   Electrostatic Analyzers

An ESA basically consists of two oppositely curved charged plates and particles passing through such plates have energies proportional to the applied voltage. ESAs have been used to measure ions and electrons in the solar wind, magnetosphere, and ionosphere. A schematic diagram of an ESA flown on the WIND and Cluster spacecraft is shown in Fig. 1.5. This "top hat" analyzer (Carlson et al. 1983) allows particles to enter at the top of the hemisphere where an opening has been cut out. The entrance aperture has uniform 360° field of view. The particles exiting the analyzer on the equatorial plane are detected by a position sensitive sensors, "channel plates."

Let the concentric spheres have a mean radius R and let the electric field E applied between the plates be directed in the radial direction. Particles traveling in circular path will pass through the plates only if the electric force just balances the centripetal force,

$$mv^2/R = qE \qquad (1.58)$$

Fig. 1.5 A schematic diagram of a symmetric spherical "top hat" ESA illustrating its focusing properties (from Lin et al., *Space Sci. Rev.*, **71**, 125, 1995)

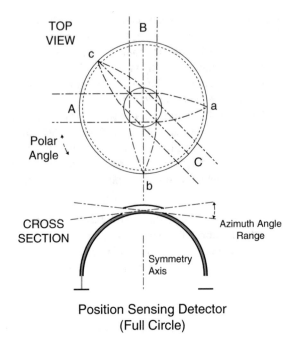

Equation (1.58) can be rewritten as

$$mv^2/2q = ER/2 \qquad\qquad (1.59)$$

where energy/charge (left side) is related to instrument quantities on the right which are all known. ESAs measure *energy/charge* of the particle, regardless of the mass, charge, or velocity and obtain 3D information on particle fluxes in one spin of the spacecraft (typically in a few seconds).

All particle detectors measure *flux* of particles. Consider a detector immersed in a plasma of density n where the particles move with a mean velocity $\mathbf{v}$. The total particle flux entering the aperture is $n\mathbf{v}$. If A is the effective area of the entrance aperture, the counting rate of a detector is $C/\Delta t = n\mathbf{v}A$ where C is the total counts recorded by the detector and Δt is the accumulation time. Here C represents dead time corrected counts. Note that the detector measures the *product* $(n\mathbf{v})$, and not n or $\mathbf{v}$.

The energy/charge (E/q) spectrum is obtained by counting the particles over a small energy range ΔE. The instrument then obtains a *differential count-rate* $C_i/\Delta t = n_i \mathbf{v}_i A \Delta E_i$ where the subindex i represents the step that counts the particle in the energy range ΔE_i. ESAs are designed to measure particles over a limited energy range, usually a few eV to about 40 keV/charge. To obtain an E/q spectrum over the entire energy range that an ESA has been designed to measure, particles of different energies are selected by varying the high voltage applied between the plates. The number of energy steps for a typical ESA is 32 or 64.

The count-rate $C/\Delta t$ depends directly on the size of the aperture A which is an important variable when designing a detector (part of the geometrical factor of the instrument). Depending on the science requirements, a detector may be designed with a small entrance aperture giving a narrow field of view (FOV). However, in small entrance apertures, the entering flux may be too small to obtain a statistically meaningful number of counts within the time interval of interest. The entrance aperture could be made larger to obtain a larger count-rate at the expense of losing angular resolution. However, if the aperture is made too large, the particle flux could become so large the detector will saturate. Saturation could be the result of the electronics (preamplifier, for example) not having sufficient time resolution to resolve all of the counts entering the aperture. The output of the data will become flat after reaching the maximum rate the electronics can handle and the data will be lost. In detector design, one considers many tradeoffs and how to optimize the instrument to accomplish the science takes much thought, good judgment, and creativity.

Differential Number Flux

The relationship between the differential flux of the particles and the differential count-rate for a given energy step is

$$F_N = \frac{C}{g_v \Delta E \, \Delta t}$$
$$= \frac{C}{g_E \, E \, \Delta t} \tag{1.60}$$

where we have omitted the subindex i for clarity. Here g_v is the velocity dependent geometrical factor, ΔE is the width of the energy channel, and Δt the accumulation time. The second line is obtained from the first by multiplying the numerator and denominator by E and defining g_E as the geometrical factor which has incorporated information about the detector efficiency at different energies and angular response of the detector. This geometrical factor g_E is then independent of the energy. The differential flux F_N is measured in units of $(\text{cm}^{-2}\,\text{s}^{-1}\,\text{sr}^{-1}\,\text{eV}^{-1})$, g_E in $(\text{cm}^2\text{-sr})$, and energy E is in eV or keV. The differential number flux is usually plotted as a function of energy to determine the form of the *Energy spectra*.

Energy Flux

If we multiply Eq. (1.60) by E, we obtain the energy flux of the particles,

$$F_E = \frac{C}{g_E \Delta t} \tag{1.61}$$

The energy flux F_E is directly proportional to the detector count-rate and is considered the primary information obtained by ESAs. Energy flux is measured in units of (ergs cm^{-2} s^{-1}). In summary plots, Energy Flux is plotted as a function of time to obtain an overall behavior of the plasma.

Distribution Function

The formal definition of the flux is

$$n\mathbf{v} = \int \mathbf{v} f(\mathbf{r}, \mathbf{v}) d^3 v$$

$$= \sum \mathbf{v} f(\mathbf{r}, \mathbf{v}) \Delta v^3 \tag{1.62}$$

where the second line takes account of the fact that the number of velocity steps in ESAs is finite. We remind the reader that the energy/charge detectors measure flux of the particles and *not* the distribution function of the particles. However, the above equations can be inverted and we can study which form of the distribution function agrees best with observations.

We showed that the flux $n\mathbf{v}$ is related to the differential count-rate of the instrument ($C_i/\Delta t = n_i v_i A \Delta E_i$). This allows us to approximate Eq. (1.62) by the differential count-rate

$$C/\Delta t = g_v v f(v) v^2 \Delta v \tag{1.63}$$

where we have replaced Δv^3 with $v^2 \Delta v$. To write this equation in a more standard form, note that kinetic energy of the particle is $E = mv^2/2$, from which we obtain $v = (2E/m)^{1/2}$ and $\Delta v = \Delta E/mv$. We use these equations to evaluate $v^3 \Delta v$ in Eq. (1.63) and obtain

$$v^3 \Delta v = \frac{2}{m^2} E^2 \frac{\Delta E}{E} \tag{1.64}$$

Rewriting Eq. (1.63), we can write $f(\mathbf{v})$ as

$$f(\mathbf{v}) = \frac{C}{\Delta t \, g_v \, v^3 \Delta v} \tag{1.65}$$

where g_v is velocity dependent geometrical factor. We now use Eq. (1.64) and obtain

$$f(\mathbf{v}) = \frac{C}{\Delta t \, g_v} \frac{m^2}{2} \frac{1}{E^2} \frac{1}{\Delta E/E} \tag{1.66}$$

Define now $g_E = g_v \, \Delta E/E$ as energy independent geometrical factor. Then using $E = mv^2/2$, $f(v)$ can be written as

$$f(v) = \frac{C}{\Delta t}\frac{1}{g_E}\frac{1}{v^4} \qquad (1.67)$$

$f(v)$ can be also be written in terms of energy flux $F_E = C/g_E \Delta t$ as

$$f(v) = F_E \frac{m^2}{2E^2} \qquad (1.68)$$

The units for $f(v)$ are s^3-cm^{-6}. Since $f(v)$ is directly related to the energy flux F_E which is the primary quantity measured by ESAs, $f(v)$ can be also looked upon as a primary quantity.

The above discussion assumes that ESAs are measuring a simple plasma distribution consisting of one ion species and electrons. However, if there are multiple ion species, then the above equations have ambiguities. This is especially important for the solar wind since typically, H^+ is 95% and He^{++} is 5% but He^{++} can increase to $\sim$20% during disturbed times. This ambiguity arises because measuring energy/charge of the particles is *not* adequate to determine the velocity which is needed to obtain information on the distribution function. In the case of ESA measurements of the SW, some very restrictive assumptions must be made (see below, Sect. 1.7.4).

To measure the distribution function of multi-component plasmas requires an ESA combined with time of flight measurements or mass spectrograph so that the velocity and q/m information of the particles are obtained (Wüest et al. 2007). However, the SW measurement is experimentally difficult because of the overwhelming dominance of H^+ over He^{++} ions. Measurements must clearly separate the two species of ions since He^{++} channels are often contaminated by H^+ ions.

The velocity space in which the phase space density is plotted could be represented either in Cartesian coordinate system using GSE variables or velocities relative to the direction of the magnetic field, $V_\parallel$ vs $V_\perp$. These velocities are defined as $V_\parallel = V_{pl} \cdot b$ where V_{pl} is the bulk plasma velocity, and b is the unit vector of B. There are two directions for $V_\perp$:

$$V_{\perp_1} = b \times (V_{pl} \times b)$$
$$V_{\perp_2} = b \times V_{\perp_1} \qquad (1.69)$$

In later chapters when we discuss the behavior of plasmas in the solar wind, above the aurora and in the plasma sheet, the distribution plots will use these variables.

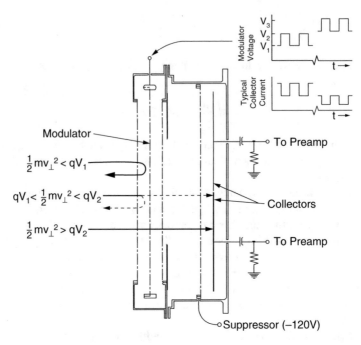

Fig. 1.6 A sketch showing how Faraday Cups work (from Bridge et al. 1960)

1.7.2 Faraday Cup and Measurements

Faraday cups have also been used to measure the SW (Bridge et al. 1960). Figure 1.6 shows a schematic cross section of a typical Faraday cup sensor (Vasyliunas 1971). It includes a series of wire mesh, planar grids, and one or more collector plates. By applying a sequence of voltages (V) to the modulator grid, only particles having energy/charge (E/q) greater than V will be able to pass through the grid and continue on to the collector plate where a measurable current is produced. In normal operation, the grid voltage V is varied between two voltages, $V1$ and $V2$ at a frequency of a few hundred Hz. Thus particles incident on the grid having E/q between $V1$ and $V2$ will produce a current on the collector plate and can be detected by a current meter. By choosing an increasing sequence of values for the modulation voltage (for example, $V2$ and $V3$), E/q spectrum of the particles can be measured.

Unlike ESAs that count the particles, Faraday cups measure the current produced by the flux of particles impinging on the detector. The relationship between the current I and the distribution function $f(\mathbf{v})$ is

$$I = eA \int \mathbf{v} f(\mathbf{v}) S(\mathbf{v}) d^3 v \qquad (1.70)$$

where $S(\mathbf{v})$ is the sensitivity function that includes the angular response, efficiency, and fraction of the aperture that intercepts the collector when projected at a given direction (like the geometrical factor). The above discussion about measuring simple plasma distributions also applies to Faraday cups. If the plasma is complex, like the solar wind, restrictive assumptions must be made about the plasma behavior.

In SW applications, the H^+ and He^{++} parameters derived from Faraday cups rely on characterizing the solar wind velocity distributions for each of the ion species with a convected bi-Maxwellian function,

$$f_i(\mathbf{v}) = \frac{n_i}{\pi^{3/2} w_{o\perp i}^2 \, w_{o\parallel i}} \exp -(w_\perp^2/w_{o\perp i}^2 + w_\parallel^2/w_{o\parallel i}^2) \qquad (1.71)$$

where w_o is thermal velocity related to temperature T by $kT = mw_o^2/2$, $\mathbf{w} = \mathbf{v} - \mathbf{V}$ where $\mathbf{V}$ is the bulk velocity, $w_\parallel = \mathbf{w} \cdot \mathbf{b}$ where $\mathbf{b}$ is the unit vector in the direction of the magnetic field $\mathbf{B}$, $w_\perp = \mathbf{w} - \mathbf{b}w_\parallel$. The subindex i denotes either H^+ or He^{++} ions. This distribution is represented in the frame of the spacecraft. The parameters that can be adjusted are the density n, bulk velocity $\mathbf{V}$ and thermal velocities, $w_{o\parallel}$ and $w_{o\perp}$. A bi-Maxwellian assumption close to the bow shock is not appropriate when there are reflected particles and particles leaking out of the magnetosheath.

Faraday cups have wide entrance apertures and do not have the angular resolution ESAs have and are not normally used to obtain 3D information. However, in principle, use of many Faraday cups pointing in different directions could obtain crude information on the arrival direction of the particles. The main advantage of Faraday cups is that they are simple, rugged, and can operate for a long time in space without deterioration or loss of gain as MCPs do.

1.7.3 Cluster Plasma Experiment

Particle instruments on numerous missions, Vela, Mariner, OGOs, IMPs, ISEE and more recently on Wind, Cluster, and Themis have shown plasmas in the vicinity of the bow shock come from several sources whose density and temperatures can vary by orders of magnitude. For example, the SW flux is cold and the flux is two orders of magnitude higher than the magnetosheath and magnetospheric fluxes (Fig. 1.7). A detector designed to measure magnetosheath fluxes would saturate in the SW and a detector designed for the SW would not count significant fluxes in the magnetosheath on dynamic time scales of interest. Measurements by saturated detectors will give incorrect density, bulk velocity, and temperature values.

Observations have also shown that plasmas on the dayside of Earth include in addition to the SW, reflected and gyrating particles from the bow shock and leakage of magnetosheath particles (Sckopke 1995; Thomsen et al. 1988). Although these particles represent a small fraction of the total population (up to 20%), they occupy different regions of the velocity space affecting the calculated values of the mean velocity and temperature. To develop models and understand the mechanisms of

Fig. 1.7 Estimates of average differential flux of plasmas observed in the various regions of space including SW, magnetosheath, magnetopause, magnetosphere, and geomagnetic tail (from Réme et al., *Space Sci. Rev.*, **79**, 303, 2001)

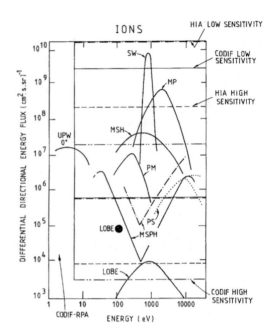

how the SW is produced, SW must be measured without contamination from the other particles, and the instruments must have sufficient energy resolution and also small gaps between the energy channels so that the shape of the distributions could be accurately determined.

1.7.4 Two Geometrical Factors

We have thus two ESAs with two different geometrical factors on Cluster (Fig. 1.8). One is designed specifically to measure the cold SW beam (g-detector) and the other the hotter magnetosheath and magnetospheric plasmas (G-detector). These analyzers have an energy resolution of $\Delta E/E \sim 16\%$. The onboard high voltage (HV) is swept with 30 contiguous energy channels and obtains data 32 or 64 times per spin (4s). Thus a two-dimensional cut through the distribution function is obtained every 1/32 or 1/64 of the spacecraft spin, giving a time resolution of 125 ms or 62.5 ms. The SW g-detector includes 8 sensors aligned along the polar (θ) direction separated by $\sim$5.625° and the SW is detected in the 8 ϕ-sectors (5.625°) in the $\sim$45° azimuthal ϕ-wedge aligned along the Sun-Earth line. The geometrical factor of the g-detector is a factor of $\sim$100 times smaller than that of the G-detector. Moreover, this instrument is microprocessor controlled and the HV sweeps only near the peak of the SW distribution. This information is stored and applied to observations on the next spin. Data from each of the 64 $\theta\phi$ directions can be examined separately. This capability allows us to identify presence of the

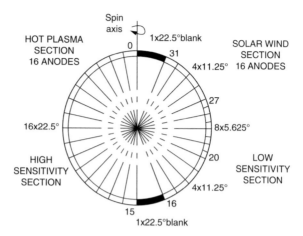

Fig. 1.8 A schematic diagram showing field of view (FOV) of two sides of the Cluster detector. The left side has 16 detectors in the 2π FOV with angular resolution $22.5°$ to measure the hot magnetosheath and magnetospheric plasmas. The right side includes 8 detectors each with an angular resolution of $5.625°$ (from Réme et al., *Space Sci. Rev.*, **79**, 303, 1997)

reflected, gyrating, and leakage of MS particles that permeate the vicinity of the bow shock which contaminate the SW measurements. Data combined from the g- and G-detector give a full 3D distribution in one spin of the spacecraft.

1.7.5 SW Measurement by g- and G-Detectors

Figure 1.9 shows a summary plot of data obtained by Cluster 1 instruments during an inbound pass on 30 January 2003. The top three panels are data from the g-detector showing the energy flux spectrogram, density (black) and temperature (red), and bulk velocity moments. The next three panels show data from the G-detector: energy flux spectrogram (top panel), density (black) and temperature (red), and bulk velocity moments. The bottom panel shows data from the magnetometer. Cluster was initially in the SW mode and the g-detector measured SW density that varied from 8–11 cc^{-1}, temperature $\sim25\,eV$ and bulk speed $\sim470\,km\,s^{-1}$. During this time interval, the G-detector was programmed to not measure particles at SW energies because the detector will saturate. A mode change occurred into a magnetospheric mode around 16 UT at a distance of $\sim15\,R_E$. Here the g-detector is turned off and the G-detector turned on to measure the full 3D distributions. The idea of the mode change was to start measuring the higher temperature plasmas from the bow shock and the magnetosheath. However, on this occasion, SC1 was still in the SW. One can see that the bulk parameters of the SW measured by the G-detector substantially differed from the values obtained earlier by the g-detector. The SW density was 4.6 cc^{-1} much lower than the earlier measurement by the g-detector, temperature $\sim86\,eV$, which was much higher than the value obtained earlier and the bulk velocity was about $520\,km\,s^{-1}$ slightly higher than the value obtained earlier. The values obtained by the G-detector are incorrect because the G-detector was saturated and

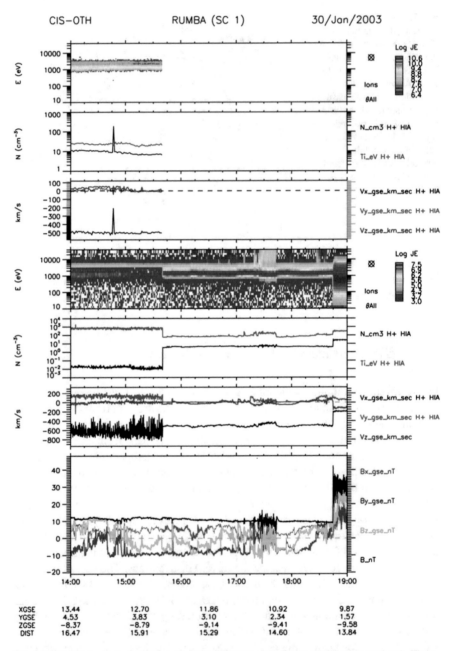

Fig. 1.9 Cluster 1 was in the SW until about 1845 UT when it entered the magnetosheath. The g-detector was measuring the behavior of the SW until about 1545 UT when the mode was changed into a magnetospheric mode to measure the hotter plasmas in the vicinity of the bow shock and magnetosheath. However, before the bow shock was crossed, the G-detector measured the cold SW beam. But the G-detector was saturated and gave incorrect estimates of the density, temperature, and bulk velocities of the SW

the obtained distributions underestimated the phase space density which has been validated examining the distribution function (not shown).

1.7.6 Caveat of Measurements by ESAs

The SW measurements that come from ESAs, which are energy per charge detectors, measure flux of particles, $n\langle v\rangle$. They do not measure n or $\langle v\rangle$ nor do they measure the charge or the mass of ions. However, we see data from these instruments plotted in velocity space identifying the ion species.

Question 1.12 How does ESA obtain information on the velocity, mass, and charge of the particles when the instrument is an energy per charge instrument and does not measure the velocity, charge, or mass?

1.7.7 Errors in Measurements

In all experimental work, one needs to worry about errors. Assuming that instrument parameters do not contribute to measurement errors (instrument has been carefully calibrated), the main error is statistical and it will come from the total number of particles counted. The error from counts can be estimated using Poisson statistics. For example, if the total number of counts obtained for a given measurement is C, the estimated error is $\sqrt{C}$. The percent (%) error for each measurement is carried through for various different quantities.

Another error can come if the measurements are made slower than the variations of some phenomenon. For example, if the sampling time to obtain the distribution function is spin period, which is 4s for Cluster, and if the phenomenon varies faster than 4s, the data will be *time aliased* and one must be careful how the data are interpreted. This can happen in the vicinity of the foreshock where one encounters a considerable amount of turbulences.

In space measurements, physical quantities are measured as a function of time at some position in space. If physical quantities are varying, one needs to assume whether the observed variation is due to time or spatial variations. This can happen, for example, when the spacecraft is crossing a boundary. If at the time of the measurement the disturbance level is fairly quiet, one can assume the variations are from spatial effects. However, one must be cautious when interpreting such data.

1.8 Examples of Plasma Distributions

As discussed above, particle instruments can measure velocity distributions of particles under restricted constraints. Below we briefly discuss the examples of different types of distribution functions frequently encountered in space plasmas.

Thermal Distribution Consider a plasma consisting of ions and electrons. If this plasma is in thermal equilibrium, the distribution function of electrons or ions has a Maxwellian form given by

$$f(\mathbf{v}) = n \left(\frac{m}{2\pi kT}\right)^{3/2} e^{-(mv^2/2kT)} \tag{1.72}$$

where m can be either electron or ion mass, n is the number density, $kT = mv_{th}^2/2$, v_{th} is the thermal velocity, and $v^2 = v_x^2 + v_y^2 + v_z^2$. If we add up all of the particles with velocities from $-\infty$ to $+\infty$, we obtain the number density in real space $\int_{-\infty}^{\infty} f(\mathbf{v})d^3v = n$. (In actual measurements, the limits are defined by instrument thresholds.)

Streaming Distribution If the Maxwellian particles are now streaming with $\mathbf{v}_o = v_x \hat{x}$, then we replace $\mathbf{v}$ in the above equation with $(\mathbf{v} - \mathbf{v}_o)$. The streaming distribution function of a Maxwellian is represented by

$$f(\mathbf{v}) = n \left(\frac{m}{2\pi kT}\right)^{3/2} e^{-m(\mathbf{v}+\mathbf{v}_o)^2/2kT} \tag{1.73}$$

This distribution is for particles in the absence of a magnetic field or they apply to distributions along the magnetic field direction. Streaming distributions are observed when plasma is streaming as in the geomagnetic tail and SW.

Bi-Maxwellian Distribution When a magnetic field is present, the mean temperatures in the direction parallel and perpendicular to **B** can be different. If the particles in the parallel and perpendicular direction are in thermal equilibrium, one can define temperatures for particles along and perpendicular to **B** as $mv_{\parallel}^2/2 = kT_{\parallel}$ and $mv_{\perp}^2/2 = kT_{\perp}$. In this case, Maxwell distribution function for particles in a magnetic field becomes

$$f(\mathbf{v}_{\parallel}, \mathbf{v}_{\perp}) = \frac{n}{T_{\parallel}^{1/2} T_{\perp}} \left(\frac{m}{2\pi k}\right) \exp\left[-\left(\frac{mv_{\parallel}^2}{2kT_{\parallel}} + \frac{mv_{\perp}^2}{2kT_{\perp}}\right)\right] \tag{1.74}$$

This distribution function is anisotropic in temperature and is called a bi-Maxwellian distribution function.

Drifting Bi-Maxwellian Distribution If there is an electric field perpendicular to **B** as often occurs in the geomagnetic tail, then the plasma will $\mathbf{E} \times \mathbf{B}$ drift and the distribution function is given by

$$f(\mathbf{v}_\parallel, \mathbf{v}_\perp) = \frac{n}{T_\parallel^{1/2} T_\perp} \left(\frac{m}{2\pi k}\right) \exp\left[-\left(\frac{m v_\parallel^2}{2kT_\parallel} + \frac{m(\mathbf{v}_\perp - \mathbf{v}_D)^2}{2kT_\perp}\right)\right] \quad (1.75)$$

where $\mathbf{v}_D = \mathbf{E} \times \mathbf{B}/B^2$. This type of distribution can describe the flowing particles in the plasma sheet.

Bi-Maxwellian Distribution Streaming Along B Similarly, if the distribution is streaming along **B** as one often sees in the aurora, we can write the distribution function for the particles streaming along **B** as

$$f(\mathbf{v}_\parallel, \mathbf{v}_\perp) = \frac{n}{T_\parallel^{1/2} T_\perp} \left(\frac{m}{2\pi k}\right) \exp\left[-\left(\frac{m(\mathbf{v}_\parallel - \mathbf{v}_{\parallel o})^2}{2kT_\parallel} + \frac{m v_\perp^2}{2kT_\perp}\right)\right] \quad (1.76)$$

Here $\mathbf{v}_{\parallel o}$ is the streaming velocity. If the distribution includes a *beam* as in auroras, it characterizes the auroral particles above the aurora and elsewhere, for example, in the boundary layer of the plasma sheet. One of the ways beams are produced is to accelerate thermal particles across potential drops along the magnetic field direction.

Cold and Ring Plasma Distribution In cold plasmas, particles do not have thermal velocities. Hence, the distribution function is represented by

$$f(\mathbf{v_o}) = n_o \delta(\mathbf{v}_o) \quad (1.77)$$

A simple modification of the cold distribution gives the ring or shell distribution,

$$f(\mathbf{v}') = n_o \delta(\mathbf{v}_o - \mathbf{v}') \quad (1.78)$$

The ring distribution involves particles gyrating around the magnetic field and applies to new born ions in the SW. A shell distribution includes particles with pitch-angles away from the magnetic field direction and the shape in velocity space is a shell. Such distributions can result from scattering particles in pitch-angles of the beam and ring distributions.

Loss Cone Distribution Loss cone distributions are important for magnetospheric dynamics because the magnetic field has mirror geometry and a loss cone is naturally developed for particles on such magnetic fields. A loss cone distribution can be represented as (Dory et al. 1965)

$$f_o(v_\parallel, v_\perp) = \frac{n_o}{\pi^{3/2} v_{th}^3 (j!)} \left(\frac{v_\perp}{v_{th}^2}\right) \exp(-v_\perp^2/v_{th}^2) \quad (1.79)$$

where n_o is the density, $v_\parallel$ and $v_\perp$ are velocities parallel and perpendicular to the magnetic field direction, v_{th} is the thermal velocity, and j is an index that measures the steepness of the loss cone feature, $j = 1, 2, 3\ldots\ldots$ Wong et al. (1985) have solved the instabilities driven by the loss cone distribution for different electron energies (30, 50, and 100 keV) and for various degrees of loss cone steepness.

Distribution Function at Boundaries Near boundaries, we encounter both magnetic field and density and or temperature gradients. We take a slab geometry and let the magnetic field be along the z-direction with a gradient along the y-direction. The equation describing this geometry is

$$B = B_o(1 - \varepsilon y)\hat{z} \tag{1.80}$$

where ε is the inverse magnetic gradient scale length. Assuming there is a balance between particle pressure and magnetic field, the relationship between ϵ and the inverse density gradient scale length $\kappa = (dn/dy)/n$ is

$$\varepsilon = \kappa(\beta_{\perp i} + \beta_{\perp e})/2 \tag{1.81}$$

where $\beta_{\perp i}$ and $\beta_{\perp e}$ are beta of the ion and electron plasmas (ratio of perpendicular particle pressure to magnetic pressure). We can incorporate the gradients into the distribution function consisting of cold plasma and a bi-Maxwellian. The result is

$$f_{jo} = n_c(\delta v_\parallel)\delta(v_\perp) + f_{jo}(v_\parallel, v_\perp)\left[1 + \kappa\left(y - \frac{v_x}{\omega_{cj}}\right)\right] \tag{1.82}$$

where

$$f_{jo}(v_\parallel, v_\perp) = \frac{n_h}{(2\pi^3)^{1/2}v_{\parallel j}v_{\perp j}^2} \exp\left(\frac{v_\parallel^2}{2v_{\parallel j}^2} - \frac{v_\perp^2}{v_{\perp j}^2}\right) \tag{1.83}$$

where n_c and n_h are the cold and hot plasma densities and $v_{\parallel j}$ and $v_{\perp j}$ are the parallel and perpendicular average velocities. The subscript j denotes the particle species. This distribution function has been applied to study how particle fluxes are modulated at the boundary of the plasma sheet (Lin and Parks 1982). The example of the distribution function at boundaries is applicable where there are density and magnetic field gradients like at the magnetopause, ring current, and plasma sheet.

Kappa Distribution Function Finally we present a nonthermal distribution that has been used in the solar wind models. Scudder (1992) has used a truncated Lorentzian (kappa distribution)

$$f_\kappa = \frac{n}{\pi^{3/2}\kappa^{3/2}v_{th}^3} \frac{\Gamma(\kappa + 1)}{\Gamma(\kappa - 1/2)}\left[1 + \frac{v^2}{\kappa v_{th}^2}\right]^{-\kappa+1} \tag{1.84}$$

Here $\Gamma(\kappa + 1)$ is the gamma function, $v_{th}^2 = [(2\kappa - 3/\kappa)(k_B T)/m]$ is the square of the thermal speed, and k_B is the Boltzmann constant. For $v^2 \gg \kappa v_{th}^2$, this distribution behaves like a power law function with a long tail and thus decreases much slower than the Maxwellian. The number flux of escaping particles for the truncated kappa distribution is

$$J_{esc}^{\kappa} = \frac{n_o v_{th}}{\pi^{1/2} \kappa^{3/2}} \frac{\Gamma(\kappa+1)}{\Gamma(\kappa-1/2)} \frac{\kappa}{\kappa-1} \left[1 + \left(\frac{v}{v_{th}}\right)^2\right] \left[1 + \frac{v^2}{\kappa v_{th}^2}\right]^{-1} \qquad (1.85)$$

With $2.62 < \kappa < 3.15$ for electrons and $2.06 < \kappa < 2.69$ for the protons, the kappa distributions can account for equal fluxes leaving the corona to produce the solar wind. This theory further suggests that the hot coronal plasma is part of the tail portion of the nonthermal distribution. This theory does not require a separate heating mechanism such as shock waves or wave-particle interactions, but it does require a source for the production and maintenance of a nonthermal distribution by the Sun. While this is probably not difficult for the dynamic Sun, the source of nonthermal distributions still needs to be identified. Candidate source mechanisms considered include the runaway electrons produced by the Fermi process that may be active on the Sun.

1.9 Concluding Remarks

We have discussed only plasma instruments that have measured particles above the SC potential. However, the thermal component is important and there is very little information about it. Recently, this point was emphasized and we recommend readers to read the article by Yau and Howarth (2016) how thermal particles can be measured. Another type of instrument we have not discussed is the neutral atom imagers. Interested readers are encouraged to read the recent article by McComas et al. (2017) which reported seven years of space observations by a neutral imager.

One of the most important results we showed in this chapter is that in equilibrium plasmas, static electric field cannot be maintained. Hence any existing electric field must come from inductive electric field. This point is important for space plasmas because magnetic field is time varying nearly all the time and in addition space plasmas are moving across interplanetary magnetic fields. Inductive electric field is thus active in all regions of space. An important area of research is to identify the details of the dynamics where and how such fields are induced.

Another important point made in this chapter is that data define the basic properties of the plasmas. We noted that in the case of SW measurements in the vicinity of Earth, there are multiple sources whose fluxes are different by several orders of magnitude and the measurements need to separate them. Thus, Cluster included two ESAs with two different geometrical factors so that each instrument provided "clean" measurements. Cluster also included a time of flight ion mass spectrometer. This instrument is very useful in the plasma sheet and the upper ionosphere for separating the different ion species but in the SW the H^+ ions overwhelm the counts over He^{++} ions and the measurements are confusing and not easily interpreted because there are spill over counts of H^+ in the He^{++} channels.

The general procedure for studying and analyzing data is to first examine the macroscopic parameters and obtain information on density, mean velocities, and temperature. If some interesting features are noted, we follow this up with examination of the distribution function. For example, suppose we see in the solar wind or plasma sheet a sudden increase of the mean velocity $\langle \mathbf{v} \rangle$. To obtain information on what contributed to the increase of the mean velocity, we examine the distribution function. The detailed features in the distribution will reveal what physical changes occurred that increased the mean velocity.

One should be aware that the macroscopic parameters could sometimes give misleading information. For example, consider a situation when we see counter streaming ions flowing along the magnetic field direction. If the intensities and energies of the streaming ions are nearly equal by chance, the mean velocity will vanish. This would indicate there is no interesting physics going on. On the contrary, counter streaming beams are important because they are unstable and could excite waves and heat particles. Multiple population spread over the velocity space will yield a high temperature, which may be incorrect, because integration in velocity is done without taking into account of the two beams. The important point is that a complete picture of space plasma behavior is obtained only by examining the primary measured quantity, which is the distribution function.

Another example of ambiguity that arises in particle measurements can come from the neighborhood of sharp plasma boundaries with dimensions around an ion Larmor radius. At sharp boundaries there is a large gradient across it because particle density on one side is very different from the other side. This situation will result in a velocity space distribution that can be non-gyrotropic, which will yield a large velocity moment even though the plasma is stationary. This important problem has been studied by Lee et al. (2004) and Wilber et al. (2004) who reexamined an event that Runov et al. (2003) have studied to illustrate how confusion can arise only looking at moments. An important lesson learned is that non-gyrotropic distribution can come from remote sensing of the boundary by a detector on the spacecraft immersed in a thin current sheet and that taking velocity moments of such nonuniform distribution will yield a large bulk velocity even when the plasma may be stationary.

1.10   Solutions to Questions in This Chapter

Solution 1.1 Heating is an irreversible process and the Vlasov equation which is time symmetric and reversible cannot explain this phenomenon. Heating involves spreading particles over a larger region of the velocity space. Scattering can be accomplished by either wave-particle interaction or physical collision which requires the Boltzmann collision term, which is absent in the Vlasov equation.

Solution 1.2 She/he deduces that inside the plasma, the charges are responding to the force $\mathbf{F} = q\mathbf{E}$ acting on them and that this electric field inside the plasma is in the opposite direction of the applied field. The charges in the blob will move until the force of the electric field is exactly equal to the applied force so that the *total force on the particles vanishes*. This observer learns that inside an equilibrium plasma a charge imbalance cannot be maintained and the force (therefore the electric field) vanishes. Hence studying electric fields in space must include the inductive effects.

Solution 1.3 That $\mathbf{E}$ is parallel to the force $\mathbf{F}_E$ and the charge q is accelerated in the direction of $\mathbf{E}$. If there are many point charges q_i, the electric field at $\mathbf{r}$ produced by *all of the particles* is $\mathbf{E} = \Sigma q_i(\mathbf{r} - \mathbf{r}')/|\mathbf{r} - \mathbf{r}'|$. A set of discrete point charges in a small volume can be described by the charge density, $\rho(\mathbf{r}) = \Sigma q_i \delta(\mathbf{r} - \mathbf{r}_i)$. The electric field is then given by $\mathbf{E}(\mathbf{r}) = \int d^3\mathbf{r}' \rho(\mathbf{r}')(\mathbf{r} - \mathbf{r}')/|\mathbf{r} - \mathbf{r}'|^3$.

Solution 1.4 The direction of the magnetic field is defined as the direction in which the charge q would move when it experiences no magnetic force. For a continuous distribution of current, we use Biot-Savat's law to compute the magnetic field, $\mathbf{B}(\mathbf{r}) = \int d^3\mathbf{r}' \mathbf{J}(\mathbf{r}')(\mathbf{r} - \mathbf{r}')/|\mathbf{r} - \mathbf{r}'|^3$ where $\mathbf{J}(\mathbf{r}) = \Sigma nq\delta(\mathbf{v} - \mathbf{v_k})$ is the current density. The current $\mathbf{J}$ vanishes in a coordinate frame moving with the charge and therefore $\mathbf{B}$ also vanishes in that frame. But the charge q is still there and so is the electric field $\mathbf{E}$. We can thus look at the electric field $\mathbf{E}$ as a fundamental quantity and the magnetic field $\mathbf{B}$ a consequence of the charges in motion.

Solution 1.5 The right side of the equation shows that the $\mathcal{EMF}$ is induced by changing either the magnetic flux in time, motion, or the shape of the contour. Since $d\Phi/dt = \partial\Phi/\partial t + (\mathbf{V} \cdot \nabla)\Phi$, Faraday's law indicates that $\mathcal{EMF}$ is induced even in steady state if the magnetic field is not uniform and there is a gradient of the magnetic flux along $\mathbf{V}$.

Solution 1.6 The only field present in the S'-frame is the static magnetic field $\mathbf{B}'$ which is stronger than $\mathbf{B}$ by a factor γ and pointing in the same direction as $\mathbf{B}$. The fourth equation in Eq. (1.33) requires that for $\mathbf{B}'_\perp$ to be *real*, $E/B < 1$ or $|\mathbf{E}| < |\mathbf{B}|$. The motion of the particle in S' is described by $d\mathbf{p}'/dt' = q\mathbf{v}' \times \mathbf{B}'$ ($\mathbf{E} = 0$) and thus consists of only the gyration motion around the field $\mathbf{B}'$ with the angular frequency $\omega_c' = (q\mathbf{B}'/\gamma m_o)$. In the S frame, however, this gyration motion is superposed on a uniform $\mathbf{E} \times \mathbf{B}$ drift, $\mathbf{V} = (\mathbf{E} \times \mathbf{B})/B^2$.

Solution 1.7 This problem is important because in magnetic neutral regions, the magnitude of the magnetic field is $|\mathbf{B}| \sim 0$. However, the magnitude of the electric field is finite. Since in the particle S'-frame $d\mathbf{p}'/dt' = q\mathbf{E}'$, the particle is now accelerated continuously in the direction of the electric field $\mathbf{E}'$. This situation is different from the case when $|\mathbf{E}| < |\mathbf{B}|$ where the motion includes only the $\mathbf{E} \times \mathbf{B}$ drift and particles are *not* accelerated.

Solution 1.8 An observe sees the charges are *not* moving, thus $\mathbf{F} = q\mathbf{V} \times \mathbf{B} = 0$. However, since the principle of relativity guarantees the physics will be the same in both frames, the displacement of the charges in the plasma blob must be observed in both S- and S'-frames. Although there is no applied electric field in the S frame

($\mathbf{E} = 0$), special theory of relativity shows that electric field is *induced* in the S' frame ($\mathbf{E}' \neq 0$). Using the relationships of the electric and magnetic fields in the two frames for $|\mathbf{V}| << c$, we see that $\mathbf{E}' \approx \mathbf{E} + \mathbf{V} \times \mathbf{B}$ and $\mathbf{B}' \approx \mathbf{B}$. Thus we can deduce that the components of the *induced electric field* in the S' frame are $E'_y = -V_x B_z$, $B'_z = B_z$ and $E'_y = E'_z = B'_x = B'_y = 0$. There is only the electric force $q\mathbf{E}'$ in the S' frame ($q\mathbf{V} \times \mathbf{B}' = 0$) and this force gives rise to the same charge separation as required.

Solution 1.9 Equation (1.37) can be rewritten by dividing through by V^2/c^2 and putting it in the standard quadric form. The solution of this equation is

$$\left(\frac{V}{c}\right) = \left(\frac{1}{2}\right) \frac{1}{|\mathbf{E} \times \mathbf{B}|} \left[(E^2/c^2 + B^2) \pm \sqrt{(E^2/c^2 + B^2)^2 + 4(\mathbf{E} \cdot \mathbf{B})} \right]$$
(1.86)

The quantities inside the square root sign involve the invariants of the Lorentz transformation for the electric and magnetic fields. Note also that since $\mathbf{E} \cdot \mathbf{B} = 0$ in the S-frame, the velocities are either $V/c = E/B$ or $V/c = B/E$ depending on the magnitude of $\mathbf{E}$ relative to $\mathbf{B}$. The final point to note is that in the case $|\mathbf{E}| = |\mathbf{B}|$, the velocity of the moving frame of reference for the two fields is $V = c$ (Evangelidis 2004).

Solution 1.10 A fundamentally important result obtained by these experiments is that the seat of the $\mathcal{EMF}$ is found in the moving conductor. The $\mathcal{EMF}$ was independent of the rotation of the magnetic field. Without a moving conductor, an $\mathcal{EMF}$ is not produced. The conclusion is that the $\mathcal{EMF}$ is not set up in the stationary part of the unipolar induction experiment. These experiments also established that it is meaningless to talk about field line motion. Readers interested in the behavior of rotating planetary dipoles are recommended to read Chapter 3 of Parks (2004).

Solution 1.11 The unknowns of the 0th order equation $\partial n/\partial t + \nabla \cdot n\mathbf{U} = 0$ are density n and flow speed $\mathbf{U}$. There are four unknowns and only three equations. Taking higher moments does not solve the problem because each order introduces new unknowns. MHD description of space plasma requires all of the velocity moments be computed, which is not practical. In practice only a finite number of moment equations are computed. To obtain closure to the equations, other assumptions must be made about the nature of the plasma. For example, one can assume that the plasma obeys an adiabatic equation of state or that the fluid is an electrical conductor and uses Ohm's law to supplement the conservation equations. However, adiabatic fluids do not allow heat flux to flow which cannot explain heat flux carried by the SW electrons and while Ohm's law works for ordinary conductors, there are no accepted models of conductivity for collisionless plasmas.

Solution 1.12 To answer this question, we discuss what assumptions are made about ESA measurements of particles. First note that the equations for energy per charge of H^+ and $He^{++}(\alpha's)$ in the spacecraft frame are

$$(E/q)_+ = m_+(v_+ + V_{sw})^2/2q_+$$

and

$$(E/q)_\alpha = m_\alpha(v_\alpha + V_{SW})^2/2q_\alpha = m_+(v_\alpha + V_{sw})^2/q_+$$

where m_+ is the mass of H^+, $m_\alpha = 4m_+$, and v_+ and v_α are thermal velocities of H^+ and He^{++} and $q_\alpha = 2q_+$ and V_{sw} is the bulk velocity. Assigning a velocity to the particles is only possible if interpretation of ESA data assumes

> all particles are traveling at the same mean velocity in steady-state plasma with a 'frozen-in'
> magnetic field
> (Hundhausen 1968)

Thus if one assumes that the SW beam is nearly cold (T~ 0) and H^+ and He^{++} are *traveling together*, we can let v_+ and $v_\alpha \ll V_{sw}$ and the above equations reduce to

$$(E/q)_+ = (m_+ V_{sw}^2)/2q_+$$

for H_+ and

$$(E/q)_\alpha = (m_+ V_{sw}^2)/q_+$$

for He^{++}.

These equations show that the energy per charge of He^{++} is twice that of H^+. Thus, if we interpret all particles as H^+ in the velocity space plots, we would find a beam centered at V_{sw} and identify it as H^+ and another beam of "protons" centered at $\sqrt{2}V_{sw}$, which we interpret as $\alpha's$.

The restriction that all particles travel together applies only if the thermal velocity is much less than the bulk motion. Moreover, it applies to motions only when V_{sw} is *perpendicular* to the magnetic field direction. However, the SW distributions relative to **B** can vary and the velocity of *field-aligned* particles can have *any* value and the different species *need not* travel together. If the motion in the direction parallel to the magnetic field can have any value, this could have significant impact in the interpretation of the solar wind which has He^{++} in addition to H^+.

References

Bridge, H., et al.: An instrument for the investigation of interplanetary plasma. J. Geophys. Res. **65**, 3053 (1960)

Chu, C., Ohkawa, T.: Transverse electromagnetic waves with E‖B. Phys. Rev. Lett. **48**, 837 (1982)

Carlson, C., et al.: An instrument for rapidly measuring distribution functions with high resolution. Adv. Space Res. **12**, 67 (1983)

Dory, R.A., et al.: Unstable electrostatic plasma waves propagating perpendicular to a magnetic field. Phys. Rev. Lett. **14**, 131 (1965)

Einstein, A.: On the electrodynamics of moving bodies. Annalen der Physik **322**, 891 (1920). Translated by Meghnad Saha, University of Calcutta, 1920

Evangelidis, E.A.: The structure of fields with electric and magnetic components parallel to each other. In: 31st EPS Conference on Plasma Physics London, 28 June–2 July 2004 ECA, vol. 28G, p. 2066 (2004)

Hundhausen, A.: Direct observations of solar wind particles. Space Sci. Rev. **8**, 690 (1968)

Jackson, J.D.:Classical Electrodynamics, 2nd edn. Wiley, New York (1975)

Landau, L.D., Lifschitz, E.M.: Electrodynamics of Continuous Media. Pergamon Press, Ltd., New York (1962)

Lee, E.S., Wilber, M., Parks, G.K., et al.: Modeling of remote sensing of thin current sheet. Geophys. Res. Lett. **31**, L21086 (2004)

Lin, C.S., Parks, G.K.: Modulation of energetic particle fluxes by a mixed mode of transverse and compressional waves. J. Geophys. Res. **87**, 5102 (1982)

Lin, R.P., et al.: A three-dimensional plasma and energetic particle investigation for the WIND spacecraft. Space Sci. Rev. **71**, 125 (1995)

McComas, D., et al.: Seven years of imaging the global heliosphere with IBEX. Astrophys. J. Suppl. Ser. **29**, 41 (2017)

Møller, C.: The Theory of Relativity. Oxford University Press, London (1952)

Parks, G.K.: Physics of Space Plasmas, An Introduction, 2nd edn. Westview Press, A Member of Perseus Books, Boulder (2004)

Rasio, F., et al.: Solving the Vlasov equation in general relativity. Astrophys. J. **344**, 146 (1989)

Rème, H., et al.: The cluster ion spectrometry (CIS) experiment. Space Sci. Rev. **79**, 303 (1997)

Runov, A., et al.: Current sheet structures near magnetic X-line observed by cluster. Geophys. Res. Lett. **30**, 1579 (2003)

Schiff, L.I.: A question in general relativity. Proc. Natl. Acad. Sci. USA **25**, 391 (1939)

Sckopke, N.: Ion heating at the Earth's quasi-perpendicular bow shock. Adv. Space Res. **15**, 261 (1995)

Schmidt, G.: Physics of High Temperature Plasmas, An Introduction. Academic Press, New York (1966)

Scudder, J.D.: On the causes of temperature change in inhomogeneous low-density astrophysical plasmas. Astrophys. J. **398**, 299 (1992)

Thomsen, M.F., et al.: On the origin of hot diamagnetic cavities near the Earth's bow shock. **93**(11), 311 (1988)

Tolman, R., Relativity. Thermodynamics, and Cosmology. Oxford Press (1934)

Vasyliunas, V.: Deep space plasma measurements. In: Methods of Experimental Physics, vol. 98, p. 49. Elsevier, New York (1971)

Widrow, L.M., Kaiser, N.: Using the Schrödinger equation to simulate collisionless matter. Astrophys. J. **416**, L-71 (1993)

Wilber, M., et al.: Cluster observations of velocity-restricted ion distributions near the plasma sheet. Geophys. Res. Lett. **31**, L24802 (2004)

Wüest, M., Evans, D., von Steiger, R. (eds.): Calibration of Particle Instruments in Space. International Space Sciences Institute, Bern (2007)

Wong, H.K., et al.: Electron cyclotron maser instability caused by hot electrons. Phys. Fluids **28**, 2751 (1985)

Yau, A.W., Howarth, A.: Imaging thermal plasma mass and velocity analyzer. J. Geophys. Res. **121**, 7326 (2016)

Additional Reading

Alfvén, H., Fälthammar, C.-G.: Cosmical Electrodynamics Fundamental Principles, 2nd edn. Oxford University Press, Oxford (1963)

Baumjohann, W., Treumann, R.: Basic Plasma Physics. Imperial College, London (1996)

Birdsall, C.K., Langdon, A.B.: Plasma Physics via Computer Simulation. McGraw-Hill, New York (1985)

Carlson, C.W., et al.: The electron and ion plasma experiment for fast. In: Robert, F.P. Jr. (ed.) The FAST Mission. Kluwer Academic Publishers, Dordrecht (2001)

Cao, J.B., et al.: Kinetic analysis of the energy transport of bursty bulk flows in the plasma sheet. J. Geophys. Res. **118**, 313 (2013)

Chandrasekhar, S.: Plasma Physics. The University of Chicago Press, Chicago (1960)

Chu, C.: Response. Phys. Rev. Lett. **50**, 139 (1983)

Chu, C., Ohkawa, T.: Response to comments by Zaghloul et al. on transverse electromagnetic waves. Phys. Rev. Lett. **58**, 424 (1987)

Dungey, J.W.: Cosmic Electrodynamics. Cambridge University Press, Cambridge (1958)

Feynman, R.P., Leighton, R., Sands, M.: The Feynman Lectures on Physics, Mainly Electromagnetism and Matter. Addison-Wesley, Reading (1964)

Geiss, J., et al.: Apollo I l and 12 solar wind composition experiments: fluxes of He and Neisotopes. J. Geophys. Res. **75**, 5972 (1970)

Gray, E.: Electromagnetic waves with **E** parallel to **B**. J. Phys. A Math. Gen. **25**, 5373 (1992)

Gurnett, D., Bhattacharjee, A.: Introduction to Plasma Physics, With Space and Laboratory Applications. Cambridge University Press, New York (2005)

Krall, N., Trivelpiece, A.: Principles of Plasma Physics. McGraw Hill Book Company, New York (1973)

Khare, A., Pradhan, T.: Transverse electromagnetic waves with finite energy, action and $\int \mathbf{E} \cdot \mathbf{B} d^4 x$. Phys. Rev. Lett. **49**, 1227 (1982)

Khare, A., Pradhan, T.: Response to comments by F. C. Michels. Phys. Rev. Lett. **52**, 1352 (1984)

Kivelson, M.G., Russell, C.T.: Introduction to Space Physics. Cambridge University Press, Cambridge (1995)

Lee, K.K.: Comments on transverse electromagnetic waves with $\mathbf{E} \parallel \mathbf{B}$. Phys. Rev. Lett. **50**, 138 (1983)

Longmire, C.L.: Elementary Plasma Physics. Interscience Publishers, New York (1963)

Maxwell, J.C.: A Treatise on Electricity and Magnetism, vols. 1 and 2. Dover Publications, New York (1954)

Michel, F.C.: Comment on transverse electromagnetic waves with non-zero $\mathbf{E} \cdot \mathbf{B}$. Phys. Rev. Lett. **52**, 1351 (1984)

Montgomery, D.C., Tidman, D.A.: Plasma Kinetic Theory. McGraw-Hill, New York (1964)

Nicholson, D.R.: Introduction to Plasma Theory. Wiley, New York (1983)

Paschmann, G., Daly, P.: Analysis methods for multi-spacecraft data. International Space Science Institute, Bern (2000)

Pfaff, R., J. Borovosky, D. Young, (eds.): Measurement Techniques in Space Plasmas. American Geophysical Union, Washington (1998)

Rosser, W.G.V.: An Introduction to the Theory of Relativity. Butterworths, London (1964)

Shimoda, K., et al.: Electromagnetic plane waves with parallel electric and magnetic fields $\mathbf{E} \parallel \mathbf{H}$ in free space. Am. J. Phys. **58**, 394 (1990)

Tollman, R.: Relativity, Thermodynamics and Cosmology. Oxford Press, Oxford (1934)

Zaghloul, H., et al.: Comment on "Transverse Electromagnetic Waves with E∥B. Phys. Rev. Lett. **58**, 423 (1987)

Chapter 2
Charged Particle Acceleration

2.1 Introduction

The momentum and energy of a charged particle obey the equations,

$$\frac{d\mathbf{p}}{dt} = q(\mathbf{E} + \mathbf{v} \times \mathbf{B})$$

$$\frac{d\mathcal{E}}{dt} = e\mathbf{v} \cdot \mathbf{E} \tag{2.1}$$

where the momentum $\mathbf{p} = \gamma m_o \mathbf{v}$ and $\gamma = (1 - v^2/c^2)^{-1/2}$. The kinetic energy $\mathcal{E}$ of the particle changes in time and energy is not conserved when an electric field is present along $\mathbf{v}$. Unlike a particle moving in magnetic field only, there are no simple solutions for the velocity of the particle. Understanding the motions of charged particles in $\mathbf{E}$ and $\mathbf{B}$ fields is fundamental for space plasmas.

The Lorentz equation is nonlinear because of the factor γ, which depends on $|\mathbf{v}|$. The equation is difficult to solve even with uniform and static electric and magnetic fields. An important point not considered in the introductory textbooks is that the solutions of the Lorentz equation depend on the relative amplitudes of $\mathbf{E}$ and $\mathbf{B}$. Analytical solutions obtained by Takeuchi (2002) show that particles can be accelerated in the $\mathbf{E} \times \mathbf{B}$ direction when $|\mathbf{E}| \geq |\mathbf{B}|$. The acceleration by $\mathbf{E} \times \mathbf{B}$ has not been studied extensively but it could be important near current sheet regions where the magnetic field is vanishingly small. The relativistic Lorentz equation when $|\mathbf{E}| = |\mathbf{B}|$ was solved by Landau and Liftschitz (1962) and for arbitrary $|\mathbf{E}|$ and $|\mathbf{B}|$ by Takeuchi in 2002 (Takeuchi 2002). Jackson (1975) has discussed briefly the solution for this case in the S' frame, but not in the S frame. The solution in the S-frame is important for then one can see the particle trajectories and also how the energy of the particles change. The problem examined by Landau and Liftschitz (1962) leads to a cubic equation and Cardan's method is used to solve the cubic equation.

© Springer Nature Switzerland AG 2018
G. K. Parks, *Characterizing Space Plasmas*, Astronomy and Astrophysics Library,
https://doi.org/10.1007/978-3-319-90041-4_2

The Lorentz equation has been successful in organizing the charged particles in the radiation belts. This chapter begins with a summary of the basic motion of charged particles in homogeneous electric and magnetic fields. The solutions of the Lorentz equation in uniform electric and magnetic fields are examined when $\mathbf{E} \cdot \mathbf{B} = 0$ for cases when $|\mathbf{E}| = |\mathbf{B}|$ and for arbitrary values of $|\mathbf{E}|$ and $|\mathbf{B}|$. We also discuss the dynamics of particles when $\mathbf{E}$ is parallel $\mathbf{B}$, treated briefly in Landau and Liftschitz (1962). This is followed with a review of particle motion in inhomogeneous electric and magnetic fields. The next part discusses several well-known acceleration mechanisms including the betatron and Fermi processes. To show how gamma ray bursts could be produced by relativistic electrons interacting with a sea of photons, we include a short discussion of the inverse Compton mechanism. Then a short discussion is given of electrostatic and electromagnetic waves and theory of wave-particle interaction. The chapter concludes with the theory of kilometric radiation associated with auroral particle acceleration and precipitation.

2.2 Motion in Uniform E and B Field

Consider a charge q of mass m and velocity $\mathbf{v}$ in an inertial frame S moving in constant uniform magnetic field $\mathbf{B}$ in the absence of an electric field ($\mathbf{E} = 0$). The relevant equations are $d\mathbf{p}/dt = q\mathbf{v} \times \mathbf{B}$ and $d\mathcal{E}/dt = 0$ where the symbols have already been defined. When there is no electric field, the energy of the particle remains constant in time. Since energy is conserved, $d\mathcal{E}/dt = \gamma m_o d\mathbf{v}/dt = 0$.

2.2.1 Uniform Magnetic Field

Let the uniform and time independent field $\mathbf{B} = B\mathbf{e}_3$ and $\mathbf{v} = v_x\mathbf{e}_1 + v_y\mathbf{e}_2 + v_z\mathbf{e}_3$ where the $\mathbf{e}_i$ are unit vectors. The components of the Lorentz equation are $dv_x/dt = \omega_c v_y$ and $dv_y/dt = -\omega_c v_x$ and $dv_z/dt = 0$, where

$$\omega_c = \frac{qB}{\gamma m_o} \tag{2.2}$$

is the cyclotron frequency (Larmor frequency). In the general case, the cyclotron frequency is velocity depend and different from the nonrelativistic case when $\gamma = 1$. The equation of motion then yields a second order differential equation that describes a simple harmonic motion of v_x and v_y. Define now $\mathbf{v} = \mathbf{v}_\parallel + \mathbf{v}_\perp$ where the subindices $\parallel$ and $\perp$ denote directions parallel and perpendicular to B. The magnitude of the velocity vector $|\mathbf{v}| = (v_\parallel^2 + v_\perp^2)^{1/2}$. The solution to the Lorentz equation can then be written as

$$\mathbf{v(t)} = v_\parallel\mathbf{e}_3 + \rho_c\omega_c(\mathbf{e}_1 + \mathbf{e}_2)e^{-i\omega t} \tag{2.3}$$

where $\rho_c = \gamma_\perp/qB = p_\perp/qB$ is the magnitude of the cyclotron radius and e_1 and e_2 are unit vectors orthogonal to $\mathbf{B}$. The cyclotron radius can be written generally as

$$\rho_c = \frac{\gamma m_o}{qB^2}(\mathbf{v} \times \mathbf{B}) = \frac{\mathbf{p} \times \mathbf{B}}{qB^2} \tag{2.4}$$

Note that ρ_c depends on $\mathbf{B}$ and $\mathbf{p}$. To arrive at this equation, we have used the relation $e^{1\omega t} = \cos \omega t + i \sin \omega t$ and $i = \sqrt{-1}$ and take only the real part of the solution.

Integration of the velocity yields $\mathbf{r}(t) = \mathbf{r}_o + i_c(\mathbf{e}_1 + \mathbf{e}_2)e^{-i\omega_c t}$. The trajectory of the particle is a helix with a cyclotron radius ρ_c and pitch-angle

$$\alpha = \tan^{-1}\left(\frac{\omega_c \rho_c}{v_\parallel}\right) \tag{2.5}$$

These equations show a charge article in constant uniform magnetic field $\mathbf{B}$ moves along a helical path with its axis in the direction of the magnetic field. The velocity and angular frequency are constant. In the frame moving with the particle (S' frame), the path is a circle with the radius ρ_c. Using the definition ω_c and ρ_c, we obtain the familiar equation $\alpha = v_\perp/v_\parallel$. Since $v_\parallel = v \cos \alpha$ and $v_\perp = v \sin \alpha$, we find that when $\alpha = \pi/2$, $v_\parallel = 0$ and when $\alpha = 0$, $v_\perp = 0$.

The perpendicular motion gives rise to a circulating current $I = q\omega_c/2\pi = q^2 B/\pi\gamma m_o$ and the magnetic moment of this "ring current" is $\mu = I\pi\rho_c^2 = q^2\rho_c^2 B/\gamma m_o$. The magnetic moment can also be written in terms of the total magnetic flux within the circular path as $\mu = q^2\Phi/2\pi\gamma m_o$ where $\Phi = \pi\rho_c^2 B$. A useful expression of the magnetic moment is

$$\mu = \frac{1}{2}\frac{p_\perp^2}{\gamma m_o B} \tag{2.6}$$

The magnetic moment (Eq. (2.6)) reduces to the more familiar form $\mu = mv_\perp^2/2B = W_\perp/B$ if $v \ll c$. Here $W_\perp$ is the kinetic energy of the motion perpendicular to the magnetic field and Eq. (2.6) is an adiabatic constant of motion when the magnetic field $\mathbf{B}$ is uniform or varying *slowly* in time and space.

Question 2.1 Cyclotron frequencies and pitch-angles of particles are measured by experiments on the spacecraft, which can be moving with respect to the plasma in which the spacecraft is immersed. In which frame of reference should these quantities be represented? Why?

2.2.2   Time Dependent Magnetic Field

What happens to particle motion when the magnetic field is varying in time? The motion of a particle in time dependent magnetic field must take account of both

electric and magnetic fields. There is now electric field induced by the changing
magnetic field. The total energy of the particle is not conserved in time dependent
magnetic fields (Alfvén and Fälthammar 1963).

Let us now calculate how much work is done on the charged particles by the
induced $\mathcal{EMF}$. Assume that **B** is constant in space but varying slowly in time. Set
up a coordinate system with the origin at the GC so the particles gyrate around the
GC in closed circles. The amount of work done on the particles in one gyration is

$$\Delta(\gamma mc^2) = -q \oint_c \mathbf{E} \cdot d\mathbf{s}$$

$$= \pi \rho q^2 \frac{\partial B}{\partial t} \tag{2.7}$$

where use was made of the relations $\oint \mathbf{E} \cdot d\mathbf{s} = -\partial \phi / \partial t$ and $\phi = \pi \rho^2 B$ is the
magnetic flux enclosed within the gyroradius. The average time derivative of the
energy is

$$\frac{d}{dt}(\gamma mc^2) = \frac{\Delta(\gamma mc^2)}{T_g} \tag{2.8}$$

where $T_g = \gamma m / q B$ is the gyro-period. Combining the above equations yields

$$\frac{d}{dt}(\gamma m) = \frac{p_\perp^2}{2\gamma m B} \frac{\partial B}{\partial t} \tag{2.9}$$

However, since $\gamma = (1 - v^2/c^2)^{-1/2} = (1 + p^2/m^2c^2)^{1/2}$, we can differentiate this
and obtain

$$\frac{d}{dt}(\gamma m) = \frac{1}{2\gamma m} \frac{dp_\perp^2}{dt} \tag{2.10}$$

In deriving this equation, we replaced p with $p_\perp$ since the acceleration is in the
perpendicular direction. These differential equations can be solved given initial
and boundary conditions. However, these equations also show that the $p_\perp^2/B$ is
conserved and we can use this invariant equation to see how particles can be
accelerated. This conservation equation is often invoked to accelerate particles in
the radiation belts.

Now let $\partial B/\partial t = dB/dt$ and combine Eqs. (2.9) and (2.10). We then obtain

$$\frac{d}{dt}\left(\frac{p_\perp^2}{B}\right) = 0 \tag{2.11}$$

indicating $p_\perp^2/B$ is a constant of motion, an adiabatic invariant. This equation also
implies that the magnetic flux in the particle's orbit remains constant, that is, $B\pi \rho_c^2$

is constant. If particles are tied to B and drift to some higher value of B, $p_\perp^2$ will increase. Particles can be accelerated very efficiently in radiation belts because B in dipole fields depend on r^3. For example, if particles on the equatorial plane are injected at $10R_E$ and allowed to radially drift earthward to $4R_E$ by a dawn-dusk electric field conserving this invariant, $p_\perp^2$ at $4R_E$ will have increased by 1.56×10^4 of the value $p_\perp^2$ had at $10R_E$.

Question 2.2 Describe the physical details of how the induced electric field accelerates particles in the above example and derive the equation for the momentum change.

2.2.3 Mass Dependence

The mass dependence can be important in the relativistic limit where the mass is velocity dependent. The time variation of the momentum is

$$\frac{d\mathbf{p}}{dt} = m\frac{d\mathbf{v}}{dt} + \mathbf{v}\frac{dm}{dt} \tag{2.12}$$

The second term shows if the acceleration of the particle changes the velocity of the particle, there will be a corresponding change in the mass of the particle.

Now since the total energy $\mathcal{E} = mc^2 = \mathcal{E}_{kin} + m_oc^2$ where $\mathcal{E}_{kin}$ is the kinetic energy and m_oc^2 is the rest mass energy, we see that $dm/dt = (1/c^2)d\mathcal{E}/dt = (1/c^2)d/dt(\mathcal{E}_{kin} + m_oc^2) = (1/c^2)d\mathcal{E}_{kin}/dt$. Also, since the change of the kinetic energy is equal to the work done on the particle, we have $d\mathcal{E}_{kin}/dt = \mathbf{F} \cdot \mathbf{v}$, $dm/dt = (1/c^2)\mathbf{F} \cdot \mathbf{v}$ where $\mathbf{F}$ is the force acting on the particle. Hence, the momentum conservation equation of relativistic particles is

$$\frac{d\mathbf{p}}{dt} = m\frac{d\mathbf{v}}{dt} + \mathbf{v}\frac{\mathbf{F} \cdot \mathbf{v}}{c^2} \tag{2.13}$$

Note that the force $\mathbf{F} = d\mathbf{p}/dt = m\mathbf{a}$ or acceleration $\mathbf{a}$ is not parallel to $\mathbf{v}$ as in the nonrelativistic case.

The relativistic mass effects are important in space and stellar dynamics. For example, in wave and particle interactions (see Sects. 2.6 and 2.7 below), the wave and particle are in resonance when

$$\omega + k_\| v_\| = \Omega_c \tag{2.14}$$

Here ω is the wave frequency, $k_\|$ and $v_\|$ are the wave number and particle velocity parallel to the magnetic field direction, and Ω_c is the cyclotron frequency of the particle. In the frame of the particle, the wave and particle are coupled when

$$\omega = \Omega_c$$
$$= \frac{qB}{\gamma m_o} = \frac{qB}{m_o}(1 - v^2/c^2)^{-1/2} \tag{2.15}$$

Electrons have speeds of ~0.98c at 2 MeV and the speed does not change and remains constant above this energy (protons reach 0.98c at ~4 BeV). The cyclotron radius of electrons $\rho_c = mv/qB$ therefore remains nearly constant for energies higher than 2 MeV. When electrons move in circular orbit, they emit synchrotron radiation due to centripetal acceleration and this is important for solar and astrophysical plasmas (radiation effects will not be discussed).

Question 2.3 Equation (2.15) shows that as v increases the cyclotron frequency will increase (assuming B remains constant). Discuss this problem considering a 50 keV electron. Do the readers think the relativistic effects for a 50 keV electron important?

2.3 E × B Acceleration

We examine below two possible cases when the particles are accelerated in the $|\mathbf{E} \times \mathbf{B}|$ direction. The first is when $|\mathbf{E}| = |\mathbf{B}|$ and the other is when $|\mathbf{E}| > |\mathbf{B}|$.

2.3.1 $|\mathbf{E}| = |\mathbf{B}|$

Consider now the case when $|\mathbf{E}| = |\mathbf{B}|$. The work of Landau and Liftschitz (1962) is reproduced here. In this case, particles can be accelerated by the $\mathbf{E} \times \mathbf{B}/B^2$ drift when the relativistic factor γ is included. Choose $\mathbf{B} = (0, 0, B)$ and $\mathbf{E} = (0, E, 0)$ and we let $|\mathbf{B}| = |\mathbf{E}|$. Then the components of the momentum equation become

$$\frac{dp_x}{dt} = qBv_y = qv_yE$$

$$\frac{dp_y}{dt} = -q(E - v_xB) = qE(1 - v_x)$$

$$\frac{dp_z}{dt} = 0 \qquad\qquad (2.16)$$

where $\mathbf{p} = \gamma m_o\mathbf{v}$ and we replaced B by E. The momentum p_x which is perpendicular to both $\mathbf{E}$ and $\mathbf{B}$ changes in time, hence work is being done in the $\mathbf{E} \times \mathbf{B}$ direction. We also see that p_y changes in time and here the particle is accelerated in the direction of the electric field $\mathbf{E} = (0, E, 0)$. The third equation shows $p_z =$ constant.

The total energy is $\mathcal{E}^2 = p^2c^2 + m_o^2c^4$ and differentiating $\mathcal{E}^2$ we obtain

$$\mathcal{E}\frac{d\mathcal{E}}{dt} = pc^2\frac{dp}{dt} \qquad\qquad (2.17)$$

Since $\mathbf{p} = \gamma m_0 \mathbf{v}$ and $\mathcal{E} = \gamma m_0 c^2$, eliminating γ and substituting the results in (2.17) yield

$$\frac{d}{dt}\mathcal{E} = \mathbf{v} \cdot \frac{d}{dt}\mathbf{p} \tag{2.18}$$

However, since $d\mathbf{p}/dt = q\mathbf{E}$, we can write

$$\frac{d}{dt}\mathcal{E} = q\mathbf{v} \cdot \mathbf{E}$$

$$= qv_y E$$

$$= \frac{eE}{\gamma m_o}p_y \tag{2.19}$$

where $p_y = \gamma m_0 v_y$. Integration of this equation yields

$$\mathcal{E} = \frac{eE}{\gamma m_o}p_y t \tag{2.20}$$

Use the relation $\mathcal{E} = \gamma m_o c^2$ and rewrite (2.20) as

$$\gamma^2 = \frac{eE}{m_o^2 c^2}p_y t \tag{2.21}$$

This equation will be useful when we compare it to the results of Takeuchi.
 Since the first equation of Eq. (2.16) shows $qv_y E = dp_x/dt$,

$$\frac{d}{dt}(\mathcal{E} - p_x) = 0 \tag{2.22}$$

where we used Eq. (2.25). Hence, $(\mathcal{E} - p_x)$ is constant and we let this constant be α. A useful algebraic relation is

$$(\mathcal{E} + p_x)(\mathcal{E} - p_x) = (\mathcal{E}^2 - p_x^2)$$

$$= c^2 p^2 + m_o^2 c^4 - p_x^2$$

$$= c^2 p_y^2 + c^2 p_z^2 + m_o^2 c^4$$

$$(\mathcal{E} + p_x)\alpha = c^2 p_y^2 + \mathcal{E}_z^2 \tag{2.23}$$

where we have used the relation $p^2 = p_x^2 + p_y^2 + p_z^2$ and we have defined $\mathcal{E}_z^2 = c^2 p_z^2 + m_o^2 c^4$, which is constant because p_z is constant. Now solve for $\mathcal{E}$ and obtain

$$\mathcal{E} = \frac{1}{\alpha}(c^2 p_y^2 + \mathcal{E}_z^2) - p_x$$

$$= \frac{1}{\alpha}(c^2 p_y^2 + \mathcal{E}_z^2) - (\alpha - \mathcal{E})$$

$$= \frac{\alpha}{2} + \frac{c^2 p_y^2 + \mathcal{E}_z^2}{2\alpha} \tag{2.24}$$

where the last line is obtained rearranging the second line. Note that the total energy $\mathcal{E}$ is proportional to p_y^2. We can obtain the expression for p_x in Eq. (2.23) using p_y from Eq. (2.24). This yields, letting $(\mathcal{E}_{kin} - p_x) = \alpha$,

$$p_x = -\frac{\alpha}{2} + \frac{c^2 p_y^2 + \mathcal{E}^2}{2\alpha c} \tag{2.25}$$

which has a p_y dependence. Consider now

$$\mathcal{E}\frac{dp_y}{dt} = eE(\mathcal{E} - v_x\mathcal{E})$$

$$= eE(\mathcal{E} - p_x)$$

$$= eE\alpha \tag{2.26}$$

where we have used the second equation of Eq. (2.19) and the second line follows because $eEv_x\mathcal{E} = eEv_x\gamma m_o c^2 = eEp_x c^2$. Use now Eq. (2.24) and obtain

$$eE\alpha = \left(\frac{\alpha}{2} + \frac{c^2 p_y^2 + \mathcal{E}_z^2}{2\alpha}\right)\frac{dp_y}{dt} \tag{2.27}$$

This differential equation can be solved to yield

$$2eEt = p_y^2\left(1 + \frac{\mathcal{E}_z^2}{\alpha^2}\right) + \frac{c^2}{3\alpha^2}p_y^3 \tag{2.28}$$

This is a cubic equation in p_y and we see that it increases along the electric field direction $(0, E, 0)$ without bounds as time t increases. We can solve for p_y and use it in Eq. (2.25) and study how p_x changes in time.

The trajectories of the particle calculated using the relation $dx/dt = c^2 p_x/\mathcal{E}$, $dy/dt = c^2 p_y/\mathcal{E}$ and $dz/dt = c^2 p_z/\mathcal{E}$. We then use Eq. (2.25) and also transform the variable using the relation $dt = \mathcal{E}dp_y/eE\alpha$, and after integration obtain

$$x = \frac{c}{2eE}\left(-1 + \frac{\mathcal{E}^2}{\alpha^2}\right)p_y + \frac{c^3}{6\alpha^2 eE}p_y^3$$

$$y = \frac{c^2}{2\alpha eE}p_y^2$$

$$z = \frac{p_z c^2}{eE\alpha}p_y \tag{2.29}$$

These equations together with Eq. (2.28) determine completely the motion of the particle in parametric form with p_y as the parameter. From Eq. (2.29), we see that $v_x = dx/dt$ increases most rapidly in the $\mathbf{E} \times \mathbf{B}$ direction (x-direction), perpendicular to both $\mathbf{E}$ and $\mathbf{B}$.

Cubic equations were known to the ancient Greeks, Babylonians, and Egyptians. In the seventh century, a Chinese mathematician and astronomer Wang Xiatong of Tang Dynasty in a mathematical treatise titled *Jigu Suanjing* solved 25 cubic equations of the form $x^3 + qx^2 + px + d = 0$. We will not solve Eq. (2.28) but anyone interested in solving this equation can use G. Cardan's formulas (see Uspensiky 1948). Cardan was an Italian algebraist who lived from 1501 to 1576. The next section summarizes his method.

Solving Cubic Equation: Cardan's Formula

We now briefly discuss the method for solving the cubic equation (Eq. (2.28)). Consider the general cubic equation of the form

$$f(x) = x^3 + ax^2 + bx + c = 0 \tag{2.30}$$

To simplify this equation, introduce new unknowns to this equation $x = y + k$ with k arbitrary and use Taylor's formula and obtain

$$f(y + k) = f(k) + f'(k)y + \frac{f''(k)}{2}y^2 + \frac{f'''(k)}{6}y^3 \tag{2.31}$$

where

$$f(k) = k^3 + ak^2 + bk + c$$
$$f'(k) = 3k^2 + 2ak + b$$
$$f''(k) = 2(3k + a)$$
$$f'''(k) = 6 \tag{2.32}$$

Let us eliminate the term involving y^2 and choose k so that

$$3k + a = 0 \tag{2.33}$$

so that $k = -a/3$. Then substituting this into the above equations shows

$$f'(-a/3) = b - a^2/3$$
$$f'(-a/3) = c - be/3 + 2a^3/27 \tag{2.34}$$

Then substitute $x = y - a/3$ into the cubic equation and transform it to

$$y^3 + py + q = 0 \tag{2.35}$$

where

$$p = b - a^3/3$$
$$q = c - ba/3 + 2a^3/27 \tag{2.36}$$

We now solve the transformed cubic equation by setting $y = u + v$. Substitution of this yields

$$u^3 + v^3 + (p + 3uv)(u = v) + q = 0 \tag{2.37}$$

with unknowns u and v. Thus unless we have another equation, the problem is indeterminate. Choose

$$3uv + p = 0 \tag{2.38}$$

Then (2.37) becomes

$$u^3 + v^3 = -q \tag{2.39}$$

The solutions to the cubic equation (2.35) can be obtained by solving

$$u^3 + v^3 = -q$$
$$u^3 v^3 = -p/27 \tag{2.40}$$

where the last equation was obtained by cubing $uv = -p/3$. Equation (2.40) gives the sum and the product of the unknowns u^3 and v^3. These unknown quantities are roots of the quadric equation

$$t^3 = qt + p^3/27 = 0 \tag{2.41}$$

whose solutions are

$$A = -q/2 + (q^2/4 + p^3/27)^{1/2}$$
$$B = -q/2 - (q^2/4 + p^3/27)^{1/2} \tag{2.42}$$

Now we can set $u^3 = A$ and $v^3 = B$. Let the three roots of u and v be denoted by $\sqrt[3]{A}$ and $\sqrt[3]{B}$. Then the three roots for u are

$$u = A^{1/3}, u = \omega A^{1/3}, u = \omega^2 A^{1/3} \tag{2.43}$$

where

$$\omega = \frac{-1 + i\sqrt{3}}{2} \tag{2.44}$$

The solutions of v can also have three values, $v = B^{1/3}$, $v = \omega B^{1/3}$, $v = \omega^2 B^{1/3}$. However, not every one of these can be associated with the three possible values of u since u and v must satisfy the relation $uv = -p/3$. If B stands for that cube root which satisfies the relation $A^{1/3} B^{1/3} = -p/3$, then the values of v that can be associated with $u's$ in (2.43) will be

$$v = B^{1/3}, v = \omega B^{1/3}, v = \omega^2 B^{1/3} \tag{2.45}$$

The solutions of (2.35) are

$$y_1 = A^{1/3} + B^{1/3}$$
$$y_2 = \omega A^{1/3} + \omega^2 B^{1/3}$$
$$y_3 = \omega^2 A^{1/3} + \omega B^{1/3} \tag{2.46}$$

These formulas are known as Cardan's formulas. We remind the readers that while $A^{1/3}$ can be chosen arbitrarily among the possible cube roots of A, $B^{1/3}$ must be chosen so that $A^{1/3} B^{1/3} = -p/3$.

2.3.2 Arbitrary $|E|$ and $|B|$

The solution for this general case involves a considerable amount of messy algebra and the physics is hidden in the mathematical details. Nevertheless, this calculation is important and we present the work of Takeuchi (2002). We will use the Gaussian units as in the original paper so that readers interested in consulting the paper can do so without confusion. Some of the missing algebra steps have been added so the derivations can be more easily followed.

Let $\mathbf{E} = (E_o, 0, 0)$ and $\mathbf{B} = (0, 0, -B_o)$. We need to solve the momentum equation $md(\gamma \mathbf{v})/dt = q(\mathbf{E} + \mathbf{v} \times \mathbf{B})/c$. With the assumed static electric and magnetic fields, the x component of the Lorentz equation is reduced to

$$\frac{d}{dt}(\gamma v_x) = (q/m_o)(E_o - v_y B_o/c)$$

$$= \frac{q B_o}{m_o c}(c E_o/B_o - v_y)$$

$$= \Omega(c\tilde{E} - v_y) \tag{2.47}$$

where $\Omega = q B_o/m_o c$ and $\tilde{E} = E_o/B_o$. Now let $\Omega dt = d\tau$ and define $\beta_x = v_x/c$, $\beta_y = v_y/c$, $\beta_z = v_z/c$. The three components of the momentum equation are

$$\frac{d}{d\tau}(\gamma \beta_x) = \tilde{E} - \beta_y$$

$$\frac{d}{d\tau}(\gamma\beta_y) = \beta_x$$

$$\frac{d}{d\tau}(\gamma\beta_z) = 0 \tag{2.48}$$

The energy equation $d(\gamma m_o c^2)/dt = q\mathbf{E} \cdot \mathbf{v}$ can be written as $d\gamma/dt = \Omega\tilde{E}\beta_x$. Hence we also have

$$\frac{d\gamma}{d\tau} = \tilde{E}\beta_x \tag{2.49}$$

The task is to solve these coupled nonlinear differential equations for various values of $\tilde{E} = E_o/B_o$ including the general cases when $\tilde{E} < 1, 0$ and > 1.

Now define new variables, $\beta_x = d\xi/d\tau$, $\beta_y = d\eta/d\tau$, $\beta_z = d\zeta/d\tau$, $\xi - \xi_o = X$, $\eta - \eta_o = Y$, $\zeta - \zeta_o = Z$. Then integration of Eqs. (2.48) and (2.49) yields

$$\gamma\beta_x = \gamma_o\beta_{xo} + \tilde{E}\tau - (\eta - \eta_o) = G$$

$$\gamma\beta_y = \gamma_o\beta_{yo} + (\xi - \xi_o) = d + X$$

$$\gamma\beta_z = \gamma_o\beta_{zo} = k$$

$$\gamma = \gamma_o + \tilde{E}(\xi - \xi_o) = \gamma_o + \tilde{E}X \tag{2.50}$$

where $x_o, y_o, \gamma_o, \beta_{xo}, \beta_{yo}, \beta_{zo}$ are integration constants. Define a modified Lorentz factor,

$$\gamma^2 - 1 = (\gamma\beta_x)^2 + (\gamma\beta_y)^2 + (\gamma\beta_z)^2 \tag{2.51}$$

We can rewrite (2.51) using (2.50) and obtain

$$(\gamma_o + \tilde{E}X)^2 - 1 = [(\gamma_o\beta_{ox}) + Y + (\tilde{E}\tau)]^2 + (\gamma_o\beta_{yo})^2 + 2\gamma_o\beta_{yo}X + (\gamma_o\beta_{zo})^2 \tag{2.52}$$

where $\gamma_o^2 - 1 = (\gamma_o\beta_{xo})^2 + (\gamma_o\beta_{yo})^2 + (\gamma_o\beta_{zo})^2$. Rearranging, we obtain

$$G^2 = (\tilde{E}^2 - 1)X^2 + 2\gamma_o(1 - \beta_{yo})X \tag{2.53}$$

The first equation of (2.50) also shows that $G^2 = (\gamma_o\beta_{xo} - Y + \tilde{E}\tau)^2$. Then, Eq. (2.53) can be rewritten as

$$G = \sqrt{(aX^2 + bX + c)} = f_1(X) \tag{2.54}$$

where $a = \tilde{E}^2 - 1, b = 2\gamma_o(\tilde{E} - \beta_{yo}), c = \gamma_o^2\beta_{xo}^2$.

The equations in (2.50) can be rewritten to yield the following differential equations,

$$\frac{\gamma \beta_y}{\gamma \beta_x} = \frac{dY}{dX} = \frac{X+d}{f_1(X)}$$

$$\frac{\gamma \beta_z}{\gamma \beta_x} = \frac{dZ}{dX} = \frac{k}{f_1(X)}$$

$$\frac{\gamma}{\gamma \beta_x} = \frac{d\tau}{dX} = \frac{\tilde{E}X + \gamma_o}{f_1(X)} \qquad (2.55)$$

where $k = \gamma_o \beta_{zo}$, $d = \gamma_o \beta_{yo}$ and $f_1(X) = G = \sqrt{(aX^2 + bX + c)}$. These differential equations have exact solutions. Equations in (2.55) can be written as

$$Y = \int \frac{X+d}{\sqrt{(aX^2 + bX + c)}} dX \qquad (2.56)$$

$$Z = \int \frac{k \, dX}{\sqrt{(aX^2 + bX + c)}} \qquad (2.57)$$

$$\tau = \int \frac{\tilde{E}X + d}{\sqrt{aX^2 + bX + C}} dX \qquad (2.58)$$

Question 2.4 Solve the above equations that involve indefinite integrals.

2.3.3 Particle Trajectories

We now examine the above equation for the three cases, when $\tilde{E} < 0$, $\tilde{E} = 0$ and $\tilde{E} > 0$ where $\tilde{E}$ is the ratio of the electric field to magnetic field, E_o/B_o.

2.3.4 $a = \tilde{E}^2 - 1 < 0$

Since $\tilde{E} = E_o/B_o$, this situation corresponds to the case $E_o < B_o$. This case corresponds to the well-known case of the $\mathbf{E} \times \mathbf{B}$ drift treated in many text books. Now examine Eq. (2.53). The particle is in the XY plane and satisfies the equation

$$(Y - \gamma_o \beta_{xo} - \tilde{E}\tau)^2 - a \left(X + \frac{b}{2a} \right)^2 = c - \frac{b^2}{4a} \qquad (2.59)$$

The first term uses the definition of $G = (Y - \gamma_o \beta_{xo} - \tilde{E}\tau)$ and the other two terms use the relation $G^2 = aX^2 + bX + c$. This equation describes an ellipse in the XY plane and the trajectory includes the $\mathbf{E} \times \mathbf{B}$ drift and pure gyration of the particles in the drift frame. The GC moves in the Y-direction with a constant velocity $V_{GC} = cE_o/B_o$. In the limit $\tilde{E} = 0$, the trajectory becomes a circle and corresponds to cyclotron motion and obeys the equation

$$(Y - \gamma_o \beta_{xo})^2 + (X + \gamma_o \beta_{yo})^2 = \gamma_o^2 (\beta_{xo}^2 + \beta_{yo}^2) \tag{2.60}$$

The definitions of a, b, and s have been used. Use of the above equations will show that the trajectory for $\tilde{E} = 0$ in the Z-direction is reduced to,

$$Z = \beta_{zo} \tau \tag{2.61}$$

2.3.5 $a = \tilde{E}^2 - 1 > 0$

This situation corresponds to when $E_o > B_o$. In this case, the electric force is stronger than the magnetic force and the particle cannot gyrate round the magnetic field and the particle moves linearly. In the first equation of Eq. (2.55), the term that dominates is $f_1(X)$ which is $G = (aX^2 + bX + c)^{1/2} \approx aX$. Hence,

$$Y \approx X/\sqrt{a}$$

$$= X/\sqrt{\tilde{E}^2 - 1}$$

$$Z = \frac{k}{\sqrt{a}} \ln |4aX|$$

$$= \frac{\gamma_o \beta_{zo}}{\sqrt{\tilde{E}^2 - 1}} \tag{2.62}$$

For the Z equation, we used the approximation $I(X) \approx (1/\sqrt{a}) \ln(2aX + 2aX)$.

2.3.6 $a = \tilde{E}^2 - 1 = 0$

This case corresponds to when the electric and magnetic fields are equal, $E_o = B_o$. Note that in this case, $G = \sqrt{bX + c} = f_2(X)$. The first two differential equations in (2.55) become

$$Y = \int \frac{X + d}{\sqrt{bX + c}} dX \tag{2.63}$$

and

$$Z = k \int \frac{dX}{\sqrt{bX + c}} \tag{2.64}$$

We use the equation corresponding to the case when $a = 0$ and obtain the trajectories of the particle as

$$Y = \frac{2X}{3b} f_2(X) + \frac{6bd - 4c}{3b^2}[f_2(X) - f_o]$$

$$Z = \frac{2k}{b}[f_2(X) - f_o] \tag{2.65}$$

where $f_2(X) = \sqrt{bX + c}$. Note also that $p = 1$ and $q = d$ in the above equations that defined the indefinite integrals.

As time goes on, X grows and for large times, the first term in (2.65) becomes larger than the second term and we can ignore the second term. Then noting that for $a = 0$, $\tilde{E} = 1$ and using $f_2(X) = (bX + c)^{1/2}$, $b = 2\gamma_o(1 - \beta_{yo})$, $c = \gamma_o^2\beta_{yo}^2$, $k = \gamma_o\beta_{zo}$, (2.65) can be simplified and rewritten as

$$Y \approx \frac{2X^{3/2}}{3\sqrt{b}}$$

$$= \frac{\sqrt{2}X^{3/2}}{3\sqrt{\gamma_o(1 - \beta_{yo})}}$$

$$Z \approx \frac{2}{k}X^{1/2}\sqrt{b}$$

$$= \beta_{zo}\left(\frac{2\gamma_o X}{1 - \beta_{yo}}\right)^{1/2} \tag{2.66}$$

The trajectories of the particle in the XY-plane are shown in Fig. 2.1. These plots are made with the parameters indicated in the figure caption. Note that for $a = 1$ and $a > 0$, particles are not gyrating because in the particle frame, $B = 0$ and the particles are only accelerated by the electrostatic field.

2.3.7 Energy Gain

First note that from Eq. (2.49) we have $d\gamma/d\tau = \tilde{E}\beta_x$ indicating for a given $\tilde{E}$, γ changes as β_x. To obtain the equation of net energy gain of the particle, we rewrite Eq. (2.49) as

$$\gamma\frac{d\gamma}{d\tau} = \frac{1}{2}\frac{d\gamma^2}{d\tau}$$

$$= \tilde{E}\gamma\beta_x \tag{2.67}$$

Fig. 2.1 Trajectories of the particles for different **E** and **B** relations in the X–Y plane. For $a < 0$, the trajectory of $\mathbf{E} \times \mathbf{B}$ drift used $\tilde{E} = 0.1$; For $a = 0$, $\tilde{E} = 1.0$; and for $a > 0$, $\tilde{E} = 3.0$. The initial values are $(X_o, Y_o) = (0, 0)$ and $(\beta_{xo}, \beta_{yo}, \beta_{zo}) = (0.3, 0.4, 0.1)$ (from Takeuchi, S., *Phys. Rev. E*, **66**, 037402, 2002)

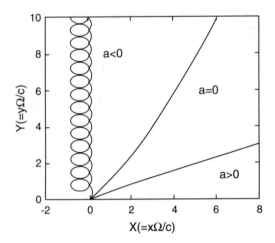

Now take the time derivative of this equation and use the equations in (2.50) and obtain

$$\frac{1}{2}\frac{d^2\gamma^2}{d\tau^2} = \tilde{E}\frac{d}{dt}\gamma\beta_x$$

$$= \tilde{E}(\tilde{E} - \beta_y) \tag{2.68}$$

Use the relations in Eq. (2.50), $\beta_y = (\gamma_o\beta_{yo} + X)/\gamma$, with $X = (\gamma - \gamma_o)/\tilde{E}$ and obtain

$$\beta_y = \frac{\tilde{E}\gamma_o\beta_{yo} + \gamma - \gamma_o}{\tilde{E}\gamma} \tag{2.69}$$

Substitution of this equation into (2.68) yields

$$\frac{1}{2}\frac{d\gamma^2}{d\tau^2} = (\tilde{E}^2 - 1) + \gamma_o\frac{(1 - \tilde{\beta}_{yo})}{\gamma}$$

$$= a + \frac{e}{\gamma} \tag{2.70}$$

where $a = (\tilde{E} - 1)$ and $e = \gamma_o(1 - \tilde{\beta}_{yo})$ are constants. Now let $\Gamma = \gamma^2$ in (2.70) reducing it to

$$\frac{1}{2}\frac{d^2\Gamma}{d\tau^2} = \left(a + \frac{2e}{\sqrt{\Gamma}}\right) \tag{2.71}$$

This differential equation has the form $d^2y/dx^2 = F(y)$ whose solution is

$$x = \pm\int\frac{dy}{\sqrt{2\int F(y)dy + c_1}} + c_2 \tag{2.72}$$

or it can be rewritten as

$$\frac{dy}{dx} = \pm [2(y)dy + c_1]^{1/2} \tag{2.73}$$

Here, $y = \Gamma$, $x = \tau$ and $F(y) = 2(a + e/\Gamma)$. Substituting these into Eq. (2.72) yields

$$\frac{d\Gamma}{dt} = \pm 4 \int \left[\left(a + \frac{e}{\Gamma} \right) d\Gamma + c_1 \right]^{1/2} \tag{2.74}$$

Squaring the above equation and integration of the right-hand side will yield,

$$\frac{1}{4} \left(\frac{d\Gamma}{d\tau} \right)^2 = a\Gamma + 2e\sqrt{\Gamma} + h \tag{2.75}$$

where h is the initial value given by

$$h = -a\Gamma_o - 2e\sqrt{\Gamma_o} + \frac{1}{4} \left(\frac{d\Gamma}{d\tau} \right)^2_{\tau=0}$$
$$= -a\gamma_o^2 - 2e\gamma_o + (\tilde{E}\gamma_o\beta_{xo})^2 \tag{2.76}$$

where the last term used Eq. (2.67). To discuss the meaning of the above Eq. (2.75), note that it can be rewritten as

$$\frac{d\tau}{d\gamma} = \frac{\gamma}{f_3(\gamma)} \tag{2.77}$$

where $f_3(\gamma) = \sqrt{a\gamma^2 + 2e\gamma + h}$.

Question 2.5 The differential equation (2.77) has the form

$$\int \frac{x\,dx}{\sqrt{a + bx + cx^2}}$$

which we have already discussed above. Obtain the solution and discuss the behavior of the solution.

The relativistic treatment of a particle when $|\mathbf{E}| = |\mathbf{B}|$ and $|\mathbf{E}| > |\mathbf{B}|$ shows particles can be accelerated in the direction of $\mathbf{E} \times \mathbf{B}$ unlike the situation when $|\mathbf{E}| < |\mathbf{B}|$. For both of these cases, the particle energy increases without bounds. Examples of particle energy gains are shown in Fig. 2.2. The $\mathbf{E} \times \mathbf{B}$ acceleration mechanism could be important at the center of Harris current sheet and at null points of magnetic field region where $|\mathbf{B}|$ approaches 0 but $|\mathbf{E}|$ can remain finite.

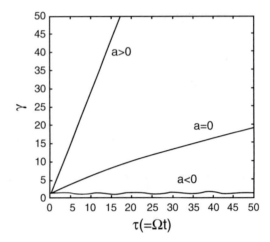

Fig. 2.2 A plot showing how the energy of the particle changes for different **E** and **B** relations. When $a = 0$ or $a > 0$, particles are accelerated. When $a < 0$, particles are not accelerated and they simply drift in the $\mathbf{E} \times \mathbf{B}$ direction (from Takeuchi, *Phys. Rev. E*, **66**, 037402, 2002)

Question 2.6 Magnetometers can measure intensities down to about 0.1 nT. What is the value of $|\mathbf{E}|$ corresponding to this intensity? Are there observations that show cases when $|\mathbf{E}| > |\mathbf{B}|$ in space plasmas? What can you say about measuring **E** and **B**?

2.4 Motion in Inhomogeneous Magnetic Field

The magnetic field in space is rarely homogeneous. Neither is the magnetic field static. Thus, the equations we have developed for the homogeneous and static **B** need to be augmented. One of the main consequences of inhomogeneous magnetic field is that the gyro-frequency of the particle motion given in Eq. (2.2) is no longer constant and it will have a spatial dependence

$$\omega_c = \frac{q}{\gamma m_o} \mathbf{B}(\mathbf{r}) \qquad (2.78)$$

If the magnetic field varies slowly in space, that is, if the distance over which **B** varies in magnitude and direction is much larger than the Larmor radius ρ_c, one can use the perturbation technique. Assume that the variation of the magnetic field is much smaller than the average magnetic field intensity inside the particle's orbit. Then the magnetic field at the particle can be expanded in a Taylor's series which, to first order, is,

$$\mathbf{B}(\mathbf{r}) = \mathbf{B}_o + (\mathbf{r} \cdot \nabla_o)\mathbf{B} + \cdots \qquad (2.79)$$

where $\mathbf{B}_o$ is the magnetic field at the GC and **r** is the vector from the gyration center to the instantaneous position of the particle. We also require that $|\mathbf{B}_o| \gg |(\mathbf{r} \cdot \nabla_o)\mathbf{B}|$, meaning that $\mathbf{B}$ at $\mathbf{r}$ is only slightly different from $\mathbf{B}_o$. The particle orbit is nearly a circle but it no longer closes on itself as it did when the magnetic field was uniform. In this limit, the expression for the gyro-frequency becomes

$$\omega_c = \frac{q}{\gamma m_o} \mathbf{B}(\mathbf{r})$$

$$\approx \omega_o \left[1 + \frac{1}{B_o} (\mathbf{r} \cdot \nabla_o)\mathbf{B} \right] \qquad (2.80)$$

where ω_o is the gyro-frequency in homogeneous magnetic field. Here ∇_o is performed relative to the center of gyration. Inserting this expression into the Lorentz equation yields

$$\frac{d\mathbf{v}_\perp}{dt} = \mathbf{v}_\perp \times \omega_c$$

$$\approx \mathbf{v}_\perp \times \omega_o \left[1 + \frac{1}{B_o} (\mathbf{r} \cdot \nabla_o)\mathbf{B} \right] \qquad (2.81)$$

The first term is the usual cyclotron motion in homogeneous $\mathbf{B}$ field. The second term is due to inhomogeneous $\mathbf{B}$ field and it can be treated as motion arising from a *perturbation*. Particle motion can be described in terms of the *guiding centers* (GC) of the particles, originally developed by H. Alfvén treated in most introductory textbooks.

The most general form of magnetic inhomogeneity $\nabla \mathbf{B}$ is a tensor with nine elements. However, with $\nabla \cdot \mathbf{B} = 0$, the sum of the diagonal terms must vanish and there remain only eight independent components (see Parks 2004). Assuming the original field $\mathbf{B}_o$ is in the z-direction, the terms $\partial B_x/\partial z$ and $\partial B_y/\partial z$ are due to the curved fields and particles traveling on these fields experience a centrifugal force. If $v_\parallel^2$ represents the square of the parallel velocity of the particle, the centrifugal force from $\partial B_y/\partial z$ is $\mathbf{F}_c = -(mv_\parallel^2/B)(\partial B_y/\partial z)\hat{\mathbf{y}}$ resulting in drift of the guiding center. Similarly, the terms $\partial B_z/\partial x$ and $\partial B_z/\partial y$ also give rise to an average force in the y direction resulting in a drift perpendicular to the gradient (Fig. 2.3).

The zeroth and first-order GC drifts due to curvature and gradient of the magnetic field are

$$\mathbf{w}_\perp = \frac{1}{\omega_c B^2} \left(v_\parallel^2 + \frac{v_\perp^2}{2} \right) \mathbf{B} \times \nabla B \qquad (2.82)$$

Fig. 2.3 A sketch showing drift of particles due to the gradient of the magnetic field

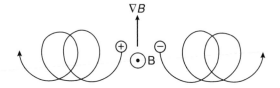

Fig. 2.4 A sketch showing mirroring of a particle in magnetic geometry like the dipole field

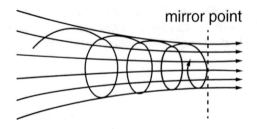

where $\omega_c = qB/\gamma m_o$ and the first and second terms in the parentheses are contributions from the curvature and gradient of the **B** field. The perpendicular motion of a charged particle in a nonuniform magnetic field **B** is described by a superposition of a gyration around a GC with the cyclotron frequency and the constant drift motion of this GC.

The diagonal terms in ∇B, $\partial B_x/\partial x$, $\partial B_y/\partial y$, $\partial B_z/\partial z$, are associated with small variations of B_o as it diverges or converges along z. A particle moving along such a magnetic field geometry will experience a radial component magnetic field and the $\mathbf{v} \times \mathbf{B}$ force will affect the motion along the z-direction (Fig. 2.4).

Consider a particle spiraling around the z-axis with a velocity $v_\perp + v_\parallel$. Since in static magnetic field the speed of the particle is constant, we can write $v_\perp^2 + v_\parallel^2 = v_o^2$ where v_o is the velocity where the magnetic field strength is B_o, a reference point on the z-axis. This reference point can be the equator of a dipole field, for example, where the strength of the magnetic field is at a minimum. Since the magnetic moment of a particle in static magnetic field under the condition being considered is constant, we can write $v_\perp^2/B = v_{\perp o}^2 B_o$ and then the motion in the parallel direction can be written as

$$v_\parallel^2 = v_o^2 - v_{\perp o}^2 \frac{B(z)}{B_o} \tag{2.83}$$

The particle will encounter stronger magnetic field as it moves away from the equator and the particle will gyrate in tighter and tighter spirals. This is accomplished at the expense of the parallel energy which is being converted to perpendicular energy. At some point, the parallel velocity will vanish and the second term in (2.83) will become larger than the first term and $v_\parallel^2$ will change sign. This is the turning point of the particle and the particle will move back in the negative z-direction. The point where the particle is reflected is called a *mirror point* and a particle bouncing back and forth between mirror points is called a *trapped particle*.

If there are other forces present perpendicular to **B**, they also contribute to the GC drift. For example, consider a nonrelativistic particle of mass m in the presence of a constant gravitation field **g**. This particle will experience a drift $(m/q)\mathbf{g} \times \mathbf{B}/B^2$ and the general GC drift equation perpendicular to **B** including the static electric field and gravity is

$$\mathbf{w}_\perp = \frac{\mathbf{E} \times \mathbf{B}}{B^2} + \frac{1}{\omega_c B^2}(v_\parallel^2 + v_\perp^2/2)\mathbf{B} \times \nabla B + \frac{m\mathbf{g}}{q B^2} \times \mathbf{B} \tag{2.84}$$

If the electric field is time dependent, we will add an inertial term that allows time variation of **E** and **B** and includes the *polarization drift* (see below).

A note of caution!. The GC is a fictitious point and its motion does *not* obey the laws of mechanics. However, the concept of the GC is useful because it allows us to study complicated motions without actually solving the exact particle equations. One studies instead the GC motion. In the reference frame moving with the GC, we know that the particles are near it and gyrating about it. For particle motions involving distances much larger than the gyroradius (as is the case in space), it is more instructive to gain knowledge about the GC motion than the details of the exact motion.

2.4.1   Inhomogeneous Electric Field

We now investigate the motion of particles in inhomogeneous electric field. This situation would arise when the electric field is not uniform in space. Let the magnetic field be uniform but the electric field is nonuniform, and for simplicity, assume that the electric field **E** is one dimensional and directed along the x-direction $(E_x, 0, 0)$ of the form

$$\mathbf{E}(y) = \hat{\mathbf{x}} E_x \cos(ky) \tag{2.85}$$

Here E_x is the amplitude and k is the wave number $k = 2\pi/\lambda$. This type of sinusoidal charge distribution can be produced by wave motions at boundaries, for example. We insert this electric field into the Lorentz equation of motion,

$$\frac{d\mathbf{p}}{dt} = q\mathbf{E}(y) + q\mathbf{v} \times \mathbf{B} \tag{2.86}$$

Assume **B** is along the z-direction. Then the velocities in the direction perpendicular to **B** are

$$\frac{dv_x}{dt} = \frac{qB}{\gamma m_o} v_y + \frac{q}{\gamma m_o} E_x(y)$$

$$\frac{dv_y}{dt} = -\frac{qB}{\gamma m_o} v_x \tag{2.87}$$

If we assume now that $\gamma \ll 1$, we can differentiate the first equation and use the second equation (similarly differentiate the second equation and use the first equation) to obtain

$$\frac{d^2 v_x}{dt} = -\omega_c^2 v_x$$

$$\frac{d^2 v_y}{dt^2} + \omega_c^2 v_y = -\omega_c^2 \frac{E_x(y)}{B} \tag{2.88}$$

Here $\omega_c = qB/m_o$ (nonrelativistic). These differential equations look identical to the ones for constant electric field except that now the electric field is dependent on y. When the electric field is constant, the motion of the particle is a superposition of the $\mathbf{E} \times \mathbf{B}$ drift on the gyration. We would still expect this to occur but now the motion will be modified because the electric field is inhomogeneous.

To see what the impact of the inhomogeneous electric field is, recall that the orbit of the particle motion for y in the absence of E is $y = y_o + r_c \cos(\omega_c t + \phi)$. Insert this into the expression of the electric field and obtain

$$\frac{d^2 v_y}{dt^2} + \omega_c^2 v_y = -\omega_c^2 \frac{E_x(y_o)}{B} \cos k(y_o + r_c \cos \omega_c t) \tag{2.89}$$

This equation is now averaged over the cyclotron motion. We will need two identities to work this out: $\cos k(y_o + r_c \cos \omega_c t) = \cos(ky_o)\cos(kr_c \cos \omega_c t) - \sin(ky_o)\sin(kr_c \cos \omega_c t)$ and $\cos k(y_o + r_c \cos \omega_c t) \approx (\cos ky_o)\left(1 - \frac{1}{2}k^2 r_c^2 \cos^2 \omega_c t\right) + \sin(ky_o)(k r_c \cos \omega_c t)$. The first is a trigonometric identity and the second a Taylor expansion of $\cos \omega_c t$ and $\sin \omega_c t$ to second order. When averaged over a cyclotron period, we find $\langle d^2 v_y/dt^2 \rangle = 0$ and $\langle \cos \omega_c t \rangle = 0$. Putting all of the pieces together, the final result we obtain is

$$\langle v_y \rangle = \frac{E_x(y_o)}{B}\left(1 - \frac{1}{4}k^2 r_c^2\right) \tag{2.90}$$

This equation can be generalized by replacing ik with ∇. Then,

$$\mathbf{W}_E = \left(1 + \frac{1}{4}r_c^2 \nabla^2\right)\frac{\mathbf{E} \times \mathbf{B}}{B^2} \tag{2.91}$$

Question 2.7 Discuss how the above equation differs from the equation when the electric field is homogeneous.

2.4.2   Time Dependent Electric Field

If the magnetic field is static and the electric field is time dependent but varying much slower than the cyclotron frequency, the GC method can be applied. The GC motion across the magnetic field is

$$\mathbf{W}_\perp = \frac{\mathbf{E} \times \mathbf{B}}{B^2} + \frac{m}{qB^2}\left(\mathbf{B} \times \frac{d}{dt}\frac{\mathbf{E} \times \mathbf{B}}{B^2}\right)$$

$$= \frac{\mathbf{E} \times \mathbf{B}}{B^2} + \frac{m}{qB^2}\frac{\partial \mathbf{E}_\perp}{\partial t} \tag{2.92}$$

where the first term is due to the presence of $\mathbf{E}_\perp$ and the second term allows for the time variation of $\mathbf{E}_\perp$.

Question 2.8 How does this equation differ relative to time independent case.

2.4.3 Inhomogeneous and Time Dependent E-Field

A more realistic case will be that the electric field is dependent on both space and time. This situation gives rise to *ponderomotive force* which is a nonlinear force that arises when time dependent electric fields are spatially varying. A variety of waves is observed above auroral altitudes and the ponderomotive force of these waves may be important for accelerating ions out into the magnetosphere during magnetic storms (Lundin and Guglielmi 2006). The ponderomotive force of large amplitude electric fields observed in the vicinity of the bow shock and the upstream region may also be important in many nonlinear interactions that are prominent in these regions.

Consider a nonrelativistic ($v \ll c$) charged particle in an oscillating electric field whose amplitude varies in space, $\mathbf{E}(\mathbf{r}, t) = \mathbf{E}_o(\mathbf{r}) \cos \omega t$. The Lorentz equation for either electron or ion is

$$m\frac{d^2\mathbf{r}}{dt^2} = q\mathbf{E}_o(\mathbf{r}) \cos \omega t \tag{2.93}$$

Let us now decompose $\ddot{\mathbf{r}} = \ddot{\mathbf{r}}_o + \ddot{\mathbf{r}}_1$ where the dots represent time derivative. Here $\ddot{\mathbf{r}}_o$ and $\ddot{\mathbf{r}}_1$ are, respectively, slow and high frequency varying components. We now assume that the spatial variation is slow so that we can Taylor expand the spatial dependence of the electric term. The above equation then becomes

$$m\ddot{\mathbf{r}} = m(\ddot{\mathbf{r}}_o + \ddot{\mathbf{r}}_1)$$
$$= q(\mathbf{E}_o + (\mathbf{r}_1 \cdot \nabla))\mathbf{E} \cos \omega t \tag{2.94}$$

where $d\mathbf{E}_o/dt$ is evaluated at r_o. We will now assume that $\ddot{\mathbf{r}}_1 \gg \ddot{\mathbf{r}}_o$ and that $\mathbf{E}_o \gg \mathbf{r}_1 d_o/dt$. The equation of motion then yields two equations associated with slow and fast variations.

$$m\ddot{\mathbf{r}}_o = q\frac{d\mathbf{E}_o}{dt} \cos \omega t \tag{2.95}$$

and

$$m\ddot{\mathbf{r}}_1 = q\mathbf{E}_o \cos \omega t \tag{2.96}$$

The solution for $\mathbf{r}_1$ is

$$\mathbf{r}_1 = -\frac{q\mathbf{E}_o}{m\omega^2} \cos \omega t \tag{2.97}$$

Use of this solution in the slow varying equation and averaging the $\langle \cos^2 \omega t \rangle$ yields

$$m\ddot{\mathbf{r}}_o = -\frac{q^2}{2m^2\omega^2}\mathbf{E}_o \cdot \nabla\mathbf{E}_o \tag{2.98}$$

The ponderomotive force can be written as

$$\mathbf{F}_{pond} = -\frac{q^2}{4m\omega^2}\nabla E_o^2 \qquad (2.99)$$

The ponderomotive force is proportional to the gradient of the square of the wave amplitude and opposite to the direction of the gradient. Ponderomotive force may be responsible for accelerating O^+ ions out of the ionosphere during magnetic storms. Readers interested in observations where the ponderomotive force was active are recommended to read Li and Temerin (1993) and Lee and Parks (1996).

2.5 Other Particle Acceleration Mechanisms

Particles in space can have very high energies. For example, in cosmic rays, particles with energies $>10^{21}$ eV are detected. Solar flares can accelerate particles to tens of BeV. Closer to Earth in the plasma sheet, ions can be accelerated to MeV energies and electrons to hundreds of keV during auroras. Terrestrial magnetic storms also accelerate particles to relativistic energies producing "killer" electrons.

There are many theories about how the particles can be accelerated. For example, the Fermi process is thought responsible for cosmic ray acceleration. Radial diffusion is thought to be responsible for the energetic radiation belt and ring current particles, and particles accelerated in flares and substorms by processes as yet not known. Fermi acceleration mechanism is often portrayed with a particle gaining energy from bouncing between two moving walls that approach each other. Radial diffusion models describe transport of particles from lower to higher intensity magnetic field regions and conservation of the first adiabatic invariant $\mu = mv^2/2B$ shows particles will be energized.

A fundamental fact is that all plasma processes involving particle acceleration require an electric field. In Chap. 1 we showed that the work done on the charged particle by the electric field is $W = \int \mathbf{F} \cdot \mathbf{v} dt = q \int (\mathbf{E} + \mathbf{v} \times \mathbf{B}) \cdot \mathbf{v} dt = q \int \mathbf{E} \cdot \mathbf{v} dt$ where W is the kinetic energy of the particle. Since the magnetic field $\mathbf{B}$ does *no* work, the kinetic energy of charged particles can only change from work done by the electric field $\mathbf{E}$. For charged particles to be accelerated, whatever the process (radial diffusion, Fermi, etc.), the mechanism must involve the electric field $\mathbf{E}$. The important task is to identify the sources of the electric fields and what mechanisms produce the electric field. Only then can we understand the physical mechanisms of how particles are accelerated in space, solar, and astrophysics plasmas.

2.5.1 *Fermi Acceleration*

Cosmic rays include ultra-relativistic energy particles, up to 10^{20} eV. As in solar-terrestrial environment, static electric fields cannot be maintained because space is filled with tenuous plasmas that have high electrical conductivity. The acceleration of cosmic rays can only be associated with inductive electric fields and/or nonstationary electric fields of electromagnetic waves.

The most popular mechanism for accelerating cosmic rays is the mechanism proposed by Fermi in 1949 (Fermi 1949). The Fermi mechanism is a stochastic mechanism in which the particles collide with plasma clouds and irregularities in the interstellar medium and galaxies with magnetic fields. The clouds move randomly with velocity V and the particles gain energy statistically. Power law distributions are expected to result. Observations show that the energy spectrum of cosmic rays has the power law form,

$$\frac{dN}{dE} = AE^{-a} \tag{2.100}$$

and thus agree with Fermi's prediction. The exponent a for energies greater than $\sim 10^9$ eV, ~ 2.6. Let us see how the Fermi mechanism works.

We want to calculate the change of energy of the particle after one collision. Assume that the mass of the cloud is much larger than the mass of the particle so that the velocity V of the cloud is unchanged when the particle collides with it. The initial energy of the particle in the moving plasma frame (S' frame) is

$$W' = \gamma(W + Vp_x) \tag{2.101}$$

where $\gamma = (1 - V^2/c^2)^{-1/2}$, $p = \gamma m_o v$ is the momentum of the particle and v is the velocity in the rest frame (S frame), and the prime (') denotes quantities measured in the moving S' frame. Let the angle between the normal of the cloud surface and the initial direction of the particle be θ so that $p_x = p\cos\theta$. The momentum measured in the S' frame is

$$p'_x = \gamma(p_x + VW/c^2) \tag{2.102}$$

It should be noted that the total energy is conserved in the collision so that $W_{before} = W_{after}$. The momentum only changes sign from p_x to $-p_x$. Transforming now back to the observer's frame of reference, S frame, we see that

$$W = \gamma(W' + Vp'_x) \tag{2.103}$$

Substitute in Eq. (2.103) the expressions for W' and p'_x and obtain

$$W = \gamma^2 W \left(1 + \frac{2vV\cos\theta}{c^2} + \frac{V^2}{c^2}\right) \tag{2.104}$$

If we assume $V/c \ll 1$, γ^2 can be expanded and we find the difference of the energy after the collision becomes

$$\Delta W = W \left[\frac{2vV \cos \theta}{c^2} + \left(\frac{V}{c} \right) \right]^2 \qquad (2.105)$$

This equation must now be averaged over θ. The probability of collision at angle θ is proportional to $(1 + V \cos \theta /c)$. Moreover, the pitch-angle of particles in the range between θ and $\theta + d\theta$ is proportional to $\sin \theta$. For simplicity, let the particle velocity $v \sim c$ and $x = \cos \theta$. In this limit, the first term in Eq. (2.105) averaged over all angles θ from 0 to π becomes

$$\left\langle \frac{2V \cos \theta}{c} \right\rangle = \frac{2V}{c} \frac{\int x[1 + (V/c)x]dx}{\int [1 + (V/c)x]dx}$$

$$= \frac{2}{3} \left(\frac{V}{c} \right)^2 \qquad (2.106)$$

Here the limit of integration for x is from -1 to $+1$. Thus, the average gain of the energy per collision is

$$\left\langle \frac{\Delta E}{E} \right\rangle = \frac{8}{3} \left(\frac{V}{c} \right)^2 \qquad (2.107)$$

This equation that shows the average energy gain in second order in V/c was derived by Fermi. It should also be apparent that this result will lead to an exponential increase in the energy of the particle since the same fractional increase occurs in every collision.

If the mean free path between the clouds along the field is λ, the time between collisions is $\lambda/c \cos \alpha$ where α is the pitch angle of the particle. To find the average time between collisions, we need to average over $\cos \alpha$, which is just $2\lambda/c$. Thus, we find that a typical rate of energy increase per collision is

$$\frac{dW}{dt} = \frac{4}{3} \left(\frac{V^2}{c\lambda} \right) W = \beta W \qquad (2.108)$$

If we now assume that the particles remain in the acceleration region for a characteristic time, the particle obeys the diffusion-like equation,

$$\frac{\partial N}{\partial t} = D\nabla^2 N + \frac{\partial}{\partial W}[b(W)n(W)] - \frac{N}{\tau_{escape}} + Q(W) \qquad (2.109)$$

where D is the diffusion coefficient constant, $b(W)$ is the energy loss term, τ_{escape} is the time it takes the particle to escape the acceleration region, and $Q(W)$ is the source term. We now assume in steady state, ignore diffusion, there are no sources and $b(W) = \beta W$. Then the above expression reduces to

$$-\frac{d}{dW}[\beta W N(w)] - \frac{N(W)}{\tau_{escape}} = 0 \qquad (2.110)$$

Differentiating and rearranging, we find

$$\frac{dN(W)}{dW} = -\left(1 + \frac{1}{\beta \tau_{escape}}\right)\frac{N(W)}{W} \qquad (2.111)$$

whose solution is

$$N(W) = AW^{-z} \qquad (2.112)$$

where $z = 1 + (\beta \tau_{escape})^{-1}$. This is the power law energy spectrum predicted by the Fermi mechanism. The readers should note that Fermi mechanism could be working for particles escaping the bow shock into interplanetary space and interacting with the SW producing higher energy particles convecting back toward the bow shock (Lourn et al. 1990; Freeman and Parks 2000).

2.5.2 Inverse Compton Effect

One of the most interesting questions in astrophysics and possibly solar physics is where the γ rays come from and how they are produced. A mechanism of particular interest is the inverse Compton effect, which can produce photons of γ ray energies by the collision of high energy electrons and ambient thermal photons. This is evidently a well-known phenomenon discovered in the laboratory. We will examine here how this mechanism works.

The interstellar space is permeated with low energy photons that have typical energies of a few eV. Before we discuss the inverse Compton effect, we need to first understand the Compton process, which is well known. The process involves scattering of a photon of fairly high energies by an electron and in the process the photon energy is shifted. We start with the conservation equations of energy and momentum,

$$E_\gamma + E_e = E_{\gamma'} + E_{e'} \qquad (2.113)$$

where the energy of the photon before and after collision is $E = h\nu$ and $E_{\gamma'} = h\nu'$ (the prime ($'$) in the subscript denotes quantities measured after collision). Here h is Planck's constant equal to 6.625×10^{-34} J s and ν is the frequency of the photon. The energy of the electron before and after collision is $\gamma m_o c^2$ and $E_{e'} = [(p_{e'}c)^2 + (m_o c^2)^2]^{1/2}$ where we have assumed the electron is initially at rest. Substitution of these quantities in (2.113) yields

$$h\nu + m_o c^2 = h\nu' + [(p_{e'}c^2)^2 + (m_o c^2)^2]^{1/2} \qquad (2.114)$$

Square (2.114) and solving for the momentum of the electron after collision yields

$$p_{e'}^2 c^2 = (h\nu - h\nu' + m_o c^2)^2 - m_o^2 c^4 \qquad (2.115)$$

Next we use the conservation of momentum equation, $\mathbf{p}_e = \mathbf{p}_\gamma - \mathbf{p}_{\gamma'}$ where $p_\gamma = h\nu/c$ and $p_{\gamma'} = h\nu'/c$. Square this equation and obtain

$$
\begin{aligned}
p_{e'}^2 &= \mathbf{p}_{e'} \cdot \mathbf{p}_{e'} \\
&= (\mathbf{p}_\gamma - \mathbf{p}_{\gamma'}) \cdot (\mathbf{p}_\gamma - \mathbf{p}_{\gamma'}) \\
&= p_\gamma^2 + p_{\gamma'}^2 - 2 p_\gamma p_{\gamma'} \cos\theta \\
&= (h\nu)^2 + (h\nu')^2 - (2h\nu)(h\nu') \cos\theta \qquad (2.116)
\end{aligned}
$$

Multiply (2.116) by c^2 and equate it to Eq. (2.114). This yields

$$\nu m_o c^2 - \nu' m_o c^2 = h\nu\nu'(1 - \cos\theta) \qquad (2.117)$$

Divide through by $\nu\nu' m_o c$ and noting that $\lambda\nu = \lambda'\nu' = c$, we obtain

$$\lambda' - \lambda = \frac{h}{m_o c}(1 - \cos\theta) \qquad (2.118)$$

where λ is the initial wavelength of the photon, and λ' is the wavelength after scattering. This equation defines the amount of wavelength shift in the Compton collision and we see that when a photon is scattered through an angle θ by a free electron, the scattered wavelength λ' is greater than the wavelength λ of the incident photon. If we substitute the values of h, m_o, and c in the above equation, the shift in wavelength can also be written as $\Delta\lambda = 0.0242(1 - \cos\theta)$ where the shift is measured in Angstroms (10^{-8} cm $= 10^{-10}$ m).

The inverse Compton process is the reverse of the Compton process where low energy photons are scattered to high energy γ rays by relativistic electrons. This effect can be observed, for example, when photons of cosmic background (CMB) move through the hot plasma that surrounds the galaxy cluster and possibly when the accelerated electrons on the Sun interact with the ambient solar photons.

Let the energy of a photon be 1 eV and let it collide head on with a 100 MeV electron in the S frame. Consider next the collision from the reference frame S' frame in which the electron is at rest before the collision. Let S' frame move with a velocity V relative to S frame where V is the speed of the electron in the S frame. The energy of the photon in the S' frame relative to S frame is

$$E' = \gamma(E - V p_x) \qquad (2.119)$$

where $\gamma = (1 - V^2/c^2)^{-1/2}$ and $p = h\nu/c$ is the momentum of the photon in the S frame. The negative sign is due to the fact that the photon is moving in the $-X$ direction. Equation (2.119) then becomes

$$E' = \gamma(h\nu + Vh\nu/c)$$
$$= \gamma(1 + V/c) \qquad (2.120)$$

For a 500 MeV electron, $\gamma = 1001$ and $V \sim c$. Hence, after collision, (2.120) shows the energy of the photon in the S' frame is $\sim$2 keV. This means in the S' frame where the electrons are at rest, the incident photons have energies $\sim$2 keV.

Apply now the transformation back to the S frame using

$$E = \gamma(E' + Vp_x)$$
$$\approx 2\,\text{MeV} \qquad (2.121)$$

which is now in the γ ray energies. Interested readers could compute the propagation direction of the scattered photon and the amount of scattering that occurs in the observer S frame.

2.6   Waves and Wave-Particle Interaction

A plasma in a magnetic field supports many different types of waves, which can be classified by the orientation of the propagation vector **k** relative to the wave electric field **E**. An electrostatic wave has **E** parallel to **k**, and an electromagnetic wave has **k** perpendicular to **E**. Both types of waves have been detected in space. For instance, plasma and "whistler mode" waves detected in Earth's magnetosphere are, respectively, electrostatic and electromagnetic waves. These waves carry much information about the plasma processes in the medium in which the waves are observed.

Waves interact with charged particles and they can be amplified or damped. Wave-particle interaction can scatter particle pitch angles and precipitate them from the magnetosphere (Oleksiy et al. 2013). Auroras are produced in this way. Waves can also grow at the expense of particles and the well-known auroral kilometric radiation (AKR) is one such example. Details of how these waves are produced requires kinetic theory. This section limits discussion of waves in the limit of quasi-linear approximation using kinetic theory. A comprehensive discussion about the waves can be found in Krall and Trivelpiece (1973).

2.6.1 Electrostatic Waves

This section will introduce how the Vlasov equation is used to study waves. Kinetic theory of waves is based on the velocity distribution function $f(\mathbf{r}, \mathbf{v}, t)$ which has six independent variables at any given time, t. Consider now a small volume in which there is equal number of electrons and ions. For simplicity, we consider only the dynamics of electrons with ions forming a fixed neutralizing background. Assume the electrons are initially in equilibrium described by the distribution function $f_o(\mathbf{v})$, for example, a Maxwellian distribution with nonzero temperature. Suppose a small perturbation removes an electron and produces a region of excess positive charge (this could simply happen because of statistical fluctuation due to finite temperature of the distribution). In response to the electric field produced by the perturbation, the electron will be pulled back. However, because of inertia, the electron will overshoot. Physically what will happen is that the electron will oscillate back and forth about the equilibrium point giving rise to electrostatic oscillation. This situation can be quantitatively described by

$$f_o(\mathbf{r}, \mathbf{v}, t) = f_o(\mathbf{v}) + f_1(\mathbf{r}, \mathbf{v}, t) \tag{2.122}$$

where f_1 is perturbed quantity and considered small. Substitution of this distribution function into the Vlasov equation and retaining terms up to first order will yield

$$\frac{\partial f_1}{\partial t} + \mathbf{v} \cdot \frac{\partial f_1}{\partial \mathbf{r}} + \frac{q}{m} \mathbf{E} \cdot \frac{\partial f_o}{\partial \mathbf{v}} = 0 \tag{2.123}$$

The $\mathbf{v} \times \mathbf{B}$ term has been ignored since we are interested only in the electrostatic oscillation. The third term is a first-order quantity because of $\mathbf{E}$ which did not exist before the perturbation. The electric field $\mathbf{E}$ is related to f_1 through the Poisson's equation

$$\epsilon_o \nabla \cdot \mathbf{E} = q \int f_1 d^3 v \tag{2.124}$$

and Faraday's equation

$$\nabla \times \mathbf{E} = 0 \tag{2.125}$$

where we have set the right side to zero for electrostatic oscillations. The three equations given above can be solved self-consistently for the unknowns, f_1 and $\mathbf{E}$, given the initial and boundary conditions.

We follow the standard procedure and let $f_1 \sim e^{i(\mathbf{k} \cdot \mathbf{r} - \omega t)}$ and $\mathbf{E} \sim e^{i(\mathbf{k} \cdot \mathbf{r} - \omega t)}$. Substitution of these relations into (2.123)–(2.125) reduces them to

$$f_1 = \frac{q}{m} \frac{i}{(\mathbf{k} \cdot \mathbf{v} - \omega)} \mathbf{E} \cdot \frac{\partial f_o}{\partial \mathbf{v}}$$

$$i\mathbf{k} \cdot \mathbf{E} = \frac{q}{\epsilon_o} \int f_1 d^3v$$

$$\mathbf{k} \times \mathbf{E} = 0 \tag{2.126}$$

Let $\mathbf{E} = E(\mathbf{k}/k)$ and use this equation in the first equation and integration of the resulting equations over all velocity space yields

$$\int f_1 d^3v = \frac{iqE}{m} \frac{\mathbf{k}}{k} \cdot \int \frac{\partial f_o}{\partial \mathbf{v}} \frac{d^3v}{(\mathbf{k} \cdot \mathbf{v} - \omega)} \tag{2.127}$$

Replacing the left side of this equation with the second equation in (2.126) then yields

$$1 = \frac{q^2}{m\epsilon_o} \frac{\mathbf{k}}{k^2} \cdot \int \frac{\partial f_o}{\partial \mathbf{v}} \frac{d^3v}{(\mathbf{k} \cdot \mathbf{v} - \omega)} \tag{2.128}$$

This equation shows the relation between ω and $\mathbf{k}$ and is called a dispersion relation. Note that the denominator vanishes when $\mathbf{k} \cdot \mathbf{v} = \omega$, when the electrons move with exactly the phase velocity of the wave. This occurs when the electrons are in resonance with the waves and the integral is not defined as formulated. The integral can be solved only if there is a prescription how to integrate around the singularity. This was first solved by Landau and Liftschitz (1962) and leads to Landau damping (see, for example, Krall and Trivelpiece 1973).

We ignore this difficulty here and integrate (2.128) by parts. Since f_o falls off rapidly with increasing velocities, the surface integral will vanish and we obtain

$$1 = \frac{q^2}{m\epsilon_o} \int \frac{f_o(\mathbf{v})}{(\mathbf{k} \cdot \mathbf{v} - \omega)^2} d^3v \tag{2.129}$$

Suppose the plasma is cold and consists of only electrons. Then $f_o(\mathbf{v}) = \delta(\mathbf{v})$ and (2.129) yields

$$1 = \frac{n_o q^2}{m\epsilon_o} \frac{1}{\omega^2}$$

$$= \frac{\omega_p^2}{\omega^2} \tag{2.130}$$

and we find that the oscillation frequency is simply the plasma frequency of electrons. These plasma oscillations occur locally and do not propagate.

Suppose we now introduce a small temperature to the electrons. This means that the electrons will have some velocities but they will be considered small so that $f_o(\mathbf{v})$ is not altered much. This approximation turns out to be valid for only small values of $\mathbf{k}$ and thus one can use $\mathbf{k} \cdot \mathbf{v}/\omega$ as an expansion parameter. We then find that (2.129) becomes

$$1 = \frac{q^2}{m\epsilon_o \omega^2} \int \left[1 + 2\frac{\mathbf{k} \cdot \mathbf{v}}{\omega} + 3\left(\frac{\mathbf{k} \cdot \mathbf{v}}{\omega}\right)^2 + \cdots \right] f_o(\mathbf{v}) d^3v \tag{2.131}$$

Given $f_o(\mathbf{v})$, each of these terms can be evaluated. The first term is just the total density, n_o. If $f_o(\mathbf{v})$ is an even function like the Maxwellian, the second term vanishes. The next term gives the $\mathbf{k} \cdot \mathbf{v}$ averaged over $f_o(\mathbf{v})$. Hence,

$$1 = \frac{\omega_p^2}{\omega^2}\left[1 + 3\frac{\langle(\mathbf{k}\cdot\mathbf{v})\rangle^2}{\omega^2}\right] \tag{2.132}$$

where $\omega_p^2 = nq^2/m\epsilon_o$. For electrostatic waves with finite temperature plasmas, k is along v and for a thermal distribution, $\langle v^2\rangle = \kappa T/m$. Also, since the thermal correction is assumed to be small, we can let $\omega_p^2 \approx \omega^2$. Then this equation reduces to

$$\omega^2 = \omega_p^2 + \frac{3\kappa T}{m}k^2 \tag{2.133}$$

where κ is the Boltzmann constant. This dispersion relation describes Langmuir waves. The second term is a correction term due to finite temperature.

2.6.2 Electromagnetic Waves

We now discuss how the electromagnetic dispersion relation can be obtained from the Vlasov equation. We begin with the wave equation

$$\nabla \times (\nabla \times \mathbf{E}) = -\mu_o\frac{\partial \mathbf{J}}{\partial t} - \frac{1}{c^2}\frac{\partial^2 \mathbf{E}}{\partial t^2} \tag{2.134}$$

which can be Fourier expanded to yield

$$-\mathbf{k} \times (\mathbf{k} \times \mathbf{E}) = -\mu_o\left(\frac{\partial \mathbf{J}}{\partial t} + \epsilon_o\frac{\partial^2 \mathbf{E}}{\partial t^2}\right) \tag{2.135}$$

The waves here have space and time dependence, $e^{-i(\mathbf{k}\cdot\mathbf{r}-\omega t)}$ and there will be longitudinal (parallel to $\mathbf{k}$) and transverse (perpendicular to $\mathbf{k}$) components. Let $\mathbf{E} = \mathbf{E}_\parallel + \mathbf{E}_\perp$ and $\mathbf{J} = \mathbf{J}_\parallel + \mathbf{J}_\perp$ where $\parallel$ and $\perp$ refer to directions relative to $\mathbf{k}$. Then, (2.135) splits into two equations

$$-\mathbf{k}\mathbf{k}E_\parallel + \mathbf{k}\mathbf{k}E_\parallel = -\mu_o\left(\frac{\partial \mathbf{J}_\parallel}{\partial t} + \epsilon_o\frac{\partial^2 \mathbf{E}_\parallel}{\partial t^2}\right) = 0$$

$$k^2\mathbf{E}_\perp = -\mu_o\left(\frac{\partial \mathbf{J}_\perp}{\partial t} + \epsilon_o\frac{\partial^2 \mathbf{E}_\perp}{\partial t^2}\right) \tag{2.136}$$

The first equation describes the longitudinal electrostatic oscillations while the second equation describes the transverse waves.

Let us examine linearly polarized electromagnetic waves and let the wave vector **k** point in the z-direction and $\mathbf{E}_\perp$ in the x-direction. This allows the second equation of (2.136) to be rewritten as

$$\left(k^2 - \frac{\omega^2}{c^2}\right) E_\perp = iq\mu_o\omega \int f_1 v_x d^3v \tag{2.137}$$

where the right-hand side is the current density **J** expressed in terms of the perturbed distribution function, f_1. The linearized Vlasov and Maxwell equations become

$$i(\mathbf{k}\cdot\mathbf{v} - \omega)f_1 + \frac{q}{m}(\mathbf{E} + \mathbf{v}\times\mathbf{B})\frac{\partial f_o}{\partial\mathbf{v}} = 0$$

$$\mathbf{k}\times\mathbf{E} = \omega\mathbf{B} \tag{2.138}$$

The second equation shows that the transverse electric and magnetic fields are coupled. Insert the second equation into the first to eliminate **B** and obtain

$$\begin{aligned}
f_1 &= i\frac{q}{m}\frac{[\mathbf{E} + \omega^{-1}\mathbf{v}\times(\mathbf{k}\times\mathbf{E})]}{(\mathbf{k}\cdot\mathbf{v} - \omega)}\cdot\frac{\partial f_o}{\partial\mathbf{v}} \\
&= i\frac{q}{m}\frac{E(\partial f_o/\partial v_x) + (Ek/\omega)[v_x\partial f_o/\partial v_z - v_z\partial f_o/\partial v_x]}{(kv_z - \omega)}
\end{aligned} \tag{2.139}$$

This equation shows the dependence of the perturbed distribution in terms of the electric field of the wave. We now insert this equation into (2.137) and obtain the dispersion relation,

$$\omega^2 - k^2c^2 = \frac{q^2}{m\epsilon_o}\int\left[\frac{(\omega - kv_z)\partial f_o/\partial v_x}{kv_z - \omega} + \frac{kv_x^2\partial f_o/\partial v_z}{kv_z - \omega}\right]d^3v \tag{2.140}$$

for transverse electromagnetic waves, except note that the integral diverges if there are particles with a velocity $v_z = \omega/k$. These resonant particles can be dealt with as in the electrostatic case using the Landau prescription for contour integration in the complex plane. There will be the usual poles and the integral must be integrated around these poles. It turns out that the damping here is less important than it was for the electrostatic oscillations. This is because for electromagnetic waves, $\omega/k > c$ for cold plasmas and one does not have many resonant particles approaching c.

On the other hand, a large class of nonthermal distributions can feed energy into the disturbance and give rise to complex ω. An example of this can be seen by integrating (2.140) by parts, assuming there are no resonant particles. This yields

$$\omega^2 - k^2c^2 = \omega_p^2 + \frac{q^2k^2}{m\epsilon_o}\int\frac{v_x^2 f_o d^3v}{(kv_z - \omega)^2} \tag{2.141}$$

Consider the anisotropic distribution function $f_o(\mathbf{v}) = \delta(v_z)\phi(v_x, v_y)$ where $\phi(v_x, v_y)$ is an arbitrary function perpendicular to the propagation direction. Insert this into Eq. (2.141) and obtain the dispersion relation

$$\omega^2 - k^2 c^2 - \omega_p^2 \left(1 + \frac{k^2 \langle v_x^2 \rangle}{\omega^2}\right) = 0 \tag{2.142}$$

which can be rewritten as

$$\omega^4 - (k^2 c^2 + \omega_p^2)\omega^2 - k^2 \omega_p^2 \langle v_x^2 \rangle = 0 \tag{2.143}$$

The solution of this quadratic equation is

$$\omega^2 = \frac{1}{2}\{k^2 c^2 + \omega_p^2 \pm [(k^2 c^2 + \omega_p^2)^2 + 4k^2 \omega_p^2 \langle v_x^2 \rangle]^{1/2}\} \tag{2.144}$$

One of the ω's is always negative for all values of k, hence this solution corresponds to the growing mode. These are unstable *lightwaves*. This instability disappears when $\langle v_x^2 \rangle$ approaches zero. We can estimate the growth rate in the limit $k^2 \to \infty$. In this case

$$\omega^2 = -\frac{k^2 \langle v_x^2 \rangle \omega_p^2}{(k^2 c^2 + \omega_p^2)} \to -\frac{\langle v_x^2 \rangle}{c^2}\omega_p^2 \tag{2.145}$$

We can ignore ω_p^2 in the denominator since $k^2 c^2$ is larger than ω_p^2. The maximum growth rate is $(\langle v_x \rangle / c)\omega_p$. This is much smaller than the growth rate of electrostatic instabilities which is of the order of ω_p. As a rule, electrostatic instabilities always have larger growth rates and is more important.

2.7 Cyclotron Resonance Theory

Magnetospheric particle distributions are not uniform in phase space and there is free energy to excite electromagnetic instabilities. An example is the loss cone distribution discussed in Chap. 1 which is unstable to generation of cyclotron waves (Kennel and Petschek 1966). These waves will interact with electrons and ions, and can violate the first invariant, scattering the pitch-angles of the particles. The cyclotron interaction mechanism is a major mechanism by which ions and electrons are lost from the magnetosphere.

Understanding wave particle interaction and precipitation requires solving the Vlasov equation that includes the ambient magnetic field. This is laborious and difficult and we shall refer the reader to the many excellent advanced plasma physics books that treat this problem (for example, Krall and Trivelpiece 1973). Briefly, the results show that complex frequencies indicating growing or damping modes will appear when energy is exchanged between the waves and particles. A particularly interesting process is the energy exchange mechanism by means of a resonance interaction. This process occurs when the particle traveling along the magnetic field has a velocity $v_\parallel$ that matches the Doppler shifted frequency $(\omega_c - \omega)/k$ of the wave, where ω_c is the cyclotron frequency. The particle is acted on by a force that is varying with the gyrating frequency and they can couple when the two frequencies are equal (as in driven oscillators).

The resonance interaction process requires examining the transverse waves traveling along the magnetic field direction. As before, one starts with a perturbation of the periodic form in space and time $e^{-i(\mathbf{k}\cdot\mathbf{r}-\omega t)}$. This leads to the linearized equation that takes the form

$$-i(\mathbf{k}\cdot\mathbf{v}-\omega)f + \frac{q}{m}(\mathbf{v}\times\mathbf{B_0})\cdot\frac{\partial f}{\partial\mathbf{v}} = -\frac{q}{m}(\mathbf{E}+\mathbf{v}\times\mathbf{B})\cdot\frac{\partial f_o}{\partial\mathbf{v}} \qquad (2.146)$$

Here, $\mathbf{B}_o$ is the ambient field and $\mathbf{B}$ is the perturbed first-order magnetic field. This equation will yield the following form for the dispersion relation,

$$c^2 k^2 = \omega^2 + \frac{\pi q^2}{m\epsilon_o\omega^2}\int\int\frac{(\omega - kv_\parallel)\partial f_o/\partial v_\perp + kv_\perp\partial f_o/\partial v_\parallel}{(\omega - kv_\parallel)\pm\omega_c}v_\perp^2\,dv_\perp dv_\parallel \qquad (2.147)$$

The limits of integration are $(-\infty, \infty)$ for $v_\parallel$, and $(0, \infty)$ for $v_\perp$. This integral is not defined for particles in resonance, when the denominator vanishes. In this case, one must again revert to the initial value problem and use the Landau prescription to integrate around the singularities.

2.7.1 Cyclotron Interaction

Our interest is to gain insight into the electron precipitation phenomenon by means of the cyclotron resonance mechanism. For this, consider the case when the β of the plasma is small. This allows us to approximate the plasma as being cold and we can use $f_o = \delta(v_\parallel)\delta(v_\perp)/2\pi$. Substitution of this in Eq. (2.147) and integration by parts will yield

$$\frac{c^2 k^2}{\omega^2} = 1 - \frac{(\omega_p^+)^2}{\omega(\omega\mp\omega_c^+)} - \frac{(\omega_p^-)^2}{\omega(\omega\pm\omega_c^-)} \qquad (2.148)$$

Here $\omega_c^\pm$ corresponds to the ion and electron cyclotron frequency, and $\omega_p^\pm$ to electron and ion plasma frequency. This is the dispersion relation for the right- and left-hand (R) modes in the limit the plasma is cold. When $\omega_c^+ << \omega < \omega_c^-$, the ion contribution is a factor m_e/m_+ smaller and can be neglected. If also the electron plasma frequency ω_p^- is large so that $\omega_p^- >> \omega_c^-$, then the dispersion relation of the right-hand mode simplifies to

$$n^2 \approx \frac{(\omega_p^-)^2}{\omega(\omega_c^- - \omega)} \qquad (2.149)$$

Electrons will interact with these R-mode waves. The velocity integration of the dispersion relation Eq. (2.147) includes an imaginary ω when $kv_{\parallel} - \omega - \omega_c^- = 0$. This occurs when the electrons are in resonance with the waves. The velocity of the resonating particle v_R must satisfy the relation

$$kv_R = \omega - \omega_c^- \tag{2.150}$$

The waves and the electrons are traveling in opposite directions and the resonance condition requires that the Doppler shifted frequency of the wave be exactly equal to the gyrofrequency in the particle frame (essentially the guiding center frame) traveling with $v_{\parallel}$. (Actually, the wave frequency must be an integer multiple of the gyrofrequency.) The electron here is acted on by a force $q\mathbf{E}_{\perp}$ and it will be accelerated or decelerated depending on the relative phase of the gyration and the phase of the wave. If the Doppler shifted wave frequency is not an integer of the gyrofrequency, no net acceleration or deceleration will occur when averaged over a gyration.

The kinetic energy of the resonant electrons is

$$E_R = \frac{1}{2}m_e v_R^2$$
$$= \frac{1}{2}m_e \left(\frac{\omega - \omega_c^-}{k}\right)^2 \tag{2.151}$$

We now eliminate k using the dispersion relation given in (2.149) using $n^2 = c^2 k^2/\omega^2$. This yields

$$E_R = \frac{m_e c^2}{2} \frac{\omega^2}{\omega_p^2} \left(\frac{\omega_c^-}{\omega} - 1\right)^3 \tag{2.152}$$

This equation can be rewritten as

$$E_R = E_{mag} \frac{\omega_c^-}{\omega} \left(1 - \frac{\omega}{\omega_c^-}\right)^3 \tag{2.153}$$

where $E_{mag} = B^2/8\pi n$ is the magnetic energy per particle (cgs unit) and n is the total plasma density. The magnetic energy per particle is a characteristic energy for the cyclotron interaction.

The growth rate for the instability can be calculated using the Landau prescription. If only the waves that are resonant with the high energy tail of the distribution are considered and if further the assumption is made that the number of these particles is small, the growth rate likewise will be small and one can write $\omega = \omega_R + i\gamma$ where ω_R and γ are real and $\gamma/\omega_R << 1$. Then, ω_R is given by the dispersion relation and the growth rate is obtained as

$$\gamma = \pi \omega_c^- \left(1 - \frac{\omega}{\omega_c^-}\right) \eta^-(v_R) \left[A^-(v_R) - \frac{1}{\omega_c^-/\omega - 1}\right] \tag{2.154}$$

Here $\eta(v_R)$ is the number of particles in resonance and defined as

$$\eta^-(v_R) = 2\pi \frac{\omega_c^- - \omega}{k} \int_0^\infty v_\perp dv_\perp f(v_\perp, v_\parallel = v_R) \tag{2.155}$$

$\eta^-(v_R)$ can be looked at as the number of electrons in the distribution that have velocities in the range $\omega_c^- - \omega/k$. The anisotropy $A^-(v_R)$ is defined as

$$A^-(v_R) = \frac{\int_0^\infty dv_\perp \left(v_\parallel \partial f/\partial v_\perp - v_\perp \partial f/\partial v_\parallel\right) \frac{v_\perp}{v_\parallel}}{2 \int_0^\infty v_\perp dv_\perp f} \tag{2.156}$$

It is understood that the right-hand side is evaluated where $v_\parallel = v_R$. This equation can be rewritten in terms of the particle pitch-angle. Use of $\tan\alpha = -v_\perp/v_\parallel$ will reduce (2.156) to

$$A^-(v_R) = \frac{\int_0^\infty v_\perp dv_\perp \tan\alpha \, \partial f/\partial\alpha}{2 \int_0^\infty v_\perp dv_\perp f} \tag{2.157}$$

Here $\partial f/\partial\alpha$ is the gradient of the distribution with respect to the pitch-angle at constant energy. In this case the electron distribution is estimated in terms of a bi-Maxwellian

$$f(v_\parallel, v_\perp) = \frac{1}{[(2\pi)^{3/2} v_{th\parallel} v_{th\perp}^2]} e^{(-v_\parallel^2/v_{th\parallel}^2 - v_\perp^2/v_{th\perp}^2)} \tag{2.158}$$

where $E = m v_{th}^2/2 = \kappa T$, then the anisotropy is represented by

$$A^-(T_\parallel, T_\perp) = \left(\frac{T_\perp}{T_\parallel} - 1\right) \tag{2.159}$$

Since dipole fields with an atmosphere have a loss cone and after the initial bounce the particles in the loss cone are lost, there will be pitch-angle anisotropy and $T_\perp/T_\parallel$ is always > 1. Since η is positive and $\omega/\omega_c^- < 1$, the growth rate will be positive if

$$A^- > \frac{1}{(\omega_c^-/\omega - 1)} \tag{2.160}$$

In summary, the growth rate depends on the pitch-angle anisotropy A^- and the fraction of particles η in resonance. These quantities can be measured. Cyclotron resonance theory has been applied to many classes of electron precipitation phenomena in Earth's magnetosphere.

Low frequency waves (also called ULF (ultra low frequency) waves) permeate the magnetosphere and are important for ring current dynamics (Zong et al. 2011; Takahashi 2016). These ULF waves have frequencies close to the azimuthal drift of the ions in the inner radiation belt where the ring current is. The ULF waves and ring current particles can resonate causing radial diffusion affecting transport of the ring current particles. The low frequency waves can also be affected by temperature gradients. Inhomogeniety of the plasma pressure becomes important near the boundary between cold and hot plasmas because the Alfvén speed varies significantly in such regions (Lee et al. 2004).

Question 2.9 Show that Alfvén waves can be obtained from the cyclotron wave dispersion relation in the limit of low frequency.

2.7.2 Relativistic Electron Cyclotron Interaction

The resonance condition derived above is for nonrelativistic electrons. However, with the recent discovery that electron precipitation in the Earth's magnetosphere often includes electrons with MeV energies, it is necessary to take account of relativistic effects (Thorne 1974). The resonance condition for relativistic electrons is

$$\gamma(\omega - kv_R) - \omega_c^- = 0 \tag{2.161}$$

where the relativistic factor γ is defined as

$$\gamma = \left(1 - \frac{v_R^2}{c^2}\right)^{-1/2} \tag{2.162}$$

These two equations are solved together with (2.149) and the result is

$$E_R = 2\left(\frac{\omega_c^-}{\omega}\right)\left(\frac{E_{mag}}{1+\gamma}\right)\left(1 - \gamma\frac{\omega}{\omega_c^-}\right)^2\left(1 - \frac{\omega}{\omega_c^-}\right) \tag{2.163}$$

Here E_R is the kinetic energy of relativistic electrons given by $m_e c^2(\gamma - 1)$.

2.7.3 Ion Cyclotron Interaction

Ions can interact with ion cyclotron waves and we can derive the resonance condition for the ions in the same manner as was done for electrons (Omura et al. 2010). For the Earth's magnetosphere, the ions that precipitate are generally not relativistic. The resonance condition for the nonrelativistic ions is

$$\omega - kv_R - \omega_c^+ = 0 \tag{2.164}$$

The ion dispersion relation, the left-hand (L) mode, in the limit of cold plasma approximation is

$$\frac{c^2 k^2}{\omega^2} = \frac{(\omega_p^+)^2}{\omega_c^+ (\omega_c^+ - \omega)} \tag{2.165}$$

Solving these two equations by eliminating k yields

$$E_R = E_{mag} \left(\frac{\omega_c^+}{\omega} \right)^2 \left(1 - \frac{\omega}{\omega_c^+} \right)^3 \tag{2.166}$$

Note that the low frequency ion cyclotron waves have been suggested to heat SW protons in the corona (Kim et al. 2015).

2.7.4 Anomalous Resonance Interaction

Resonance can occur between ion cyclotron waves and relativistic electrons. This anomalous resonance occurs when the ion waves and the electrons are both moving in the same direction. Since the electrons are relativistic, the electron velocity $v_\parallel$ is larger than the wave phase velocity v_{ph}. The electrons will overtake the wave and although the ion wave is a left-hand polarized wave in the plasma frame, it becomes a right-hand polarized wave when Doppler shifts into the electron frame. This effect is called anomalous Doppler shift. We use the electron resonance condition but since the ion wave frequency is much less than the electron cyclotron frequency, the resonance condition is simplified to

$$- \gamma k v_R - \omega_c^- \approx 0 \tag{2.167}$$

This equation is used to eliminate k in the ion dispersion relation in (2.149) together with the relation for γ given in (2.162). We then obtain

$$(\gamma^2 - 1) = 2 \frac{E_{mag}}{m_e c^2} \left(\frac{m_i}{m_e} \right) \left(\frac{\omega_c^+}{\omega} \right)^2 \left(1 - \frac{\omega}{\omega_c^+} \right) \tag{2.168}$$

This equation can be rewritten in terms of the kinetic energy of the electrons, $E_R = m_e c^2 (\gamma - 1)$ as

$$E_R = \left\{ \left[2 \frac{E_{mag}}{m_e c^2} \left(\frac{m_i}{m_e} \right) \left(\frac{\omega_c^+}{\omega} \right)^2 \left(1 - \frac{\omega}{\omega_c^+} \right) + 1 \right]^{1/2} - 1 \right\} m_e c^2 \tag{2.169}$$

Observations from Earth's magnetosphere indicate that most of the relativistic electron precipitation due to anomalous interaction with ion waves occurs in the local time sector centered around dusk.

2.8 Concluding Remarks

The motion of charged particles guided by the Lorentz equation shows particles are accelerated when they move in the presence of electric fields. Particles moving across current sheets are also accelerated especially in regions where the magnetic field vanishes. We discussed the work of Takeuchi (2002) who showed that when $|\mathbf{E}|$ is larger than $|\mathbf{B}|$, particles can be accelerated in the $\mathbf{E} \times \mathbf{B}/B^2$ direction. We have also reviewed the basic motions of particles in the presence of gradients of magnetic and electric fields and several types of well-known particle acceleration mechanisms. In nature, particles are often acted on by time varying fields. The sources of time varying fields are not always known but they occur in many regions of space. For example, during disturbed geomagnetic times, large amplitude micropulsations and Alfvén waves are detected in the radiation belts, solar wind, and geomagnetic tail. These time varying magnetic fields induce electric fields and particles interacting with these fields will be accelerated. Betatron acceleration combined with radial diffusion could be an important source of relativistic electrons observed in the inner radiation belts (see also Chap. 7). The perturbed plasma can also excite higher frequency waves and these waves can interact with electrons and ions accelerating and precipitating them. The $\mathbf{E} \times \mathbf{B}$ acceleration mechanism is a relatively new concept in space plasma physics and could be important in current sheets where the magnetic field is very small. Attention is also called to a new source of radiation belt electrons discovered by the MMS mission, called *microinjection* events whose source is still not known (Fennell et al. 2016). We have focused on application of the particle theory in Earth's environment. For application to other planets, see for example, Halekas et al. (2017).

2.9 Solutions to Questions in Chap. 2

Solution 2.1 To obtain the physical meaning of these quantities, they must be transformed to the rest frame of the plasma. This is especially important when the plasma bulk velocity is high as in the SW and the plasma sheet. This is necessary because the bulk motions add to the velocity vector and will alter the values of the pitch-angle and Larmor frequency calculated in the spacecraft frame.

Solution 2.2 For this problem, consider the motion of an electron in a changing magnetic field. We let the magnetic field be everywhere on a plane pointing in the vertical direction (along z), symmetric about some axis, but the strength will depend on the distance from the axis. The electron is moving on a path that is a circle of constant radius with its center at the axis of the field. An electric field $\mathbf{E}$ due to the changing magnetic field is tangent to the electron's orbit which will drive the electron around the circle. Because of the symmetry, this electric field will have the same value everywhere on the circle. If the electron's orbit has a radius $\mathbf{r}$, the line integral of $\mathbf{E}$ around the orbit is equal to the rate of change of the magnetic flux through the circle. The line integral of $\mathbf{E}$ is its magnitude times the circumference of the circle, $2\pi r$. Let B_{av} be the average magnetic field inside the circle. Then we have $E2\pi r = \partial/\partial t (B_{av}\pi r^2)$ Since we are assuming r is constant, E is proportional to the time derivative of the average magnetic field. Hence $E = (r/2)d B_{av}/dt$. For the circular orbit we have assumed, the electric force is always in the direction of its motion, so its total momentum will be increasing at the rate given by the above equation. The force $q\mathbf{E}$ will accelerate the electrons and the rate of the change of the momentum is $dp/dt = qE$. Combine the two equations and obtain $dp/dt = qr/2d B_{av}/dt$. Integration of this equation yields $p - p_o = qr/2\Delta B_{av}$ where p_o is the momentum the electron started out with, and the term on the right shows the momentum change is proportional to ΔB_{av} where ΔB_{av} is the change in the average magnetic field. In interplanetary space and the radiation belts, magnetic micropulsations are always detected during disturbed times and particles will be accelerated by the inductive electric fields. The changing magnetic fields in space are always inducing electromotive force $(\mathcal{EMF})$ and the changing magnetic flux $\Phi = \int \mathbf{B} \cdot d\mathbf{A}$ will accelerate charged particles.

Solution 2.3 For a 50 keV electron, the relativistic correction for this electron is determined from the total energy of the particle, $\mathcal{E} = mc^2 = m_oc^2/(1-v^2/c^2)^{1/2} = T + m_oc^2$. We then obtain

$$1 - \frac{v^2}{c^2} = \left(\frac{m_oc^2}{T + m_oc^2}\right)^2 = 1.312/1.440 = 0.91111$$

Here we used $m_oc^2 = 0.51$ MeV and $T = 0.05$ MeV. We thus find $(1 - v^2/c^2)^{1/2} = 0.9111$ indicating there is a 8.9% change of the cyclotron frequency. This example shows mass is even important in cyclotron resonance interactions when electrons e^- have energies only $\sim$50 keV. For a fixed frequency wave ω, the particle and wave will then get out of phase and lose resonance. However, since there is a spectrum of waves available, the particles will resonate with lower frequency waves as v increases. The readers can show that for H^+, mass becomes important when energies are greater than $\sim$40 MeV.

Solution 2.4 To solve these equations, note that

$$\int \frac{px + q}{\sqrt{ax^2 + bx + c}} dx = \frac{p}{a}\sqrt{ax^2 + bx + c}(q - bp/2a)I(x)$$

where we define the function $I(x)$ as, for $a > 0$,

$$I(x) = \frac{1}{\sqrt{a}} \ln |2ax + b + 2\sqrt{a(ax^2 + bx + c)}|$$

for $a < 0$ and $b^2 - 4ac > 0$, and if $a = 0$, $I(x)$ satisfies

$$I(x) = -\frac{1}{\sqrt{|a|}} \sin^{-1}\left(\frac{2ax + b}{\sqrt{b^2 - 4ac}}\right).$$

Then

$$\int \frac{px + q}{\sqrt{ax^2 + bx + c}}dx = \frac{2p(bx - 2c) + 6bq}{3b^2}\sqrt{bx + c}$$

When $a \neq 0$, a little algebra will show that these equations yield the exact solutions

$$Y = \frac{1}{a}[f_1(X) - f_o] + \left(d - \frac{b}{2a}\right)[I(X) - I_o]$$

$$Z = k[I(X) - I_o]$$

$$\tau = \frac{\tilde{E}}{a}[f_1(X) - f_o] + \left(\gamma_o - \frac{b\tilde{E}}{2a}\right)[I(X) - I_o]$$

where f_o and I_o are the initial values at $X = 0$. Use the second equation above and eliminate $[I(X) - I_o]$ from the first and third equations. Then multiply the resulting first equation by $-\tilde{E}$ and add to the τ equation. This eliminates the first terms involving $[f_1(X) - f_o]$. Then use the definitions $a = (\tilde{E}^2 - 1)$, $b = 2\gamma_o(\tilde{E} - \beta_{yo})$, $d = \gamma_o\beta_{yo}$ and $k = \gamma_o\beta_{zo}$ to arrive at a new relation

$$\beta_{zo}\tau = \tilde{E}\beta_{zo}Y + (1 - \tilde{E}\beta_{yo})Z$$

Solution 2.5 The solution can be written as

$$\tau = \frac{f_3(\gamma) - f_3(\gamma_o)}{a} - \frac{e[I(\gamma) - I(\gamma_o)]}{a}$$

As before, we have different situations to consider.

2.9.1 $a = \tilde{E}^2 - 1 > 0$

In the case, $E > B$, the first two terms are dominant if $a > 0$ and the last two terms if $a < 0$. Equation (2.49) shows $d\gamma/dt = \tilde{E}\beta_x = \tilde{E}d\xi/dt$. Hence $\Delta\gamma = \tilde{E}\Delta\xi$ and we see that if ξ increases, γ must also increase. When $\gamma \gg 1$, for very relativistic particles and $a > 0$, we note that $f_3(\gamma) \sim \sqrt{a}\gamma$. Then, $d\tau/d\gamma \approx \gamma/f_3(\gamma)$ and we find that

$$\gamma = (\tilde{E} - 1)^{1/2}\tau$$

2.9.2 $a = 0, E = B$

If $a = 0$, then the above equation reduces to

$$\tau = \frac{\gamma f_4(\gamma) - \gamma_o f_4(\gamma_o)}{3e} - \frac{h[f_4(\gamma) - f_4(\gamma_o)]}{3e^2}$$

where $f_4(\gamma) = \sqrt{2e\gamma + h}$. $\tilde{E} = 1$ and for $\gamma \gg 1$, $\tau \approx \gamma f_4(\gamma)/3e$ and $f_4 \approx (2e\gamma)^{1/2}$. Then we obtain

$$\gamma = [9\gamma_o(1 - \beta_{yo})/2]^{1/3} \tau^{2/3}$$

Now the γ dependence on time shown in the above equation should be the same as the above in Eq. (2.21). To see this, note that Eq. (2.21) shows that $\gamma^2 \sim p_y t$ and the earlier Eq. (2.28) shows that $p_y^3 \sim t$. Hence we see that $\gamma^2 \sim t^{4/3}$ or $\gamma \sim t^{2/3}$ which has the same time dependence as the above equation.

Solution 2.6 The relative magnitude of **E** and **B** is best examined using the Gaussian unit because in Gaussian unit, 1 statVolt/cm = 1 G. Consider, for example, that near a magnetic neutral region the magnetic field strength is 10^{-10} T. Since $1 G = 10^{-4}$ T, we have a field strength of 10^{-6} Gauss and this is equal to 10^{-6} statVolts/cm. Now, the relationship between Volts and statVolt is 1 statVolt = 300 V. Hence, we obtain 3×10^{-4} V/cm which is 0.3 mV/cm or 30 mV/m. This value is high but electric field of this magnitude or larger has been measured in Earth's radiation belt (Cattell et al. 2008). It is interesting to note that the electric field instruments can measure field magnitudes down to a few millivolt/m. The corresponding values of $|B|$ for a few millivolt/m $|E|$ is not measurable by magnetometers at this time.

Solution 2.7 The usual **E** × **B** drift in homogeneous electric field is modified by the second term in the parenthesis. This correction term is called the finite Larmor radius effect. The **E** × **B** term is now mass, energy, and charge dependent through r_c. The finite Larmor radius effect could be important across sharp boundaries such as the magnetopause. The readers should be aware that sharp boundaries could be unstable to generation of instabilities, for example, drift instability.

Solution 2.8 This second term is due to the *polarization drift* and the particles drift along the direction of the electric field with electrons and ions drifting in opposite directions. The particles pick up or lose energy depending on whether the electric field is increasing or decreasing. The readers should be aware that space plasmas in time dependent electric field can behave like a dielectric material. Maxwell's equation can then be written for a material medium and one can obtain expressions for the dielectric constants and polarization currents. This topic is generally treated in introductory text books and will not be repeated here.

Solution 2.9 For this problem, we start with the dispersion relation of circularly polarized electromagnetic waves propagating in the z-direction along the magnetic field (B_o). Equation (2.148) is relevant and repeated here for convenience.

$$\frac{c^2 k^2}{\omega^2} = 1 - \frac{(\omega_p^+)^2}{\omega(\omega \pm \omega_c^+)} - \frac{(\omega_p^-)^2}{\omega(\omega \mp \omega_c^-)}$$

where k is parallel to B and the perturbed electric field E_1 is perpendicular to B_o. Recall the $\pm$ sign refers to right- and left-handed polarized waves, $\omega_p^{\pm 2} = ne^2/m^{\pm}\epsilon_o$ is the plasma frequency where the $\pm$ on mass refers to ions and electrons,

$$\frac{c^2 k^2}{\omega^2} = 1 - \frac{ne^2}{\omega}\left[\frac{1}{m^-\omega_c^-} + \left(1 + \frac{\omega}{\omega_c^-}\right)\right]^{-1} - \left[\frac{1}{m^+\omega_c^+} + \left(1 + \frac{\omega}{\omega_c^+}\right)\right]^{-1}$$

$$= 1 + \frac{ne^2\mu_o}{B}\frac{1}{\omega_c^+}\left(1 + \frac{m^-}{m^+}\right)$$

$$= 1 + \frac{nm^+\mu_o}{B^2}$$

$$= \frac{1}{V_A^2}$$

where $V_A = B_o/\sqrt{(nm^+\mu_o)}$ is the Alfvén speed. Thus we see here that in the limit of low frequency, the dispersion relation of circularly polarized waves reduces to the dispersion relation of Alfvén waves. This discussion also shows that the Alfvén waves are *degenerate* cyclotron waves. Our calculation could have also started with only the left-hand circularly polarized waves and the conclusion will be the same. For modeling ULF waves in a compressed dipole magnetic field, see Degeling et al. (2010).

References

Alfvén, H., Fälthammar, C.-G.: Cosmical Electrodynamics, Fundamental Principles, 2nd edn. Oxford University Press, Oxford (1963)

Cattell, C., et al.: Discovery of very large amplitude whistler-mode waves in earth's radiation belts. Geophys. Res. Lett. **35**(1), L01105 (2008)

Degeling, A., et al.: Modeling ULF waves in a compressed dipole magnetic field. J. Geophys. Res. **115**, A10212 (2010)

Fennell, J., et al.: Microinjections observed by MMS FEEPS in the dusk to midnight region. Geophys. Res. Lett. **43**, 6078 (2016)

Fermi, E.: On the origin of cosmic radiation. Phys. Rev. **15**, 1165 (1949)

Freeman, T., Parks, G.K.: Fermi acceleration of supra thermal solar wind oxygen ions. J. Geophys. Res. **105**, 15715 (2000)

Halekas, J., et al.: Flows, fields, and forces in the mars-solar wind interaction. J. Geophys. Res. **122**(1), 320 (2017)

Jackson, J.D.: Classical Electrodynamics, 2nd edn. Wiley, New York (1975)

Kennel, H., Petschek, H.: Limit on stably trapped particle fluxes. J. Geophys. Res. **71**, 1 (1966)

Kim, E.H., et al.: Localization of ultra-low frequency waves in multi-ion plasmas of the planetary magnetosphere. J. Astro. Space Sci. **32**, 289 (2015)

Krall, N.A., Trivelpiece, A.W.: Principles of Plasma Physics. McGraw Hill Book Company, New York (1973)

Landau, D.D., Liftschitz, E.M.: The Classical Theory of Fields. Addison-Wesley Publishing Company, Reading (1962)

Lee, N., Parks, G.K.: Ponderomotive acceleration of ions by circularly polarized electromotive waves. Geophys. Res. Lett. **23**, 327 (1996)

Li, X., Temerin, M.: Pondoromotive effects on ion acceleration in the auroral zone. Geophys. Res. Lett. **20**, 13 (1993)

Lourn, P., et al.: Trapped electrons as a free energy source for the auroral kilometric radiation. J. Geophys. Res. **95**, 5938 (1990)

Lundin, R., Guglielmi, A.: Ponderomotive forces in cosmos. Space Sci. Rev. **127**(1), 1–116 (2006)

Oleksiy, A., et al.: Statistics of whistler mode waves in the outer radiation belt: cluster STAFF-SA measurements. J. Jeophys. Res. **118**, 3407 (2013)

Omura, Y., et al.: Theory and observation of electromagnetic ion cyclotron triggered emissions in the magnetosphere. J. Geophys. Res. **115**, 5553 (2010)

Parks, G.K.: Physics of Space Plasmas, An Introduction, 2nd edn. Westview Press, A Member of Perseus Books Group, Boulder (2004)

Takahashi, K.: ULF waves in the inner magnetosphere. In: Keiling, A., Lee, D.H., Nakariakov, V. (eds.) Low Frequency Waves in Space Plasmas. Geophysical Monograph, American Geophysical Union. Wiley, Hoboken (2016)

Takeuchi, S.: Relativistic $\mathbf{E} \times \mathbf{B}$ acceleration. Phys. Rev. E **66**, 037402 (2002)

Thorne, R.: A possible cause of dayside relativistic electron precipitation events. J. Atmos. Terres. Phys. **34**, 635 (1974)

Zong, Q., et al.: On magnetospheric response to solar wind discontinuities. J. Atmos. Sol. Terr. Phys. **73**(1), 1–4 (2011)

Additional Reading

Chen, F.F.: Introduction to Plasma Physics. Plenum Press, New York (1974)

Friedman, Y., Semon, M.: Relativistic acceleration of charged particle in uniform and mutually perpendicular electric and magnetic fields as viewed in the laboratory frame. Phys. Rev. E **72**, 026603 (2005)

Lee, D.H., Lysak, R., Song, Y.: Investigations of MHD wave coupling in a 3D numerical model: effects of temperature gradient. Adv. Space Res. **33**, 742 (2004)

Northrop, T.G.: The Adiabatic Motion of Charged Particles. Interscience Publishers, New York (1963)

Rankin, R., et al.: Self-consistent wave-particle interactions in dispersive scale long-period field-line-resonances. Geophys. Res. Lett. **4**, L23103 (2007)

Roederer, J.G.: Dynamics of Geomagnetically Trapped Radiation. Springer, New York (1970)

Rosser, W.G.V.: Introductory Relativity. Butterworths, London (1967)

Schmidt, G.: Physics of High Temperature Plasma: An Introduction. Academic, New York (1966)

Schulz, M., Lanzerotti, L.J.: Particle Diffusion in the Radiation Belts. Springer, Berlin (1974)

Uspensky, J.V.: Theory of Equations. McGraw Hill Book Company, Inc., New York (1948)

Zhou, X.Z., et al.: Charged particle behavior in the growth and damping stages of ultralow frequency waves: theory and Van Allen probes observations. J. Geophys. Res. **112**, 3254 (2016)

Zou, H.: Short-term variations of the inner radiation belt in the South Atlantic anomaly. J. Geophys. Res. **120**, 4475 (2015)

Chapter 3
Escaping Stellar Particles

3.1 Introduction

Particles from hot stars are continuously escaping. It was known from early on that were it not for such particles, the interplanetary and interstellar space would be a near perfect vacuum with an occasional cosmic ray zipping through. The escaping particles from Sun have been called solar wind (SW). The SW travels across the interplanetary magnetic field (IMF) inducing $\mathcal{EMF}$ that drives a heliospheric current sheet which then interacts with magnetized planets creating structures like magnetospheres populated by high energy particles.

Nearly all of the plasma activities observed in our heliosphere are one way or another associated with solar disturbances carried out by the SW. Because of the SW, all dynamics of solar-terrestrial plasmas occurring in the environment of Earth's magnetosphere are connected and depending on the local plasma properties, different types of processes become active and instabilities excited.

The existence of the SW was predicted from observations of cometary tails by Biermann (1960) and later measured by in situ experiments on Lunik 2 and Lunik 3 (Gringaus et al. 1961). Subsequently, Explorer 10 (Bonetti et al. 1963) verified these first measurements. Lunik and Explorer 10 observations were however for short durations and it was not until Mariner 2 that the existence of the SW was firmly established (Neugebauer and Snyder 1966). The physics of SW is one of the most important problems to understand because it can apply to collisionless plasmas that permeate the Universe.

The well-known Mariner 2 observations showed that the SW speed on any given day can vary from a few hundred $\mathrm{km\,s^{-1}}$ to more than $1000\,\mathrm{km\,s^{-1}}$ (Fig. 3.1). The different speeds of the SW were interpreted as natural variations occurring in the SW source. The high speed SW is thought to represent the ambient state and escaping from the *coronal holes* where the density is low and the plasma cool, and the magnetic field is directed radially outward. The coronal holes are mainly observed

© Springer Nature Switzerland AG 2018
G. K. Parks, *Characterizing Space Plasmas*, Astronomy and Astrophysics Library,
https://doi.org/10.1007/978-3-319-90041-4_3

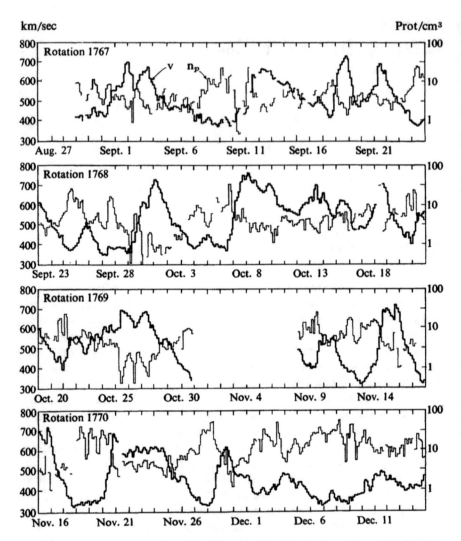

Fig. 3.1 Solar wind bulk velocity and density measured by Mariner 2 experiment. The bold line is bulk velocity and the lighter line density of protons (from Neugebauer, M. and C. Snyder, *J. Geophys. Res.*, **71**, 4469, 1966)

in the polar regions during quiet years and near the equator during more active times. The slow SW comes from closed magnetic field regions and may include the coronal streamers.

Luhman et al. (2013) have reviewed some of the features observed with the recent unusual solar cycle minimum. They showed that solar surface fields in both polar and low-to-mid-latitude active region zones routinely produce coronal magnetic fields and related solar wind sources much more complex than a dipole. They compared observations to magnetogram-based coronal field models and discussed the conditions that prevailed in the corona from the decline of cycle 23 into the rising phase of cycle 24. Their results emphasize the need for adopting new views of what is "typical" solar wind, even when the Sun is relatively inactive.

Since the early observations, we have learned that the SW can have many different sources. Mariner 2 measurements likely included solar disturbances because the measurements were made at times of very high KP indices. Thus, the high SW speed they measured could have included the Interplanetary Coronal Mass Ejections (ICMEs) and solar disturbances like co-rotating interaction regions and flares.

Even though the Mariner SW instrument was primitive by today's standards, it was able to identify the presence of helium He^{++} in the distribution function associated with the peak that appeared at twice energy per charge of the ESA measurements. The early observations established that the typical SW includes about 96% H^+ ions and 4% heavy ions with He^{++} most abundant. Although these observations showed that the SW He^{++} can increase to $\sim$20%, it was not known until much later that such increases were associated primarily with interplanetary coronal mass ejection (ICME) events.

The SW electrons have different distributions than the ions. The SW electron distribution includes four components: core ($\sim$5–50 eV), halo ($\sim$50 eV–1 keV), *strahl*, the high energy end of halo component that is field-aligned ($\geq$50 eV), and the super-halo ($\geq$1 keV) component. The narrowly field-aligned *strahl* component is the main component carrying the heat flux from the Sun. The sources of the different electron features were not known until recently, where observations have shown that they are possibly accelerated at different times (Wang et al. 2016).

This chapter will begin discussion of the basic features of the SW that have been observed since the beginning of space age. An important question still not answered is how much of the observations made at 1 AU represents the original features produced at the Sun. To learn how the SW behaves closer to the Sun, we will examine Helios data that were obtained as close as 0.3 AU (astronomical unit) from the Sun. Then SW electron observations are summarized and we discuss what could cause the electrons to behave so differently from the ions. Next we discuss some of the issues with fluid models and examine what fluid models can and cannot explain. We then present a short review of kinetic models of the SW and discuss their salient features. The chapter concludes with a discussion of electrostatic solitary waves observed in the SW which have been interpreted as *double layers* (Mangeney et al. 1999). Double layers are produced by unstable currents but where these nonlinear structures in the SW are produced is not known. However, if these double layers were produced on the Sun and have propagated out, then taking a cue from Earth's auroral observations, we *speculate* there are large-scale solar current systems similar to Earth's auroral current system to produce double layers. Like Earth's aurora, parallel electric field could be accelerating the solar particles outward producing the SW (alternatively, the double layers may be produced in the SW, in which case, the observations would imply currents strong enough are produced in the SW to excite double layers).

3.2 Observations of Solar Wind Ions

Solar wind has been measured in the vicinity of Earth since the beginning of space age (Dessler 1967). Although much is known about the SW (Cranmer 2012), we will revisit some of the important features using data from Cluster (Escoubet et al. 1997). Cluster data will confirm not only previous observations but also aspects of the measurements that have been improved. Figure 3.2 is a summary plot (spin averages) of typical bulk ion plasma data measured by the SW g-detector (top three panels) and the G-detector (next three panels) and the bottom panel shows the intensity of the magnetic field and components. The g-detector measures the cold SW beam only in the direction looking toward the Sun while the G-detector is not counting in that direction and thus the measurements in the SW for this detector are meaningless. The spacecraft was initially in the magnetosheath and crossed the bow shock into the SW around 0645 UT as can be deduced from the discontinuities observed in the magnetic field (bottom panel) and the bulk quantities, the energy flux (top panel), macroscopic density (black), and temperature (red) mean velocities computed from moments of the 3D distribution function. It then returned back into the magnetosheath at ∼0708 UT.

An important observation is the measurement of the plasma beams in the magnetosheath. Fluid shock theories predict that the SW beam is thermalized in the downstream magnetosheath. But clearly, the plasma in the magnetosheath is not a thermalized distribution because we see discrete beams (red). What produced these beams in the magnetosheath? Are they the directly transmitted SW beam suggested previously in the ISEE data (Sckopke 1995)? What more can we learn from these data? These questions will be discussed further below.

Question 3.1 Review and discuss the main differences in the type of data obtained by the two detectors (g- and G-detectors) on Cluster.

Question 3.2 Many of the well-known bulk features of SW plasma in the neighborhood of the bow shock are shown in Fig. 3.2. However, interesting questions arise about the detailed features present in the data. For example, the SW beam is clearly seen as the red line in the energy flux spectrogram covering a narrow energy range (top panel). What are the low intensity counts (blue) scattered slightly above and below the SW beam that appear in the energy flux? Are they part of the SW? Can their source be identified and the particles removed in the calculation of the SW density, velocity, and temperature? Why don't they show up in the bulk parameters?

3.2.1 Solar Wind Ion Beams

The distribution function contains the most basic information about the particles measured by plasma instruments. Figure 3.3 shows a two-dimensional (2D) cut of the three-dimensional (3D) distribution function $f(\mathbf{r}, \mathbf{v})$ obtained during one spin period of the spacecraft (4 s) displayed in the spacecraft frame of reference

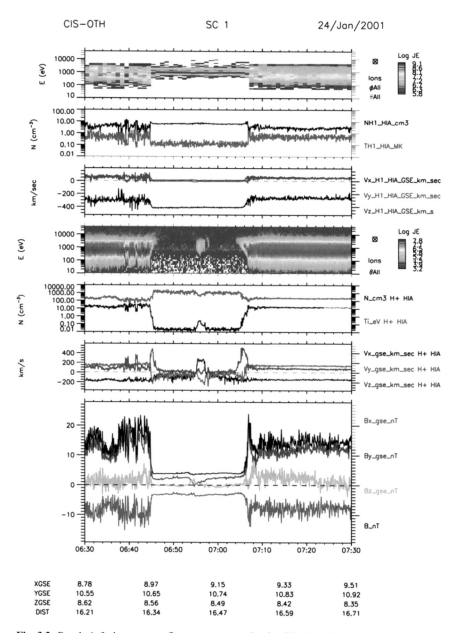

Fig. 3.2 Panels 1–3 show energy flux spectrogram, density (black) and temperature (red), mean velocities from the g-detector, panels 4–6 same bulk parameters from the G-detector and the bottom panel shows the magnetic field intensity and components (the geometrical factor of g detector is smaller by a factor of ~50). Data are shown in the spacecraft frame of reference. V_x, V_y, V_z are in GSE (Geocentric Solar Ecliptic) coordinates. See text for further explanations

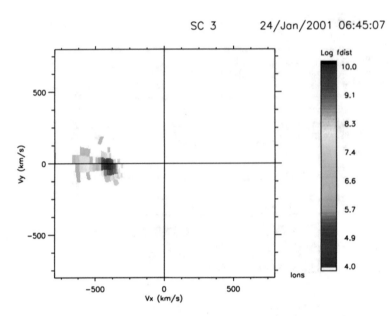

Fig. 3.3 2D cuts of 3D $f(\mathbf{v})$ in spacecraft frame (S-frame) of the SW beam in V_y and V_x plane. The velocity scale is $\pm 800\,\mathrm{km\,s^{-1}}$. The SW H$^+$ is the intense beam and the weaker beam (yellow) is He^{++}. The color bar on the right defines the density of particles in units of (s^3 km^{-6}). This plot includes bins with greater than 1 count

(S-frame). This plot shows the SW phase space density in V_y and V_x plane (GSE coordinates). A key feature of the SW is that the particles occupy a small region of the velocity space. When the particles all have nearly the same velocity, the distribution is called a *beam*. The intense red beam at V_x approximately $-408\,\mathrm{km\,s^{-1}}$ is H$^+$ and the less intense beam (orange) at $-567\,\mathrm{km\,s^{-1}}$ is He^{++}. The He^{++} beam is at ~ 1.39 times the velocity where the H$^+$ beam is observed. In the 2D velocity space, we can also see the weak fluxes of particles that appear scattered about the main SW beam.

Fundamental questions about the SW distributions measured at 1 AU include what mechanisms on the Sun can accelerate the particles to form the beams (Tu et al. 2002). What caused the displacement of the beam relative to the inertial frame? How much of the observations reflects the original properties of the SW at the Sun? Have the original SW distributions been modified in transit toward Earth? How was the temperature anisotropy produced? These questions will be discussed further below. Wave-particle interactions could pitch-angle scatter the particles and produce such anisotropy (Marsch 2006). How does one validate this suggestion?

Question 3.3 Where is the spacecraft instrument located in Fig. 3.3?

Question 3.4 The SW beam observed at 1 AU is always coming from V_x direction regardless of the pitch-angle. Pitch-angles from near 0° to 90° have been observed but the SW beam is coming from $+V_x$ direction. Provide a physical explanation of these observations.

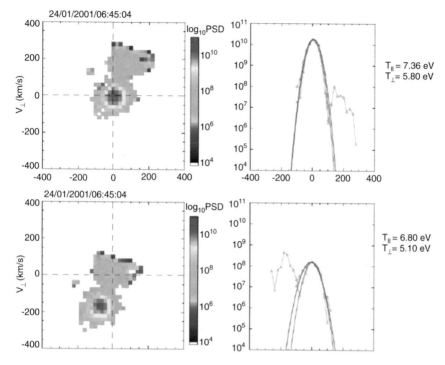

Fig. 3.4 Temperature estimates of H$^+$ and He^{++} beams in the plasma frame of reference (S'-frame). The top 2 panels show the 2D and 1D cuts of the H$^+$ beams in the SW (0645:06 UT) and the bottom two panels He^{++} beams. In the 1D distributions, the data points are circles and the solid lines are Maxwellian fits. Red is $T_\parallel$ and blue is $T_\perp$. Errors of temperature measurements are about 5–10% for H$^+$ and 10–20% for the He^{++} ions. Velocity scales are ± 400 km s^{-1} (from Parks et al., *Astrophys. J. Lett.*, **825**, L27, 2016)

3.2.2 Temperature of SW Ion Beams

The left panels of Fig. 3.4 show 2D cuts of 3D phase space density of the SW in $V_\parallel$ and $V_\perp$ plane in the SW-frame of reference (S'-frame) where the $\parallel$ and $\perp$ refer to directions relative to the magnetic field. The right panels show 1D cuts of the distribution in directions parallel and perpendicular to the magnetic field. Estimations of $T_\parallel$ (red) and $T_\perp$ (blue) from 1D distributions (right) assume the shapes of the distributions are Maxwellian. The computed SW temperatures of the beams in the parallel and perpendicular directions for H$^+$ were $T_\parallel$ ~7.4 eV and $T_\perp$ ~5.8 eV and for He^{++} beam $T_\parallel$ ~ 6.8 eV and $T_\perp$ ~5.3 eV. Although SW temperatures have been measured since 1974 (Feldman et al. 1974), Cluster observations are one of the cleanest measurements ever made to date.

Note that anisotropic distributions in temperature have also been observed on Wind and Helios (Kasper et al. 2003; Liu et al. 2006; Marsch 2016; He et al. 2015; Tu and Marsch 1997) that often indicate parallel temperatures are higher than the perpendicular temperatures.

Question 3.5 Discuss how the plots in Fig. 3.4 were obtained. In particular how one identified the H^+ and He^{++} ions.

Question 3.6 The computed temperatures of both ions are nearly the same (within errors). What does this mean? But the distributions are anisotropic with the parallel temperatures slightly higher than the perpendicular temperatures especially for He^{++}. Discuss what this could mean.

3.2.3 Solar Wind Ions Measured by Helios

Information about the SW closer to the Sun has been obtained by *Helios* spacecraft, a two spacecraft mission which was designed to measure the SW behavior to distances of ~0.3 AU. Figure 3.5 shows a summary plot of SW macroscopic parameters from Helios 2. Shown are the mean ion velocity V, density N, and temperature T obtained during four consecutive solar rotations from distances of ~1–0.3 AU. The plots are arranged so that the distance to the Sun decreases from top to bottom. The recurrent structure is clearly seen in the SW during these rotations. We can also possibly see steepening of the SW velocity profiles as the spacecraft approached closer to the Sun. The proton number densities are the highest just before the onset of high speed flows associated with the compression regions produced by the interacting fast and slow streams.

3.2.4 Multi-Streaming Ion Beams

Helios observations showed the behavior of the SW beams with distance from the Sun. Figure 3.6 shows examples of 2D cuts of 3D distributions of H^+ measured by Helios. These plots have removed He^{++} ions from the measured distributions. The three columns are for different SW speeds increasing from left to right, with $|V|$ on the left column ~350 km s^{-1}, middle column ~500 km s^{-1}, and right column ~750 km s^{-1}. The heliocentric distances where measurements are made are top row ~0.95 AU, second row ~0.7 AU, third row ~0.5 AU, and fourth row ~0.3 AU.

Note the distributions here are shown in the SW frame (S'-frame), obtained by transforming the measured SW on the SC frame (S-frame) by the amount the beam is shifted from the spacecraft (unfortunately the original data observed on the spacecraft frame are not shown in the article). In these plots, the center of H^+ beam is at the origin and the spacecraft (S-frame) is shifted away from the origin. This analysis a priori assumes that the solar corona is expanding with a constant flow velocity and the coronal particles are in thermal equilibrium represented by the "core" distribution.

The distributions are complicated and they often include nonthermal tails and secondary peaks aligned along the magnetic field direction (dashed lines). The

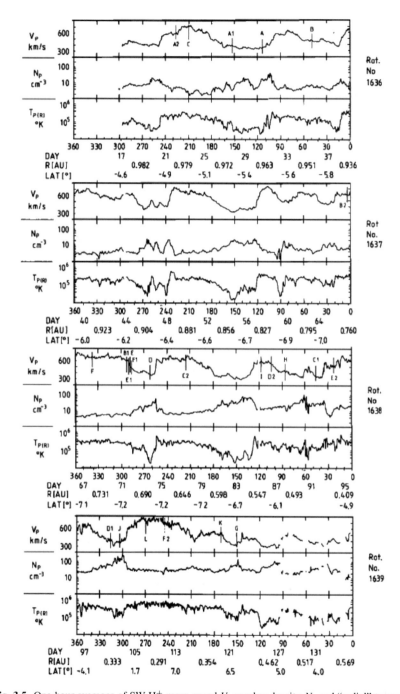

Fig. 3.5 One-hour averages of SW H$^+$ mean speed V, number density N, and "radial" temperature T versus Carrington longitude for four successive solar rotations (January1 7, 1976 and May 3, 1976). Time of measurement, day, heliocentric distance, and solar latitude are given on the bottom. Data shown come from radial distances between 0.3 and 1 AU measured by an ESA (from Marsch et al., *Astron. Astrophys.*, **164**, 77, 1986)

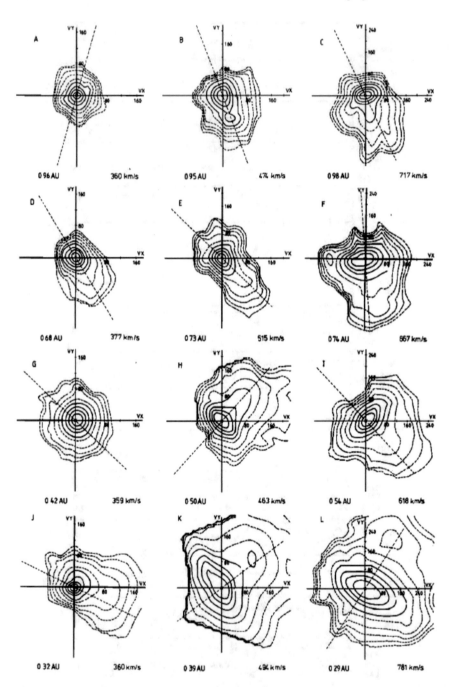

Fig. 3.6 2D cuts of 3D ion velocity distributions plotted in the SW frame. V_x and V_y are speeds in GSE coordinates. The dashed line shows the direction of the B-field (from Marsch et al., *J. Geophys. Res.*, **87**, 52, 1982a)

"core" distribution is also anisotropic with $T_\perp > T_\parallel$. Note also that the SW always has a field-aligned component for moderate and fast SW speeds (second and third columns). Near the Sun, even the slow SW is field-aligned (bottom left). The two isotropic distributions (a) and (g) were obtained at the sector boundaries. Readers interested in further details of these plots are recommended to consult the original article (Marsch et al. 1982a,b). Unfortunately, because the original data measured on the spacecraft frame are not available, it was not possible to verify the final results shown (note that from an experimental point of view, the practice is always to present first the quantities that have been measured before the original data are manipulated).

He^{++} Distributions

The He^{++} ions have similar features as H$^+$ ions. Examples of He^{++} distributions from Helios are shown in Fig. 3.7. As with the H$^+$ distributions, the He^{++} distributions are plotted in the moving SW frame (S'-frame) by subtracting the measured mean SW velocity assuming He^{++} ions are traveling together with the H$^+$ ions. The He^{++} velocity distributions show they also have nonthermal components and secondary peaks aligned along the magnetic field direction.

The mechanisms that produce the nonthermal particles, anisotropy, and the secondary peaks are not currently known. However, the field-aligned pitch-angle distributions are consistent with particles being accelerated by $\mathbf{E}_\parallel$. Our suggestion is that field-aligned distributions observed by *Helios* could have been accelerated along the magnetic field direction by a field-aligned potential drop. But where this could be occurring on the Sun is not known and remains an important problem still to be solved.

3.3   Observations of Solar Wind Electrons

The ions carry most of the momentum and the electron contribution is much smaller. However, the electrons carry the current and have complex distributions with high energy tails important for understanding the basic SW mechanisms. The SW electrons are also measured by ESAs and the energy range is typically from a few eV to a few keV (low energy threshold determined by photoelectrons and spacecraft potential). We have known since the early measurements of 1D electron velocity distributions that the SW (Fig. 3.8) electrons consist of an isotropic core component that is nearly a Maxwellian and an isotropic "halo" component with a field-aligned high energy component called "strahl." A more recent discovery shows there is in addition a "super-halo" component whose particle energies can exceed $\sim$100 keV (Lin 1997; Wang et al. 2013b).

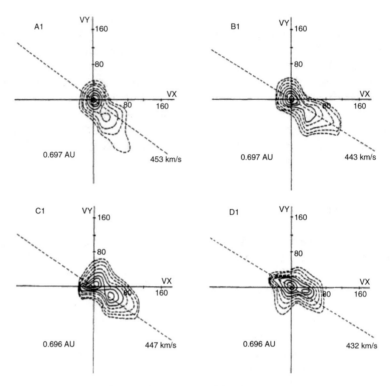

Fig. 3.7 2D cuts of 3D velocity distributions of He^{++} plotted in the SW frame. V_x and V_y are speeds in GSE coordinates. The dashed line shows the direction of the B-field (from Marsch et al., *J. Geophys. Res.*, **87**, 52, 1982b)

3.3.1   Core Electrons

Generally, electrons from ∼5 to 50 eV form the *core electrons* whose temperature T_e is ∼5 eV. The core electrons have isotropic pitch angle distributions although the temperatures parallel and perpendicular to **B** can be different. The speed of a 5 eV electron is about 4000 km s^{-1}, ten times larger than the bulk velocity of the ions. The electron bulk velocity is not usually computed. Instead, the mean speed of the electrons is assumed to be the same as the ions. Are there any difficulties in determining bulk velocities of electrons? What are they? Can the electrons have different bulk velocities than the ions?

3.3.2   Halo Electrons

Electrons from ∼50 eV to 1 keV form the *halo* component. They generally have nearly isotropic pitch-angles but include a field-aligned nonthermal high energy tail (*strahl* electrons). Figure 3.9 shows examples of the angular distribution of

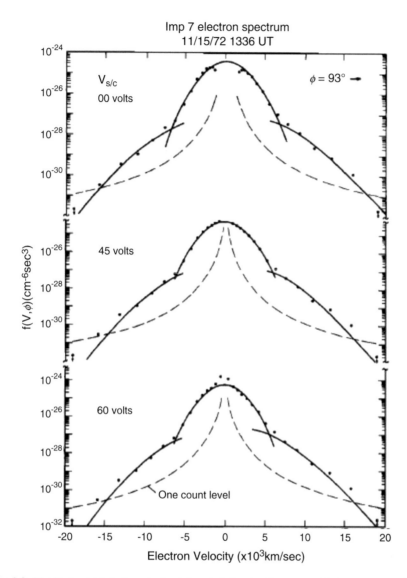

Fig. 3.8 Distributions of electrons in the solar wind showing the *core* and *halo* components for three different spacecraft potentials. The solid line is a bi-Maxwellian fit to the data (solid points). The velocity scale is 20×10^3 km s^{-1}. The measurements come from IMP 7 (from Feldmann et al., *J. Geophys. Res.*, **80**, 4181, 1975)

an electron beam observed in the fast solar wind from an early Explorer mission. The full width at half maximum (FWHM) of the peaks decreases with increasing energy, from 35° at 86 eV to 15° at 250 eV. These field-aligned nonthermal electrons are a possible source of heat flux and are observed mostly with fast SW. High energy halo electrons may also be important for producing the electrostatic potential needed to produce a fast solar wind (see Sect. 3.5 below, *Kinetic models*). The

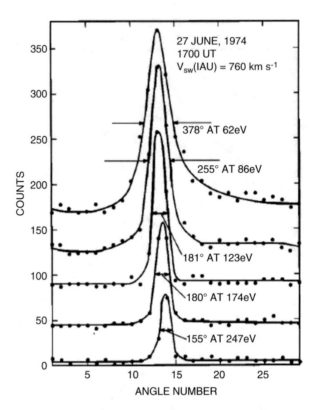

Fig. 3.9 Electron beams observed at various energies in the fast solar wind. Note that the high energy tail is field-aligned. The beam is more field-aligned at higher energies (from Feldmann et al., *J. Geophys. Res.*, **83**, 5285, 1978)

dependence of the narrowing pitch-angles as the energy increases has been verified by Wind observations (Fitzenreiter et al. 1998). However, Ulysses observations show contradicting features (Hammond et al. 1996). Readers should check to see if Ulysses data may have been time-aliased because the observations were made with 60 s time resolution magnetic field data. The issue of the pitch-angle dependence against the electron energy is important and warrants further studies.

The field-aligned SW electrons have received much attention in the observational and theoretical communities because they have been interpreted as the escaping solar coronal electrons along the magnetic field direction carrying heat flux. The distribution functions are anisotropic and skewed with respect to the magnetic field direction. The electron temperatures vary from 0.7×10^5 to 2×10^5 K with the temperature along the magnetic field $T_\parallel$ larger than the temperature in the perpendicular directions $T_\perp$ by a factor of 1.1–1.2. The heat flux along **B** is directed away from the Sun and varies from 5×10^{-3} to 2×10^{-2} ergs cm^{-2} s^{-1}. The readers should consider if the observed heat flux can be described by Spitzer and Härm (1953) transport model.

Fig. 3.10 Quiet-time solar wind electron VDF spectra from ~9 eV to 200 keV measured by WIND (blue) and STEREO A (black) and B (red) on 2007 December 6. The solid lines represent a fit by a sum of Maxwellian and Kappa function to the solar wind core and halo distributions and a power-law fit to the super-halo component (from Wang et al., *AIP Conf. Proc.*, **1539**, 229, 2013b)

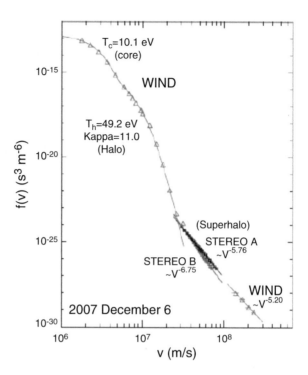

Question 3.7 If the particles leaving the Sun conserve the first adiabatic invariant, $p_\perp^2/B$, one would then expect the distributions to become narrowly field-aligned at 1 AU. Quantify this statement. How will wave-particle interaction affect this measurement?

3.3.3 Super-Halo Electrons

The SW measurements of electron distributions have now been extended using three separate detectors (Wang et al. 2013b). These measurements (Fig. 3.10) show electrons from a few eV to ≥ 100 keV are included in the "super-halo" component. The source of the high energy electrons is not known but this component is seen even during a quiet Sun suggesting that electrons of these energies are continuously escaping the Sun if they are of solar origin. There are many suggested theories about how the SW electrons are accelerated (Yoon et al. 2012a,b).

The statistical properties of 2–20 keV super-halo electrons in the solar wind have been measured by the STEREO/STE instrument during 2007 March through 2009 March during solar minimum. They show the quiet-time super-halo electrons have a nearly isotropic angular distribution and a power-law spectrum. The integrated density of the super-halo electrons at 2–20 keV varies from 10^{-8} to 10^{-6} cm^{-3}, which is about 10^{-9} to 10^{-6} of the solar wind density. The power-law spectrum

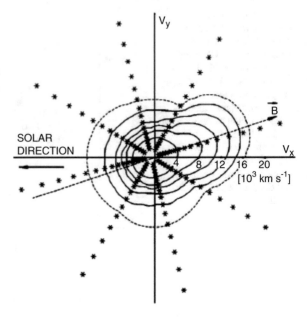

Fig. 3.11 2D electron distribution from Helios plotted in the moving SW frame. The core and halo components are evident (from Pilipp et al., *J. Geophys. Res.*, **92**, 1103, 1978)

shows no correlation with solar wind protons. These quiet-time super-halo electrons are present even in the absence of any solar activity: no active regions, flares, microflares, and type III radio bursts, suggesting that they may be accelerated by wave-particle interactions in the interplanetary medium or by a nonthermal process possibly related to the solar wind acceleration. Why the distributions of SW ions and electrons are so different is not understood.

3.3.4   Solar Wind Electrons Measured by Helios

Figure 3.11 shows an example of Helios electron distribution in 2D near the Sun. Here one can see the distributions are field-aligned. The distributions show core and halo components. The core consists of the bulk of the electrons with energies less than $\sim$50–100 eV and is nearly isotropic (or moderately anisotropic). The halo is about 5% of the total electron density and is anisotropic accelerated possibly by an electrical potential of $\sim$1–4 kV. Helios instrument did not measure electrons of energies responsible for the super-halo component.

3.3.5   Acceleration of Electrons

The question of how the SW electrons are accelerated is one of the most important questions in solar physics. Recent investigation of the injection timing of electron 3He-rich SEP (solar energetic particle) events (Wang et al. 2016) has found that the low-energy electron injection starts ~9 min before the coronal release of type III radio burst; the high-energy electron injection starts ~8 min after the type III burst, and the injection of ~MeV/nucleon, 3He-rich ions, begins ~1 h later, after the associated CME event was at altitudes of ~2–10 $R_\odot$. This work also showed that the selected electron 3He-rich SEP events have a remarkable one-to-one association with fast west-limb CME events.

Furthermore, it was shown that in the interplanetary medium, low- and high-energy electrons propagate differently, with more scattering occurring at higher energies. The authors suggest that the scattering is caused by the interaction with waves and turbulences at scales larger than the gyroradius of thermal protons in the solar wind. Although a transition occurs at energies where the electron injection delays are observed, the results show that scattering is not enough to produce these delays. Based on these results, a new coherent picture of electron 3He-rich SEP events has been suggested to fill the gap in knowledge about a fundamental process in acceleration mechanism.

This phenomenological model suggests that at the Sun, the low-energy electrons are accelerated first in jets that may appear as CME events high in the corona, and these low-energy electrons generate type III radio bursts. The high-energy electrons are accelerated subsequently at distances $\geq 1\,R_\odot$ by CMEs acting on the seed electrons provided by the initial low-energy injection. The MeV/nucleon 3He-rich ions are accelerated even later by a selective resonance mechanism with electron-beam generated waves and by fast, narrow CMEs. In the interplanetary medium within 1 AU, both low- and high-energy electrons often propagate nearly scatter-free, but the high-energy electrons experience more scattering than the low-energy electrons due to waves and turbulences generated by solar wind ions. This model provides the first look of a mechanism consistent with observations.

Super-halo electrons could originate from nonthermal processes related to the solar wind source or could form in the interplanetary medium due to effects arising from particle propagation. Interchange reconnection may be responsible for the super-halo component (Wang et al. 2016). In this picture, some of the accelerated electrons travel upward along open magnetic field lines into the interplanetary space to form the solar wind super-halo population, while the rest of the electrons propagate downward into the lower atmosphere. Both upward and downward traveling electrons collide with the solar atmosphere and generate the quiet-Sun high-energy X-ray emissions via the bremsstrahlung process.

3.4 Solar Wind Models

The current SW models are basically of two types. The SW is treated as either fluid or as particles. Both fluid and kinetic models assume the SW originates in the solar corona where the temperature is about 10^6 K, much hotter than the photospheric temperature of ~6000 K. What can heat the coronal plasma to a million degrees is not explained. Neither is anything known about the thermal and nonthermal transport mechanisms (Hunten 1982). Both types of models assume the time scales are much longer than the ion cyclotron frequency ($\omega \ll \Omega_c$) and spatial scales larger than the ion cyclotron radius ($r \gg R_\perp$). These models have been discussed in numerous review papers and conference proceedings (Lemaire and Scherer 1971; Hundhousen 1968; Meyer-Vernet 1999; Tam and Chang 1999; Maksimovic et al. 2001; Cranmer 2002; Echim et al. 2011). We briefly summarize what is known and discuss new material not covered in the review articles.

3.4.1 Fluid Models

Magnetohydrodynamic (MHD) fluid theory and concepts have been used since the early days to obtain a large-scale picture of space plasmas. The main advantage of the MHD fluid theory is that the dynamics can be described using only a few macroscopic parameters. However, as already mentioned in Chap. 1, MHD conservations equations, derived from moments of the collisionless Boltzmann equation, are incomplete because the procedure of taking moments always leaves us with more unknowns than the number of equations, hence they do not form a closed set. The fluid SW models are based on the original model proposed by Parker (1958). The model assumes SW flow is in steady state (no time dependence) and the flow equations are developed for the radial direction. There are also more complicated fluid models, for example, two fluid models that use the Chew-Goldberger-Low equations (Chew et al. 1956).

The basic equations used in the fluid model are continuity of mass and momentum, ideal gas assumption, and adiabatic equation of state. Fluid SW models have been discussed extensively in introductory space plasma physics textbooks. Hence our presentation is short, noting only the main points.

$$\frac{d}{dr}(r^2 \rho_m U) = 0$$

$$\rho_m \frac{d}{dr} U = -\frac{dp}{dr} + -\rho_m \frac{GM_\odot}{r^2}$$

$$p = nk_B(T_e + T_i) = 2nk_B T$$

$$p = p_o \left(\frac{\rho_m}{\rho_{mo}}\right)^{5/3} \tag{3.1}$$

where $\rho_m = nm$ is the mass density, n and m are number density and mass of protons, U is the flow velocity, and p is the pressure which is related to the temperature T. Assuming thermal equilibrium, $T_e = T_i$. These equations can be solved to yield

$$\left(\frac{U^2}{C_s^2} - 1\right)\frac{dU}{U} = \left(2 - \frac{GM_\odot}{C_s^2 r}\right)\frac{dr}{r} \tag{3.2}$$

where $C_s = (dp/d\rho_m)^{1/2}$ is the sound speed, G is the gravitational constant, $M_\odot$ is the mass of the Sun, and r is distance measured from the center of Sun. Integration of this equation yields a solution U that satisfies

$$\frac{3}{8}\frac{1}{v_{th}^2}(U^2 - U_c^2) - \frac{1}{2}\ln\frac{U}{U_c} = \ln\frac{r}{r_c} + \frac{r_c}{r} - 1 \qquad (3.3)$$

where U_c is an integration constant and represents the solar wind speed at $r = r_c = GM_\odot m/4kT$ and $v_{th}^2 = 3kT/m$. Note that U_c and r_c are free parameters. U_c is also called the critical speed and for the model used here it is the speed of sound, C_s. Equation (3.3) can be rewritten as

$$\frac{U^2}{U_C^2} - 2\ln\left(\frac{U}{U_c}\right) = 4\ln\left(\frac{r_c}{r}\right) + 4\left(\frac{r_c}{r}\right) - 3 \qquad (3.4)$$

This equation has a family of solutions for different values of U_c and r_c. Equation (3.4) represents one of four possible solutions to the differential equation (3.2). While all four solutions are mathematically acceptable, physical arguments eliminate the three solutions and only the one shown is satisfactory for the SW problem (one of the three solutions is Chamberlain's solution which corresponds to "solar breeze"). The solution equation (3.4) starts with $U < C_s$ near the base of the solar corona, reaches C_s at the critical distance r_c, and continues to increase beyond r_c (Parks 2004).

3.4.2 Issues with Fluid Models

The SW in the fluid model starts out at the base of the solar corona. The flow speed is fairly small there but evolves to a super-sonic speed shortly after leaving the base. The fluid SW is driven by the available heat energy at the Sun and the important issue is to understand how this heat is transported outward against the gravitational binding energy (Meyer-Vernet 2007).

Assuming that there are only protons p^+ and the flow velocity is V_{SW} at large distances (1 AU), we can write the energy balance equation (ignore mass of electrons m_e)

$$\frac{V_{SW}^2}{2} \approx \frac{5k_B T_o}{m_p} - \frac{M_\odot G}{r_o} + \frac{Q_o}{n_o m_p V_o} \qquad (3.5)$$

where the subindex (o) refers to parameters measured at the coronal base. Thus, r_o is the distance on the Sun where the solar wind originates, T_o is the temperature, n_o is density of protons (or electrons), m_p is the proton mass, $M_\odot$ is the mass of the Sun, and G is the Universal gravitational constant. The three terms on the right-hand side are enthalpy per unit mass due to both protons and electrons (heat content), gravitational binding energy, and heat flux per unit mass. Since the bulk flow speed is assumed small at the source, the bulk kinetic energy of the flow has been neglected.

If we use $V_{SW} = 4 \times 10^5 \, \mathrm{m\,s^{-1}}$, the numerical value of the left side is $\approx 1.6 \times 10^{11} \, \mathrm{J\,kg^{-1}}$. To estimate the values on the right side, we use $T = 2 \times 10^6 \, \mathrm{K}$, $r_o \sim r_\odot \approx 7 \times 10^8 \, \mathrm{m}$, and $M_\odot \approx 2 \times 10^{30} \, \mathrm{kg}$. This yields $\sim 0.8 \times 10^{11} \, \mathrm{J\,kg^{-1}}$ for the enthalpy and $\approx 2 \times 10^{11} \, \mathrm{J\,kg^{-1}}$ for the gravitational binding energy. The available enthalpy is therefore *not* sufficient to overcome the Sun's gravitation energy indicating the heat flux term is important.

To evaluate the heat flux term in (3.5), we note that fluid plasma is collisional and the applicable heat conduction equation is

$$Q = -\kappa_o \frac{dT}{dr} \tag{3.6}$$

A fluid is made up of electrons and ions but since electrons have much higher thermal speed than the protons, heat is mainly transported by the electrons. For a fluid with a density n, the heat capacity is $3nk_B/2$ per unit volume, where k_B is Boltzmann's constant. The thermal conductivity is

$$\kappa_o \approx 3kn \times v_{\mathrm{th}}^e \times \lambda_c \tag{3.7}$$

where the thermal velocity of the electron is $v_{\mathrm{th}}^e = \sqrt{2kT/m_e}$, the collisional mean free path is $\lambda_c \approx 1/\pi n r_c^2$, and πr_c^2 is the Coulomb cross section for charge particle interaction. The expression of collisional mean free path is $\lambda_c \sim 3 \times 10^7 T^2/n$, given by Spitzer and Harm (1953). Assume that there are no heat losses at the source. We can then use

$$\frac{d}{dr}\left[r^2 \kappa_o \frac{dT}{dr} \right] = 0 \tag{3.8}$$

to deduce how the temperature T varies as a function of the distance away from the source r. Let the heat conductivity be described by $\kappa_o \sim T^{5/2}$ and set the boundary condition that $T \to 0$ at large distances. Equation (3.8) then shows $T \sim r^{-2/7}$. Substitution of this relationship into (3.6) shows that the heat flux at the coronal base reduces to

$$Q_o \approx 3.7 \times 10^7 k^{3/2} m e^{-1/2} T_o^{7/2} \tag{3.9}$$

We now need to estimate the value of $n_o V_o$ in (3.5). This is obtained using the equation of conservation of the number flux. At the Earth's orbit ($\sim 215 r_\odot$), $n \sim 5 \times 10^6 \, \mathrm{m^{-3}}$, $V \sim 400 \times 10^3 \mathrm{m\,s^{-1}}$, yielding $\sim 2 \times 10^{12}$ particles $\mathrm{m^{-2}\,s^{-1}}$. The number flux at the Sun is $n_o V_o \sim 2 \times 10^{12} \times (215)^2$. Substitution of all of these numbers into (3.5) assuming $T = 2 \times 10^6 \, \mathrm{K}$ yields $Q_o \sim 2 \times 10^{11} \, \mathrm{J\,kg^{-1}}$. This value just balances the gravitational binding energy. The left side of Eq. (3.5) $V_{SW}^2/2$ must be accounted for by the enthalpy term, $\sim 0.8 \times 10^{11}$ J kg^{-1}, which will produce a terminal velocity of 400 km s^{-1}.

While this calculation shows there is enough enthalpy to account for the observed high velocity of the SW, the enthalpy term is extremely sensitive to the temperature T because heat flux varies as $T^{7/2}$ (Eq. (3.9)). This means that if the temperature is smaller by 15%, the right-hand side of (Eq. (3.5)) becomes *negative*! This fact is important because fluid models cannot explain how the fastest SW comes from the coldest regions of the solar corona where the temperature is only $\sim 10^5$ K or less. The thermal conductivity here falls short by an order of magnitude of the value needed to drive the solar wind. In the original model Parker assumed uniform T requiring ∞ heat conductivity.

Additional energy sources have been sought to save the fluid model. For example, suggestions have been made that micro flares and Alfvén waves could provide the additional energy. However, these are perturbations and it is not known how they could provide the needed energy in the right place at the right time to accelerate the solar wind. The general conclusion is that thermally driven SW fluid models do not have the capability to give quantitatively accurate estimates of the terminal SW velocity. Moreover, there are many other features of the SW observations that are not consistent with the fluid models.

3.5 Kinetic Models of the SW

Nearly all kinetic models assume that the SW plasma consists of H$^+$ ions and an equal number of electrons e$^-$ so there is charge neutrality. In the real SW, H$^+$ ions are typically 95% and the other 5% is heavier ions with He^{++} typically 4% which can increase to 20% during ICMEs. Nevertheless, we will assume as most theories and models do that SW consists of only H$^+$ ions and an equal number of electrons.

Theoretical treatment of kinetic models uses the distribution function $f(\mathbf{r}, \mathbf{v})$. Fortunately, the velocity distribution function is now routinely measured in situ by instruments carried on satellites. Kinetic models use the Boltzmann equation,

$$\partial f/\partial t + \mathbf{v} \cdot \partial f/\partial \mathbf{r} + \mathbf{a} \cdot \partial f/\partial \mathbf{v} = (\partial f/\partial t)_c \qquad (3.10)$$

where $\mathbf{a}$ is acceleration $\mathbf{a} = (q/m)(\mathbf{E} + \mathbf{v} \times \mathbf{B}) + \mathbf{g}$ and $\mathbf{g}$ is solar gravitation. The term on the right is due to Coulomb collision between particles and it vanishes if the plasma is assumed collisionless. Inclusion of the Coulomb collision term will require solving the Fokker-Planck equation. There are two equations, one for electrons and another for ions. If He^{++} is included, a third equation will be needed. The electric $\mathbf{E}$ and magnetic $\mathbf{B}$ fields in the Boltzmann equation are obtained from Maxwell equations, $\nabla \times \mathbf{E} = -\partial \mathbf{B}/\partial t$, $\nabla \cdot \mathbf{D} = \sum e \int f d^3v$, $\nabla \times \mathbf{H} = \partial \mathbf{D}/\partial t + \sum e \int \mathbf{v} f d^3 v$, and $\nabla \cdot \mathbf{B} = 0$. where the charge and current densities require summing over the different particle species. The general problem of the SW is extremely complicated and has not been solved yet in entirety (Zouganelis et al. 2003; Yoon et al. 2012a,b). A recent review paper has discussed possible instabilities driven by anisotropies observed in the SW (Yoon 2017).

3.5.1 Electrostatic Potential of the Atmosphere

One of the first particle models of how the particles escaped the solar atmosphere was proposed by Chamberlain (1960). The model assumed solar plasma was in *hydrostatic equilibrium* and used the electric field model of Pannekoek (1922) and independently derived by Rosseland (1924). The Pannekoek-Rosseland model treats free electrons and ions as ideal gases. They assume there is no charge density from external sources, hence $\rho_{ext} = 0$. The Boltzmann distributions for the number of free electrons and ions at height $\mathbf{r}$ are given by

$$N_e(r) = N_{eo}e^{(-m_e\phi_g+e\phi_E/kT)}$$
$$N_i(r) = N_{io}e^{(-m_i\phi_g-Ze\phi_E/kT)} \tag{3.11}$$

where N_{eo} and N_{io} are densities at a reference height r_o, $\phi_g = GM_\odot/R$ is the gravitational potential, $M_\odot$ is the solar mass, G is Universal gravitational constant, R is radius of the Sun, ϕ_E is electrostatic potential, and the subindices e and i refer to electrons and ions. Let the charge number $Z = 1$. Then the potentials are obtained by solving the Poisson equation,

$$\nabla^2\phi_E = -\rho_c/\epsilon_o$$
$$\nabla^2\phi_g = -4\pi G\rho_m \tag{3.12}$$

where $\rho_c = ZeN_i - eN_e$ and $\rho_m = m_iN_i + m_eN_e$ are the charge and mass densities. These equations are satisfied when

$$\phi_E = -\frac{m_i - m_e}{e(Z+1)}\phi_g \tag{3.13}$$

For solar plasma consisting of H^+ and e^-, $q^+ = -q^- = |e|$ and $n^+ = n^- = n$ since the plasma is neutral and in thermal equilibrium. Let us now evaluate the charge density,

$$\rho_c = e(N_i - N_e)$$
$$= e[N_{io}e^{(-m_i\phi_g-e\phi_E)/kT} - N_{eo}e^{(-m_e\phi_g+e\phi_E)/kT}]$$
$$= -\frac{en}{kT}(m_i - m_e)\phi_g + \frac{2ne^2}{kT}\phi_E \tag{3.14}$$

where the last line comes from expanding the exponential terms. The mass of the electrons in the first term can be ignored since $m_- \ll m_+$. Now use the relation

$$\phi_g = \frac{GM_\odot m_i}{r} = -\frac{m_i g R^2}{r} \tag{3.15}$$

where $g = -GM_\odot / R^2$. One of the Poisson equations then becomes

$$\nabla^2 \phi = \frac{enm_i g R^2}{\epsilon_o k T r} + \frac{2ne^2}{\epsilon_o k T} \phi_E$$

$$= \frac{1}{\lambda^2} \phi - \frac{m_i g R^2}{e\lambda_D^2} \mathbf{r} \tag{3.16}$$

where $\mathbf{r}$ is measured from the center of the Sun and the second line used the relation $\lambda_D^2 = kT\epsilon_o / nq^2$ where λ_D is the deBye length. The solution of this differential equation is

$$\phi = \frac{1}{4\pi} \int \frac{e^{-|\mathbf{r}-\mathbf{r}'|}}{|\mathbf{r}-\mathbf{r}'|/\lambda_D} d\mathbf{r}' \frac{m_i g R^2}{e\lambda_D^2 |\mathbf{r}'|} \tag{3.17}$$

Now since λ_D / r is very small because $r > R \gg \lambda_D$, the integrand is finite only in a small volume about λ_D^3 centered about r and it is reduced to $4\pi\delta(\mathbf{r}-\mathbf{r}')$. Hence,

$$\mathbf{E} = -\nabla\phi = \frac{m_i g R^2}{2er^2}\hat{\mathbf{r}} = \frac{m_i g}{2e}\hat{\mathbf{r}} \tag{3.18}$$

where we let $r^2 \approx R^2$. The electric field of the solar corona under gravity is due to pile-up of charges and since the ions are heavier, the electric field points outward. This equation shows that the vertical electric field reduces the force $q\mathbf{E}$ on each ion to one half its gravitational value as if the ions are lighter in the plasma atmosphere than in neutral gas and it provides a downward force on the electrons equal to the upward force on the ions. This downward force on the electrons is much larger than the gravitational force on the electron, as if the electrons are half as heavy as the ions.

Question 3.8 A small charge imbalance induces an electric field $\mathbf{E}$ pointing outward. The total force on any parcel of the atmosphere in equilibrium must vanish since the force balance requires that $F_g - q\mathbf{E} = q\mathbf{E}$ where the left side is force on electrons and it balances the force on ions, right side. What is the magnitude of the induced electric field and the amount of estimated charge imbalance?

3.5.2 Flux of Escaping Particles

To calculate the number of charged particles escaping the solar atmosphere to produce the SW, the book by Jeans (1925) is an excellent source of reference. Jeans was one of the first to consider how the planetary and stellar atmospheres evolved in space and time. The calculations he made can be expanded by adding the electromagnetic forces applied to charged particles.

Consider a sphere of radius r_o below which the plasma is collision dominated and above which the plasma is collisionless. This distance has been called *exobase* and located where the collisional mean free path is just larger than the scale height (Lemaire and Scherer 1973). The particles at r_o are gravitationally bound except for the particles that have kinetic energies that exceed the binding potential energy.

If collisions are ignored for $r > r_o$, the total energy of the particles is conserved and we can write

$$\frac{1}{2}mv^2 + m\phi_g(r) + Ze\phi_E(r) = \frac{1}{2}mv_o^2 + m\phi_g(r_o) + Ze\phi_E(r_o) \tag{3.19}$$

where v is the velocity of the particle at r ($r > r_o$), $\phi_g = -GM_\odot/r$ is the gravitational potential, ϕ_E is the electrostatic potential, and Ze the electric charge. The escape velocity at r_o is then

$$v_{\text{esc}}(r_o) = \left[-2\phi_g(r_o) - \frac{2Ze\phi_E(r_o)}{m} \right]^{1/2} \tag{3.20}$$

For electrons, the gravitational term can be neglected and the escape velocity is simplified to

$$v_{\text{esc}}^e(r_o) \simeq \left[\frac{2e}{m_e}\phi_E(r_o) \right]^{1/2} \tag{3.21}$$

Note that the flux of particles leaving the corona is crucially dependent on the model of the electrostatic potential $\phi_E(r)$.

To compute the particle fluxes escaping the atmosphere, assume a static atmosphere and that the particles are in thermal equilibrium allowing us to use the Boltzmann distribution. The flux of particles $(\text{cm}^2\text{-s})^{-1}$ leaving the sphere in the outward direction with a velocity $> v_{\text{esc}}$ is

$$\begin{aligned} J_{\text{esc}} &= n_o \left(\frac{m}{2\pi k_B T} \right)^{3/2} \int_{v_{\text{esc}}}^{\infty} e^{m(v_x^2+v_y^2+v_z^2)/2kT} v_x dv_x dv_y dv_z \\ &= \frac{n_o}{2\sqrt{\pi}} v_{\text{th}} \left(1 + \frac{v_{\text{esc}}^2}{v_{\text{th}}^2} \right) e^{-v_{\text{esc}}^2/v_{\text{th}}^2} \end{aligned} \tag{3.22}$$

where n_o is the density of particles at $r = r_o$ and v_x is the positive velocity component normal to the sphere. The integration is taken for all values of v_x, v_y and v_z and $v_x^2 + v_y^2 + v_z^2 > v_{esc}^2$. Once a particle escapes the Sun, it is in a hyperbolic trajectory and thus is permanently lost from the solar (stellar) atmosphere. We have used spherical coordinates with the integration limits θ $(0, \pi/2)$, ϕ $(0, 2\pi)$, and v (v_{esc}, ∞).

When this equation is applied to the solar atmosphere with a temperature 10^6 K, one finds that nearly 50% of electrons will escape and $\ll 1\%$ of protons will. This will leave the solar corona with a huge net positive electrical charge and consequently we need to consider now how to deal with this serious problem.

Question 3.9 Solve the equation to estimate the fluxes of the escaping electrons.

Question 3.10 Solve the equation to estimate the fluxes of the escaping protons.

3.5.3 Zero Net Current Requirement

Charge neutrality condition is not sufficient to produce a SW model. The next step is to find the conditions for an equal number of electrons and ions to escape, requiring a zero net current. The charge neutrality requires $n_e(r_o) = n_i(r_o)$ and zero net current requires $J_{esc}^e = J_{esc}^i$. Use of the equations derived above will then yield

$$\left(1 + \frac{v_{esc}^2}{v_{th}^2}\right) e^{-v_{esc}^2/v_{th}^2} = 2 \left(\frac{v_{th}^i}{v_{th}^e}\right)$$

$$= 2 \left(\frac{m_e}{m_i}\right)^{1/2} \tag{3.23}$$

The above equation shows that the kinetic SW model is a natural consequence of the solar coronal plasma maintaining quasi-neutrality of charges and zero net current.

Question 3.11 Estimate the value of the electrostatic potential for a solar coronal temperature of 10^6 K.

There are other solar wind models. For example, some kinetic models predict that the presence of the non-Maxwellian high energy tail can increase the solar wind speed and may account for the fast solar wind. Other models have suggested that cyclotron resonance heating occurring at the source may account for the bulk acceleration of the solar wind. Another model predicts that the inclusion of the spiral interplanetary magnetic field can change the plasma temperature. Finally, a few papers have attempted to model the high speed solar wind from collision-dominated lower-coronal heights into the collisionless interplanetary space using the Fokker-Planck collision operator to describe the Coulomb collisions of electrons (Echim et al. 2011). The readers are encouraged to discuss how the existing models can be further improved, focusing on what parameters should be measured. Solar magnetic fields have been ignored in most models. How serious is this?

3.6 Heuristic Interpretation of the Solar Wind

Our understanding of the behavior of the SW has come largely from macroscopic variables calculated from the velocity moments of the distribution function $f(\mathbf{r}, \mathbf{v})$. Fluid models ignore two important features about the SW that need to be discussed further. The first is how to interpret the beam distribution and the second is the meaning of the beam displaced from the measurement platform (see Fig. 3.3). These features are not discussed by any models but as shown below we think they are important.

3.6.1 Fluid Interpretation

Both fluid and kinetic models assume a priori that the Sun's corona is hot and is expanding into space. The expansion velocity at 1 AU is obtained by calculating the velocity moment of the measured SW distribution. The mean velocities can vary from a few hundred $\mathrm{km\,s^{-1}}$ to more than $1000\,\mathrm{km\,s^{-1}}$. For the example shown (Fig. 3.3), the mean velocity of the SW was V_x equal to $-400\,\mathrm{km\,s^{-1}}$. In this calculation, the measured distribution did not include many particles from the reflected and gyrating population but it included some He^{++}. The SW can be represented analytically by a flowing Maxwellian,

$$f(\mathbf{v}) = C e^{-(\mathbf{v}-\mathbf{U})^2/v_{th}^2} \tag{3.24}$$

where $C = n_o/(\pi^{3/2} v_{th}^3)$, v_{th} is thermal velocity, and $\mathbf{U} = \int \mathbf{v} f(\mathbf{v}) d^3 v$ is the mean velocity of the distribution measured by a stationary observer (spacecraft). Fluid models interpret the SW beam as representing the thermal distribution of the expanding coronal atmosphere flowing out with a velocity $\mathbf{U}$. The range of SW speeds and temperature observed would then imply how the expansion velocity and thermal temperatures of the coronal atmosphere vary for different times.

3.6.2 Particle Interpretation

Another possible interpretation is to recognize that the Sun has a magnetic field and the coronal plasma is magnetized. The SW interpretation must then take into account how charged particles interact with electric and magnetic fields. The magnetic field on the Sun is very complex and depending on the solar cycle, there are many small scale transient structures in addition to the general dipole field (Fig. 3.12). We will not consider the complex transient magnetic fields. Instead, for our purpose, we will assume that the Sun has only a general dipole field.

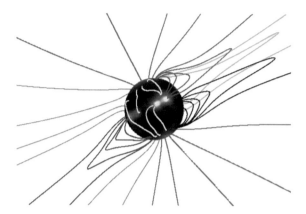

Fig. 3.12 A computer simulation of general solar magnetic field. The Sun's magnetic field is complex but includes a dipole field of about 1 Gauss. The Sun's plasma in the corona is active and produces currents which in turn generate smaller scale transient magnetic fields. Fast SW is observed to escape the corona along the polar magnetic fields. The slower SW originates in the equatorial regions where the fields are modified dipole shaped because of currents generated by the plasmas. The colors represent intensity of the magnetic field with yellow on the surface most intense (Courtesy of NASA)

Consider now the particles on the Sun's equatorial plane at coronal altitudes. The particles are trapped and moving in the dipole field executing the gyrating motion, longitudinal and drift motions as trapped particles do in Earth's radiation belts. To produce a SW, the guiding centers of these particle must move away from the Sun by crossing the dipole field. How can this be done? We know the answer because the Lorentz equation has already shown us how this can be achieved. Let the particles be nonrelativistic. These equatorial particles will cross the magnetic field when they experience an electric field perpendicular to the magnetic field direction.

For simplicity, we let the magnetic field on the equator be given $\mathbf{B} = (0, 0, B)$ and the electric field by $\mathbf{E} = (0, E, 0)$. The two fields are perpendicular to each other. Lorentz equation then shows in the particle frame (S' frame), an observer will see the particle gyrating about the magnetic field while in S frame the particle is gyrating but also drifting in the $\mathbf{E} \times \mathbf{B}$ direction moving away from the Sun. The $\mathbf{E} \times \mathbf{B}$ drift is charge independent so electrons and ions will both drift out together maintaining charge neutrality in the plasma. The shift of the SW beam in $V_{\perp}$ can then be interpreted as the velocity of the S' frame whose velocity is $(\mathbf{E} \times \mathbf{B})/B^2$.

We have thus far discussed observations at 1 AU. These observations have raised questions about how much of the features measured at 1 AU represent the original properties of the SW. Has the SW been modified in transit from Sun to Earth? Is the first magnetic invariant conserved? Where is the temperature anisotropy observed at 1 AU produced? If the SW has been modified in transit one needs to study what mechanisms are possibly working in transit to modify the distributions, which is a different question if the observed features have been produced at the source. Some theoretical interpretation of the microphysics of the solar wind assumes that the distribution function of the measured SW retains features from the origin and thus the observed SW reveals information about the plasma state in the coronal source region (Marsch 2006). But this question needs to be further studied.

Question 3.12 Estimate the value of the electric field at the position of the particle for it to drift across the solar magnetic field at coronal heights. Can this perpendicular **E** field produce a SW beam?

3.6.3   Speculative Model of the Solar Wind

In many ways the SW electron distribution (see Fig. 3.10) resembles the electron distributions above Earth's aurora. On Earth, well-formed beams are observed accelerated by the potential drop. The auroral beams show $\Delta\phi$ is typically a few hundred eV to ~keV. However, the beam is always accompanied by lower and higher energy electrons that are nearly isotropic.

For Earth, the lower energy electrons have been interpreted as secondaries produced by the electron beam propagating through the atmosphere (Evans 1974). Simulation results of field aligned keV beams going through the Earth's atmosphere have reproduced features observed in the data giving strong support to the secondary electron interpretation. The low energy electrons could also come from redistribution of unstable beam electrons. The higher energy electrons have been interpreted as electrons from the plasma sheet.

Question 3.13 Suppose one assumes the solar coronal electrons have gone through a potential drop producing a beam. Suggest how the other electron components might be produced on the Sun.

Plasma theories suggest that particle acceleration can come from the energy flow in field-aligned currents that couple the lower regions to the more distant regions. Electromagnetic energy is converted to particle kinetic energy so that $\mathbf{j} \cdot \mathbf{E} > 0$. One possible mechanism in Earth's aurora is the double layer which is a positive-negative charge separation that is maintained by the flow of ions and electrons into the double layer.

Speculating once again that similar mechanisms are active at solar coronal heights, we suggest that plasma motions at coronal heights induce electric field and particles are accelerated outward along the magnetic field direction. There is no physical reason why such a plasma process is not occurring on the Sun. This model would suggest that the solar coronal H^+ beams measured by spacecraft experiments at $400\,\mathrm{km\,s^{-1}}$ to more than $1000\,\mathrm{km\,s^{-1}}$ represent the amount of equivalent potential drops the ions have gone through. The readers are encouraged to discuss if it is possible to have a current system on the Sun like the one on Earth. What can one do to support or dispel this suggestion?

3.7 Electrostatic Solitary Waves

Figure 3.13 shows three examples of ESWs detected by an experiment on Wind at L_1. We see here uni- and bi-polar pulses lasting for a few milli-seconds. Noting their similarities to the structures observed in Earth's auroras, these structures have been interpreted as double layers (DL) by Mangeney et al. (1999). The DLs have dimensions of a few tens of Debye lengths and are aligned along the magnetic field direction, as are the ESWs observed in the auroral ionosphere, although the SWDL amplitudes are much smaller. Double layers can transport charges and currents efficiently over long distances retaining their shape and velocity. A theory of double layers is given in Chap. 6. Excitation of ESWs has also been modeled by PIC simulations (Choi et al. 2012).

3.7.1 Double Layers

Hannes Alfvén in 1958 suggested that auroral field-aligned currents are sufficiently intense to produce space-charge regions where charge neutrality is not maintained and high potential difference could be developed along the magnetic field direction.

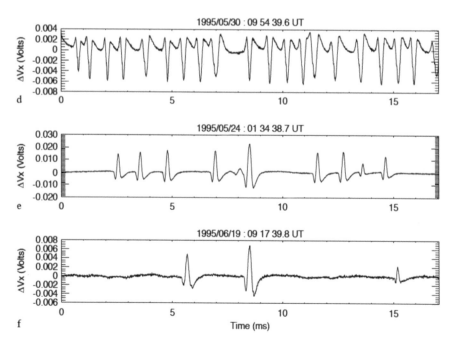

Fig. 3.13 Examples of electrostatic solitary waves in the SW. Time here is in milli-second (from Mangeney et al., *Ann. Geophys.*, **17**, 439, 1999)

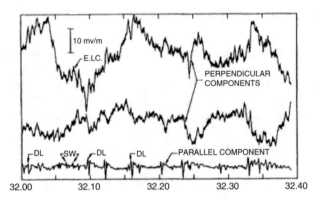

Fig. 3.14 Shown are examples of perpendicular (top two) and parallel (bottom) components of the electric field measured over the aurora on the S3-3 satellite at an altitude of ~6000 km. The "spikes" in the parallel component are double layers (DL) and solitary waves (SW). The total time interval represents 0.4 s (from Temerin et al., *Phys. Rev. Lett.*, **48**, 1175, 1982)

Such space-charge regions are most likely associated with *double layers* (DL) which have been observed above the aurora with field-aligned currents and beams. Double layers were first discovered in laboratory plasma experiments. The first detection of DL in space occurred over the auroral zone in 1977.

An example of DL detected with downward going electron beams observed over the aurora is shown in Fig. 3.14. The top two curves show measurements of electric field perpendicular to the magnetic field $\mathbf{E}_\perp$ direction, and the bottom curve electric field parallel to the magnetic field $\mathbf{E}_\parallel$. The fast variations observed in the $\mathbf{E}_\perp$ curves are electrostatic ion cyclotron (EIC) waves. The "spikes" labeled DL in the $\mathbf{E}_\parallel$ component are double layers and have durations ~2–20 ms. The spikes labeled SW are *solitary waves* consisting of two opposite polarities of approximately equal magnitudes (SW is also used to label solar wind, but there should not be any confusion here). Both DLs and SWs have magnitudes $|\mathbf{E}_\parallel| \sim 15 \times 10^{-3}$ V m^{-1}. There are no corresponding spikes of $\mathbf{E}_\parallel$ in the $\mathbf{E}_\perp$ curves. These DLs were observed in the evening sector of the auroral zone where upward field-aligned current $|\mathbf{J}_\parallel| \sim 10^{-7}$ A m^{-2} flowed and 0.5 keV ion beams were moving upward and 0.5 keV electrons moving downward.

Although DLs have dimensions of only a few to few tens of Debye lengths they are stable and can travel a long distance carrying the potential and retaining their shapes. On Earth, DLs produced at auroral heights travel along the geomagnetic field and have been seen even in the distant regions of the plasma sheet boundary layer (Chap. 6). The DLs may be the source of large-scale $\mathbf{E}_\parallel$. While this problem remains unsolved, one suggestion is that $\mathbf{E}_\parallel$ may come from many double layers that are distributed along the geomagnetic field.

3.8   Concluding Remarks

We have used old and new observations and asked questions that still remain unanswered. We have given a brief review of the possible mechanisms proposed by various models. The current models have not discussed quantitatively what mechanisms could accelerate electrons and ions differently.

The energy per charge detectors do not distinguish the different ion species. We discussed that H^+ is distinguished from He^{++} because one assumes that all particles in the SW travel together carrying frozen in magnetic field. But as noted, this assumption is only valid for SW particles traveling perpendicular to the magnetic field and does not apply for SW particles traveling along the magnetic field direction. We will thus argue that even though SW has been studied since the beginning of space age, instruments have not measured SW unambiguously. Correct solar wind measurements must be made by experiments designed to measure energy/charge followed by mass/charge. The design of the instrument must take note of the fact that 95% of the SW flux is H^+ and one must prevent "spill over" of H^+ into He^{++} channel so the He^{++} data are not contaminated.

Mass composition experiments have been flown to identify m/q of the ions detected. However, because the H^+ is so dominant, it is difficult to measure separately He^{++} ions without contamination from H^+ ions. No solar wind experiment has provided clean separation of H^+ and He^{++} ions. The MMS mission uses a new instrument with a novel technique to measure H^+ and He^{++} but as of this writing we have not seen the data to ascertain the measurements of He^{++} are clean and not contaminated by the abundant H^+. Note that plasma instruments flown on the future ESA's THOR mission will include mass ion analyzers that have been designed to sort out the different ion species.

3.9   Solutions to Questions in Chap. 3

Solution 3.1 There are two detectors on Cluster with different geometrical factors, g and G. The g-detector includes 8 sensors aligned along the polar (θ) direction separated by $\sim5.625°$ and the SW is detected only in the $\sim45°$ azimuthal ϕ-wedge aligned along the Sun-Earth line divided into 8 smaller ϕ sectors ($5.625°$). A microprocessor controls the HV and sweeps only near the peak of the SW distribution. Data from each of the 64 $\theta - \phi$ directions can be examined separately. The different panels represent examples of the output of the 8-θ detectors from the different azimuthal ϕ-directions (the time resolution between the panels is 62.5 ms). The g-detector measures counts only from the X_{GSE} direction (toward the Sun) for particles with energies near the peak energy of the SW. The peak energy is determined from data obtained from the previous spin which is stored and subsequently used. Consequently, the measurements do not give information on the more energetic particles. The G-detector measures the full 3D distributions from a few eV to about 30 keV/charge but omits measurements in the X GSE direction where the SW is normally found because the detector will saturate by the intense SW beam. Hence, in the upstream SW, it is programmed to not measure the SW but measures reflected and gyrating populations and particles from all other directions. The combined data from the two detectors give complete information on the 3D distributions from the SW to the downstream MS.

Solution 3.2 These counts are background counts contributed by cosmic ray particles and they are not part of the SW. To be sure, one needs to check the distribution function plots to see whether they are part of the SW distribution. The bulk parameters do not show these counts because in the process of integration they have been averaged out.

Solution 3.3 The spacecraft is the observer frame of reference, S-frame. The SC in velocity space is at the origin.

Solution 3.4 The SW is always coming from the $+V_x$ direction (GSE coordinate system) regardless of the SW bulk parameters and direction of the magnetic field. To explain this observation, compute the kinetic energy of the bulk motion and convince yourself that it is generally larger than the thermal energy and magnetic energy. Since the energy of the bulk motion dominates and the Sun is in the $+X$ direction, the SW will thus always be observed coming from the X-direction.

Solution 3.5 The temperature of the SW beams is calculated in the rest frame of the plasma. To achieve this, the velocity is shifted until the beam is located at the origin ($< v >= 0$). Moreover, a constraint has been used so that the peak of the distributions $f(v_\parallel)$ and $f(\perp)$ will be the same and the peak of the distributions is as close as possible to the center. If the cuts along $v_\parallel$ and $v_\perp$ are somewhat off from the center then $f(v_\parallel)$ or $f(v_\perp)$ do not go through the center, the calculated temperature will be affected. We let the readers decide if it will be slightly lower or higher. He^{++} is identified by ESAs assuming that ESAs are measuring H$^+$ and as discussed in Chap. 1, the He^{++} appears at twice the energy/charge of H$^+$. So we conclude these ions are He^{++} ions.

Solution 3.6 The same temperature suggests that thermal equilibrium has been achieved. The anisotropy results in the SW could be telling us the mechanisms for thermalization in the two directions are not the same. The readers are urged to come up with other possible reasons.

Solution 3.7 The pitch-angles will become narrower because the intensity of the solar magnetic field will be reduced considerably and one would expect the pitch-angles to become field aligned as the particles move to smaller magnetic field values. For example, if the magnetic field at the corona is 10^4 nT and the magnetic field where the distributions is measured at 1 AU is 10^8 nT, the pitch-angle will be reduced to about 0.5°. The observed distributions will be field-aligned and narrow, but will cover larger pitch-angle ranges because of inadequate pitch-angle resolution of the instrument. However, it is also possible that the original pitch-angle distributions could have been increased by wave particle interactions that can scatter the original pitch-angles to larger values.

Solution 3.8 The estimated charge imbalance can be computed from $\nabla \cdot \mathbf{E} = \rho/\epsilon_o = q\delta n/\epsilon_o$ yields $\delta n = \epsilon_o m_p M_\odot G/q^2 r^3 = 8.85 \times 10^{-12} \times 1.67 \times 10^{-27} 2 \times 10^{30}/(1.6 \times 10^{19})^2 (3 \times 7 \times 10^8)^3 \approx 10^{-6} \mathrm{m}^{-3}$ if we let $r = 3r_\odot$. The electron density at $3r_\odot$ is $\approx 8.86 \times 10^{20} \mathrm{m}^{-3}$, hence $\delta n/n = 10^{-6}/10^{21} \approx 10^{-27}$, which is a very small number. Use $E \approx F_g/2q = m_p M_\odot G/2qr^2$ to estimate the electric field. Putting in the numbers, we obtain $|\mathbf{E}| \sim 6.6 \times 10^{11} \mathrm{V\,m}^{-1}$. This is a huge number indicating that even a very small charge imbalance will induce a huge electric field. Note that this calculation does not take into account of collective interaction, in which case charge imbalance occurs only inside a DeBye sphere. Some models assume that this electric field can drive the SW out but this assumption is not obvious. The readers are recommended to discuss if any of the future SW missions have experiments that measure these fields and can verify the predictions.

Solution 3.9 For this problem, we use spherical coordinates in Eq. (3.22) and let $d^3 v = v^2 \sin\theta d\theta d\phi dv$ where the integration limits for θ are $(0, \pi/2)$, ϕ $(0, 2\pi)$ and v (v_{esc}, ∞). Then,

$$J_{esc}^e = \frac{n_{eo}}{\pi^{3/2} v_{th}^3} \int v^3 e^{v^2/v_{th}^2} dv \sin\theta \cos\theta d\theta d\phi$$

$$= n_o \pi \left(\frac{m}{2\pi kT}\right)^{3/2} \int v^3 e^{v^2/v_{th}^2} dv$$

where $v_{th} = \sqrt{2kT/m}$ and the second line follows since $\int \sin\theta \cos\theta d\theta = 1/2$. Let $u = v_{th}^2$ and integrate this equation by parts and obtain

$$J_{esc}^e(r_o) = \frac{n_o(r_o)}{2\sqrt{\pi}} v_{th}^e \left(1 + \frac{v_{esc}^2}{v_{th}^2}\right) e^{-v_{esc}^2/v_{th}^2}$$

Note that since $v_{esc}^2/v_{th}^2 = e\phi_E/k_B T$ (Eq. (3.21)), the escaping electron fluxes depend strongly on the electrostatic potential. The electrons that can escape into space have thermal velocities larger than v_{esc} and are thus dependent on the shape of the electron distribution function.

Solution 3.10 The potential at distance r is $-GM_\odot m_i/r + e\phi_E(r)$ and is positive and monotonically decreasing as r increases. The ion speed increases with r and thus in this model all ions can escape. The ion velocity distribution at height r_o is assumed Maxwellian,

$$f_{io}(v) = \frac{n_{io}}{\pi^{3/2} v_{th}^3} e^{-(v^2/v_{th}^2)}$$

where $v_{th} = (2k_B T/m_i)^{1/2}$. Similar to electron calculation, we integrate the above equation with integration limits $0 < \theta < \pi/2$. Since all ions can escape, the velocity limits are $0 < v < \infty$. Since there are no particles for $\pi/2 < \theta < \pi$, the density of ions at r_o will just be half of the value of the integral over the entire velocity space. Hence, $n_i(r_o) = n_{io}/2$. The calculate flux of escaping ions is thus given by

$$J_{esc}^i(r_o) = \frac{n_p(r_o)}{\sqrt{\pi}} v_{th}^i$$

where we have used the relation $n_i(r_o) = n_{io}/2$. Note that the escaping flux of ions does not depend on the electrostatic potential ϕ_E because the model assumes that all ions are escaping at the exobase, different from the electrons.

Solution 3.11 In the above equation, $v_{esc}^2/v_{th}^2 = e\phi_{Eo}/k_B T_{eo}$. The mass ratio $m_e/m_i \approx 5.4 \times 10^{-4}$ giving $v_{th}^i/v_{th}^e \approx 5$ indicating $e\phi_E(r_o) \approx 5k_B T_o$ and the electrostatic potential is directly related to the coronal plasma temperature. A coronal temperature of 10^6 K yields $e\phi_E \approx 490$ V. This value is approaching the value of the SW beam flowing at 400 km s^{-1} but cannot explain the fast SW.

Solution 3.12 To answer the first question use the SW speed of 400 km s^{-1} and a magnetic field of 10^{-4} T. This will yield a value of 0.4 V m^{-1}. To answer the second question, we note that perpendicular electric field that is energy independent cannot produce a beam. To speculate how the beam is produced which is shifted from the origin by 400 km s^{-1} in velocity space (Fig. 4.3), note that particles from Earth's aurora have shown us that one of the simplest ways to produce a beam along **B** is to let the particles go through a quasi-electrostatic potential drop $\Delta\phi_E$. A Maxwellian distribution of the particles entering one side of the potential will come out of the other side having the form

$$f(\mathbf{v}) = Ce^{-(W-q\Delta\phi)/k_B T}$$

where W is initial thermal energy of the particle before entering the potential region, q is charge, $\Delta\phi$ is the potential drop, k_B is Boltzmann's constant, and T is the temperature. (Pitch-angle scattering and heating in the acceleration process have been ignored.) A possible particle picture of the SW suggested here is that the SW beam is produced because the coronal thermal particles have gone through a potential drop parallel to the magnetic field direction. The value of the potential drop is given by the streaming velocity which is just exactly how much the beam is shifted from the origin. For the SW, this shift is typically $\sim$1 keV (inferred from the 400 km s^{-1} SW shift from the origin), but it could be several times larger since beam velocities to $\sim$1000 km s^{-1} have been observed during disturbed solar wind conditions. We can also speculate that the potential structures are on magnetic fields located close to the polar region. These magnetic fields extend far out into space ("open" field line region) and the accelerated beam could travel out to 1 AU without crossing the dipole magnetic field.

Solution 3.13 One possible suggestion is that this beam could be the *strahl* component. As on Earth, the lower energy core electrons are secondaries produced by the beam as it propagates through the solar coronal atmosphere. The nearly isotropic halo electrons could come from part of the beam electrons that have been redistributed to lower energies by instabilities. The source of the nonthermal super-halo electrons is not explained by this simple potential drop model. However, one could speculate that the super-halo electrons consist of the nonthermal field-aligned halo electrons that have been accelerated to "run away" energies by $E_\parallel$ followed by pitch-angle scattering. This would predict that super-halo electrons will become nearly isotropic, but this speculative model cannot be verified easily with the existing observations.

References

Biermann, L.: Observed Dynamical processes in interplanetary space. In: Clauser, F.H. (ed.) Plasma Dynamics. Addison-Wesley, Reading, MA (1960)

Bonetti, A., et al.: Explorer X plasma measurements. J. Geophys. Res. **68**, 4017 (1963)

Chamberlain, J.: Interplanetary gas, 2, expansion of a model solar corona. Astrophys. J. **131**, 47 (1960)

Chew, G.F., Goldberger, M.L., Low, F.E.: The Boltzmann equation and the one-fluid hydromagnetic equations in the absence of particle collisions. Proc. R. Soc. **A236**, 112 (1956)

Choi, C.-R., et al.: A study of solitary wave trains generated by injection of a blob into plasmas. Phys. Plasmas **19**, 102903 (2012)

Cranmer, S.R.: Coronal holes and the high speed solar wind. Space Sci. Rev. **101**, 229 (2002)

Cranmer, S.R.: Self consistent models of the solar wind. Space Sci. Rev. **172**, 145 (2012)

Dessler, A.: Solar wind and interplanetary magnetic field. Rev. Geophys. **5**, 1 (1967)

Echim, M., et al.: A review of solar wind modeling: kinetic and fluid aspects. Surv. Geophys. **32**, 1 (2011)

Escoubet, P., et al.: The Cluster and Phoenix Missions. Kluwer Academic, Dordrecht (1997)

Evans, D.: Precipitating electron fluxes formed by a magnetic field aligned potential difference. J. Geophys. Res. **79**, 2853 (1974)

Feldman, W., et al.: The solar wind He^{2+} to H+ temperature ratio. J. Geophys. Res. **79**, 2319 (1974)

Feldman, W., et al.: Solar wind electrons. J. Geophys. Res. **80**, 4181 (1975)

Feldman, W., et al.: Characteristic electron variations across simple high-speed solar wind streams. J. Geophys. Res. **83**, 5285 (1978)

Fitzenreiter, R., et al.: Observations of electron velocity distribution functions in the solar wind by the WIND spacecraft: high angular resolution Strahl measurements. Geophys. Res. Lett. **25**, 249 (1998)

Gringaus, K., et al.: Some results of experiments in interplanetary space by means of charged particle traps on soviet space probes. Space Res. **2**, 539 (1961)

Hammond, C., et al.: Variation of electron strahl width in the high-speed solar wind: Ulysses observations. Astron. Astrophys. **316**, 350 (1996)

He, J., et al.: Evidence of landau and cyclotron resonance between protons and kinetic waves in solar wind turbulence. Astrophys. J. Lett. **800**, L31 (2015)

Hundhousen, A.: Direct observations of solar wind particles. Space Sci. Rev. **8**, 690 (1968)

Hunten, D.: Thermal and nonthermal escape mechanisms for terrestrial bodies. Planet. Space Sci. **30**, 773 (1982)

Jeans, J.H.: The Dynamical Theory of Gases. Cambridge University Press, Cambridge (1925)

Kasper, J.C., et al.: Solar wind temperature anisotropies. In: Velli, M., Bruno, R., Malara, F. (eds.) Proceedings of the Tenth International Solar Wind Conference, vol. 538. American Institute of Physics, Melville (2003)

Lemaire, J., Scherer, M.: Kinetic models of the solar wind. J. Geophys. Res. **76**, 7479 (1971)

Lemaire, J., Scherer, M.: Kinetic models of the solar wind and polar wind. Rev. Geophys. Space Phys. **11**, 427 (1973)

Lin, R.P.: Energetic particles in the solar wind and at the Sun. AIP Conf. Proc. **385**, 25 (1997)

Liu, Y., et al.: Thermodynamic structure of collision-dominated expanding plasma: heating of interplanetary coronal mass ejections. J. Geophys. Res. **111**, A01102 (2006)

Luhman, J., et al.: Solar origins of solar wind properties during the cycle 23 solar minimum and rising phase of cycle 24. J. Adv. Res. **4**, 221 (2013)

Maksimovic, M., et al.: On the exospheric approach for the solar wind acceleration. Astrophys. Space Sci. **277**, 181 (2001)

Mangeney, A., et al.: WIND observations of coherent electrostatic waves in the solar wind. Ann. Geophys. **17**, 439 (1999)

Marsch, E., et al.: Solar wind proton: three dimensional velocity distributions and derived plasma parameters measured between 0.3 AU and 1 AU. J. Geophys. Res. **87**, 52 (1982a)

Marsch, E., et al.: Solar wind helium ions: observations of the Helios solar probes between 0.3 and 1 AU. J. Geophys. Res. **87**, 31 (1982b)

Marsch, E., et al.: Acceleration potential and angular momentum of undamped MHD-waves in stellar winds. Astro. Astrophys. **164**, 77 (1986)

Marsch, E.: Kinetic physics of the solar corona and solar wind. Living Rev. Sol. Phys. **3**, 1 (2006)

Marsch, E.: Diffusion in velocity space of solar wind protons exposed to parallel and oblique plasma waves. AIP Conf. Proc. **1539**, 243 (2016)

Meyer-Vernet, N.: How does the solar wind blow? A simple kinetic model. Eur. J. Phys. **20**, 167 (1999)

Meyer-Vernet, N.: Basics of the Solar Wind. Cambridge University Press, Cambridge (2007)

Neugebauer, M., Snyder, C.: Mariner 2 observations of the solar wind, 1. Average properties. J. Geophys. Res. **71**, 4469 (1966)

Pannekoek, A.: Ionization in stellar atmospheres. Bull. Astron. Inst. Neth. **1**, 107 (1922)

Parker, E.N.: Dynamics of the interplanetary gas and magnetic fields. Astrophys. J. **128**, 664 (1958)

Parks, G.K., Lee, E.S., Fu, S.Y., et al.: Transport of solar wind H^+ and He^{++} ions across Earth's bow shock. Astrophys. J. Lett. **825**, L27 (2016)

Pilipp, W., et al.: Characteristics of electron velocity distribution functions in the solar wind derived from the Helios plasma experiment. J. Geophys. Res., **92**, 1075 (1978)

Rosseland, S.: Note on the absorption of radiation within a star. Mon. Not. R. Astron. Soc. **84**, 525 (1924)

Sckopke, N.: Ion heating at the Earth's quasi-perpendicular bow shock. Adv. Space Res. **15**, 261 (1995)

Spitzer, L., Härm, R.: Transport phenomena in a completely ionized gas. Phys. Rev. **89**, 977 (1953)

Tam, S., Chang, T.: Kinetic evolution and acceleration of the solar wind. Geophys. Res. Lett. **26**, 3189 (1999)

Temerin, M., et al.: Observations of double layers and solitary waves in the auroral plasma. Phys. Rev. Lett. **48**, 1175 (1982)

Tu, C.-Y., Marsch, E.: Two-fluid model for heating of the solar corona and acceleration of the solar wind by high-frequency Alfvén waves. Sol. Phys. **171**, 363 (1997)

Tu, C.-Y., Wang, L.H., Marsch, E.: Formation of the proton beam distribution in high-speed solar wind. J. Geophys. Res. **107**, 1291 (2002)

Wang, L., et al.: Quiet time solar wind super halo electrons at solar minimum. AIP Conf. Proc. **1539**, 299 (2013)

Wang, L.H., et al.: The injection of ten electron/3He-rich SEP events. Astron. Astrophys. **585**, A119 (2016)

Yoon, P., et al.: Asymmetric solar wind electron distributions. Astrophys. J. **755**, 112 (2012a)

Yoon, P., et al.: Langmuir turbulence and suprathermal electrons. Space Sci. Rev. **173**, 459 (2012b)

Yoon, P.: Kinetic instabilities in the solar wind driven by temperature anisotropies. Rev. Mod. Plasma Phys. **1**, 4 (2017)

Zouganelis, I., et al.: A new exospheric model of the solar wind acceleration: the transonic solutions. In: *Solar Wind*, **10**, 315 (2003)

Additional Reading

Asbridge J.R., et al.: Helium and hydrogen velocity differences in the solar wind. J. Geophys. Res. **81**, 2719 (1976)

Barnes, A., et al.: Solar wind heating. Cosmic Electrodyn. **3**, 254 (1972)

Formisano, V., Palmiotto, F., Moreno, G.: α-particle observations in the solar wind. Solar Phys. **15**, 479 (1970)

Gaelzer, R., et al.: Asymmetric solar wind electron super thermal distributions. Astrophys. J. **677**, 676 (2008)

Geiss, J., Eberhardt, P., et al., Apollo I 1 and 12 solar wind composition experiments: fluxes of He and Ne isotopes. J. Geophys. Res. **75**, 5972 (1970)

He, J., et al.: Proton heating in solar wind compressible turbulence with collisions between counter-propagating waves. Astrophys. J. Lett. **813**, L30 (2015)

He, J. et al.: Sunward propagating Alfvén waves in association with sunward drifting proton beams in the solar wind. Astrophys. J. **805**, 176 (2015)

Hollweg, J.: Some physical processes in the solar wind. Rev. Geophys. Space Phys. **16**, 689 (1974)

Kim, S., et al.: Asymptotic theory of solar wind electrons Astrophys. J. **806**, article id. 32 (2015)

Maksimovic, M., et al.: Radial evolution of the electron distribution functions in the fast solar wind between 0.3 and 1.5 AU. J. Geophys. Res. **110**, A9104 (2005)

Meyer-Vernet, N., Issautier, K.: Electron temperature in the solar wind: generic radial variation from kinetic collisionless models. J. Geophys. Res. **103**, 29705 (1998)

Montgomery, M.: Solar-wind electrons Vela 4 measurements. J. Geophys Res. **73**, 4999 (1968)

Ogilvie K., et al.: Electron energy flux in the solar wind. J. Geophys. Res. **76**, 8165 (1971)

Parks, G.K.: Physics of Space Plasmas: An Introduction, 2nd edn. Perseus Book Company, New York (2004)

Pezzi, O. Solar Wind Collisional Heating, J. Plasma Phys., **83**, 555830301 (2017)

Pierrard, V., Lemaire, J.: Electron velocity distribution functions from the solar wind to the corona. J. Geophys. Res. **104**, 17021 (1999)

Pierrard, V., Lamy, H., Lemaire, J.: Exospheric distributions of minor ions in the solar wind. J. Geophys. Res. **109**, A02118 (2004)

Rème, H., et al.: First multi-spacecraft ion measurements in and near the Earth's magnetosphere with the identical cluster ion spectrometry (CIS) experiment. Ann. Geophys. **19**, 1303 (2001)

Richardson, J., et al.: Pressure pulses at Voyager 2: drivers of interstellar transients? Astrophys. J. **834**, 190 (2017)

Robbins, D., et al.: Helium in the solar wind. J. Geophys. Res. **75**, 1178 (1970)

Savoini, P., et al.: Under and over-adiabatic electrons through a perpendicular collisionless shock: theory versus simulations. Ann. Geophys. **23**, 3685 (2005)

Shi, Q.Q., et al.: Solar wind entry into the high-latitude terrestrial magnetosphere during geomagnetically quiet times. Nature (2013). https://doi.org/10.1038/ncomms2476

Tao, J., et al.: Quiet-time suprathermal (0.1–1.5 keV) electrons in the solar wind. Astrophys. J. **820**, 22 (2016)

Tu, C.-Y., The damping of interplanetary Alfvénic fluctuations and the heating of the solar wind. J. Geophys. Res. **93**, 7 (1988)

Tu, C.-Y., Marsch, E.: MHD structures, waves and turbulence in the solar wind: observations and theories. Space Sci. Rev. **73**, 1 (1995)

Tu, C.-Y., Marsch, E.: Wave dissipation by ion cyclotron resonance in the solar corona. Astron. Astrophys. **368**, 1071 (2001)

Tu, C.-Y., Marsch, E., Wang, L.H.: Cyclotron-resonant diffusion regulating the core and beam of solar wind proton distributions. In: Proceedings of the Tenth International Solar Wind Conference, AIP Conference Proceedings, vol. 679, p. 389 (2003)

Vocks, C.: A kinetic model for ions in the solar corona including wave-particle interactions and Coulomb collisions. Astrophys. J. **568**, 1017 (2002)

Volkov, A.N.: On the hydrodynamics model of thermal escape from planetary atmospheres and its comparison with kinetic simulation. Mon. Not. R. Astron. Soc. **459**, 2030 (2016)

Wang, L., et al.: Pitch-angle distributions and temporal variations of 0.3–300 keV solar impulsive electron events. Astrophys. J. **727**, 121 (2011)

Wang, L., et al.: Simulation of energetic neutral atoms from solar energetic particles. Astrophys. J. Lett. **793**, L37 (2014)

Wang, L.H., et al.: Solar wind 20–200 keV superhalo electrons at quiet times. Astrophys. J. Lett. **803**, L2 (2015)

Yang, L.: Proton heating in solar wind compressible turbulence with collisions between counter-propagating waves. Astrophys. J. Lett., **811**, L8, 2015.

Zong, Q., et al.: Fast acceleration of inner magnetospheric hydrogen and oxygen ions by shock induced ULF waves. J. Geophys. Res.,**117**, A11206 (2012)

Zouganelis, I., et al.: Acceleration of weakly collisional solar-type winds. Astrophys. J. Lett. **626**, L117 (2005)

Chapter 4
Collisionless Shocks

4.1 Introduction

When a flowing fluid encounters an obstacle in its path, a boundary forms. If the fluid flows faster than the local sound speed, a shock wave forms. Ordinary fluid shocks are produced by effects of compression and the supersonic flow becomes subsonic in the downstream region by converting the ordered flow energy into disordered thermal energy. The thickness of the shock transition region is of the order of a collision mean free path. Collisions play a fundamental role.

The possible existence of shock waves in collisionless space plasmas was suggested by Gold (1955) to explain the prompt rise time of sudden commencements during geomagnetic storms. At that time, the rise time was thought due to time delays of different energy particles from solar flares arriving on Earth compressing against the geomagnetic field. But the observed rise times of sudden commencements were faster than the expected rise times produced by particles of observed energy spectra in solar flares.

The concept of collisionless shock waves was not accepted immediately and serious debates followed as to whether shocks could form without collisions. However, when the solar wind (SW) was discovered and the velocity moments yielded super-Alfvénic flow speeds, the magnetic discontinuity in front of Earth was accepted as evidence of a collisionless shock. The Earth's bow shock was discovered in 1963 (Sonnet et al. 1963; Ness et al. 1964). The theoretical debates ended quickly without a clear resolution of the fundamental issues about shock formation and entropy generation in collisionless plasmas.

Shock processes in fluids involve irreversible dynamics and the shock boundary represents a transition between two regions of local thermodynamic equilibrium (Krall 1997). The picture of ordinary fluid shocks has been carried over to describing the formation of Earth's collisionless bow shock. Analogous to ordinary shocks, the hotter plasma downstream of the bow shock has been interpreted to represent the thermalized SW (Kennel et al. 1985). However, the mean free path of the SW is

© Springer Nature Switzerland AG 2018

G. K. Parks, *Characterizing Space Plasmas*, Astronomy and Astrophysics Library,

https://doi.org/10.1007/978-3-319-90041-4_4

about an astronomical unit (AU), seven orders of magnitude larger than the bow shock transition width. Even though Earth's bow shock has now been studied for more than 50 years, the question of how the SW could thermalize on ion Larmor radius scales without collisions is still not answered and has been the subject of many experimental and theoretical studies.

One important difference between the bow shock observations and the predictions of fluid theories is that the bow shock includes a foreshock region consisting of suprathermal particles. Two possible sources of the foreshock particles are the SW particles reflected from the bow shock and particles leaking out of the magnetosheath. The fraction of particles reflected from the bow shock can be as high as ~20% and the remaining 80% that is directly transmitted to the downstream region is not immediately thermalized (Sckopke 1995). The reflected particles travel back along the interplanetary magnetic field (IMF) direction and can perturb the incident SW producing waves, and instabilities affecting the bow shock dynamics, structure, heating, and acceleration of particles to high energies.

Another important observation of the upstream feature is that the amplitudes of the excited waves can grow by processes as yet unidentified. These waves subsequently couple to the SW and excite transient nonlinear structures with density depleted regions and steepened edges that resemble shock waves. The upstream particles and transient structures convect Earthward with the SW, further affecting the structure and dynamics of the bow shock (Burgess et al. 2012). The microphysical processes of the upstream nonlinear structures must be treated as an integral part of the bow shock physics regulating the large-scale dynamics.

This chapter begins with a brief discussion of what we know about the bow shock from both early and recent observations. Our goal is to supplement the material covered in review articles about theories, simulation, and observations of Earth's bow shock (Wu 1982; Krall 1997; Lembège et al. 2004; Bale et al. 2005; Eastwood et al. 2005; Burgess et al. 2012; Treuman 2009; Wilson 2016; Parks et al. 2017). Moreover, two textbooks have been published recently on the basic physics of collisionless shocks (Balogh and Treumann 2013; Burgess and Scholer 2015). Recent results from some of the experiments on Cluster on the bow shock and other regions of the magnetosphere are given in Goldstein et al. (2015).

We use Cluster data to show some of the new features. For example, we show that the mean speed of plasma in the downstream region can remain super-Alfvénic, a feature that is not a part of any shock models. Moreover, we show examples of the SW beams in the downstream MS that can retain many of the original upstream SW features indicating that the SW sometimes does *not* interact crossing the bow shock. To further understand how the bow shock interacts with the SW, we have calculated the Boltzmann's entropy across the bow shock using Cluster data. We find that the entropy change is finite and is due to the presence of the multiple plasma distributions in the vicinity of the bow shock. A simple model shows that this entropy behavior can be explained by the Vlasov theory. An example of ICME shock is introduced to show how the scale may affect the dynamics. We conclude this chapter presenting observations of the nonlinear structures in the upstream region and discussing some of the unresolved issues.

4.2 Observations of Earth's Bow Shock

Observations from early on have noted that many features about the SW interaction with Earth's bow shock departed from predictions of shock theories. For example, HEOS-1 (Formisano et al. 1970) showed that while thermalization of H^+ generally occurs across the shock boundary, sometimes the SW He^{++} ions are found with unchanged energy spectra downstream of the shock (magnetosheath). AMPTE observations reported He^{++} and O^{+6} ions in the magnetosheath had shell-like distributions (Fuselier et al. 1995) indicating that heavy ions had not thermalized. These studies were based on data accumulated over times much longer than the SW variations, and the results were very likely affected by spatial and temporal averaging. We now show Cluster data and validate some of the earlier observations as well as show new features not previously known.

4.2.1 Super-Alfvénic Flow in the Magnetosheath

To illustrate the basic features of the bow shock, we begin with a SW event observed on Cluster that was discussed in the previous chapter (see Fig. 3.2) except now we focus on the time interval when the spacecraft crossed the bow shock into the downstream MS region. Data from both g- and G-detectors are shown. Let us go over this plot in detail and point out some of the important features.

Cluster was outbound on 24 January 2001 and crossed the bow shock into the SW at ~0645 UT. The top four panels of Fig. 4.1 show data obtained by the g-detector that was designed to measure accurately the 3D SW distributions. The bow shock crossing can be identified by the discontinuities observed of particle fluxes or other macroscopic parameters such as density n, temperature T, and mean velocity V. The red narrow line in the energy flux spectrogram is from H^+ in the SW (top panel, after 0645 UT). He^{++} is the lighter green line above it at twice the energy per charge of H^+. The SW density (panel 2) was ~7 cc^{-1} (black), temperature ~0.13×10^6 K (red) corresponding to 11.2 eV. The mean velocity of the SW was mainly in $V_x = -408$ km s^{-1} (panel 3). The magnetic field intensity $|\mathbf{B}|$ in the SW was ~4 nT, $B_x = -3$ nT, $B_y = +2.5$ nT, and $B_z \sim 0$ nT. The Alfvén Mach number M_{V_A} of the SW was ~7 and $\theta_{BN} \sim 70°$. This bow shock is a quasi-perpendicular super-critical shock. Our observations of many events indicate that the bulk flow speed in the MS is super-Alfvénic most of the time and rarely sub-Alfvénic. The results simply interpreted imply that the SW beams have not been thermalized contributing to high velocity moments.

A very important feature is the appearance of beams in the MS as observed by the g-detector. These beams are SW beams that have penetrated into the MS but now have slightly lower energies than the original SW beam. These beams are seen deep in the MS and raise important questions about the interaction of the SW beams with the bow shock.

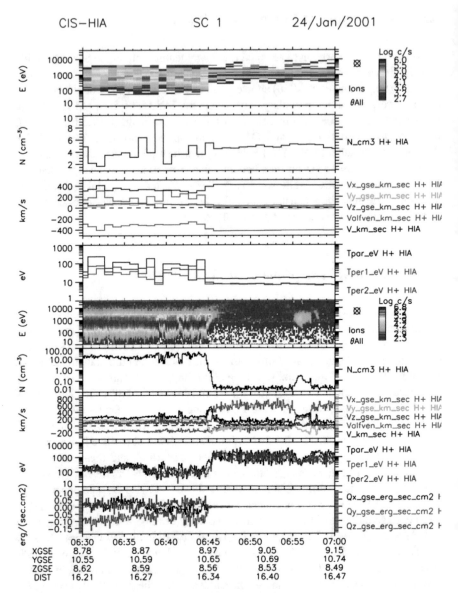

Fig. 4.1 Summary plot showing data obtained by SC1 (*S*-frame) from downstream MS to upstream SW. The narrow red line in the energy flux spectrogram measured by g-detector (top panel) at ∼1 keV is SW H$^+$ and the green line above it at approximately twice the energy charge^{-1} is He^{++}. MS particles cover a broad energy range, ∼100 eV to several keV. The bow shock was crossed around ∼0645 UT. The second panel shows density (black) and temperature (red), the third panel components of velocity moments in GSE including Alfvén speed (magenta). The 3D bulk moments were computed at spin resolution and are transmitted every minute (telemetry limitations). The bottom four panels show data from the G-detector which bypasses the SW when the detector is pointing toward the Sun. The measurements include only the particles above the SW energy (from Parks et al., *Astrophys. J. Lett.*, **825**, 27, 2016)

Question 4.1 Discuss the features shown in Fig. 4.1 relating them to the capabilities of the g-detector. Explain why the spectrograms of the bow shock crossing observed by the g-detector are sharp whereas the G-detector shows a dispersion in energy.

4.2.2 Distribution Function of SW in the Magnetosheath

We now show the details of the distribution functions of the penetrated SW beams in the MS and examine how they evolved. Figure 4.2 summarizes the behavior of the distribution function from 0632:04 UT to 0646:07 UT. Here the velocity is the total velocity defined by $v = (v_x^2 + v_y^2 + v_z^2)^{1/2}$. The velocity moments of

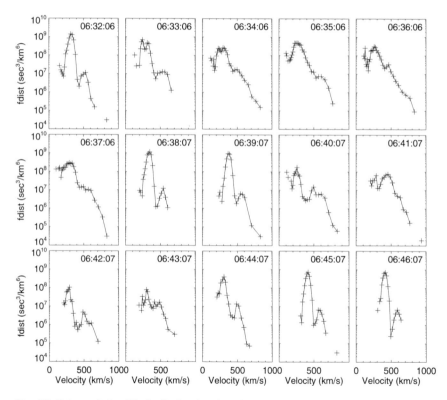

Fig. 4.2 Spin resolution 3D distribution function of ions detected on Cluster 3 from 0632:04 in the MS to 0646:07 UT in the SW in the spacecraft frame of reference. Cluster was in a mode that transmitted the distribution function about every minute. The vertical axis is the distribution function calculated by summing all of the counts from the different θ detectors (θ goes from +90° to −90° with +90° along the spacecraft spin axis) in the 45° ϕ wedge in the azimuthal direction. The main beam is H+ and the weaker beam is He++ (from Parks et al., *Astrophys. J. Lett.*, **825**, 27, 2016)

the bulk parameters shown in Fig. 4.1 come from these distributions. The behavior shows the interaction across the bow shock is very dynamic and can change from moment to moment. We see cases that show broadening of the H^+ and He^{++} suggesting that the beams have been heated (0637:06 UT), while others (0632:06 UT; 0638:07, 0638:07) show the H^+ and He^{++} beams remained almost the same as in the upstream SW (0646:07 UT) suggesting that the SW beams hardly interacted with the bow shock. Note that He^{++} ions do not appear to slow down as much as H^+ ions (0632:06 UT).

Question 4.2 Compare the different panels in Fig. 4.2 and discuss some of the important features not mentioned above.

4.2.3 $\theta - \phi$ *Distribution of SW Beams in the MS*

Another perspective of the SW crossing the bow shock is to examine the behavior in $\theta - \phi$ directions (Fig. 4.3).

The SW beams were also observed on SC3 which was a few hundred km from SC1. While the details were different, SC3 also show some beams deep in the MS with similar features as in the SW. The observed features all indicate that the physics of SW interaction at the bow shock is different from the theoretical models.

The distributions of the SW before (0645:05 UT) (left two columns) and after (0632:06 UT) (right two columns) obtained by Cluster as the SC crossed the bow shock into the downstream magnetosheath. These plots come from the g-detector and there are eight panels in each for different θ and ϕ angles. The eight θ detectors are separated by 22.5° and consist of a detector pointing north of the ecliptic plane $+90°$, to the one pointing to the south, $-90°$. The SW is normally found close to the equatorial plane (0°) whose data are shown on the fourth and fifth panels. The H^+ beam is located around 400 km s^{-1} and the lighter intense beam H^{++} beam around 600 km s^{-1}. The SW beams were also observed on SC3 which was a few hundred km from SC1. While the details were different, SC3 also shows some beams deep in the MS with similar features as in the SW. The observed features all indicate that the physics of SW interaction at the bow shock is different from the theoretical models.

Question 4.3 Discuss and compare the features in Fig. 4.3 observed in the upstream and downstream of the bow shock.

4.2.4 *SW Temperature Upstream and Downstream*

We now discuss how we compute the temperature of the individual SW beam. The SW temperature using the spin resolution (4 s) 3D distribution functions obtained by the g-detector is shown in Fig. 4.4. During 0630–0645 UT, there were ∼18 SW beams in the MS. The temperatures (actually kT) of H^+ and He^{++} beams

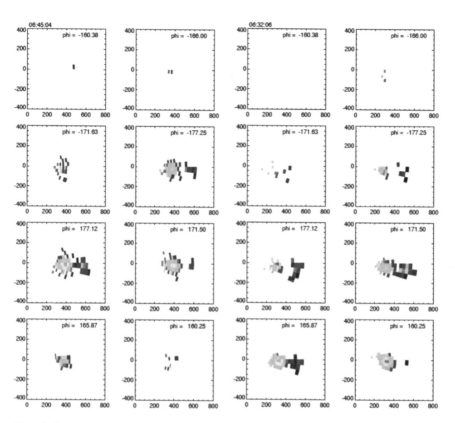

Fig. 4.3 2D cuts of 3D θ and ϕ distributions of the SW beam in the upstream and downstream MS displayed in the spacecraft frame. Velocity is in GSE relative to the spacecraft (observer frame). The main SW beam H$^+$ is detected between $\phi = 172°$ (left two panels). The Sun's direction is $\phi = 180°$. The two right panels show the SW beam in the MS at $\sim160°$, indicating the SW beam was deflected crossing the bow shock. The maximum scale of V is ±400 km s^{-1} for the Y-axis and 800 km s^{-1} for the X-axis

in the MS have been determined for the distributions shown in Fig. 4.3. The left panels of Fig. 4.4 show 2D cuts of the 3D SW distributions in the upstream and downstream MS region in the plasma frame of reference (S'-frame). The temperature is determined assuming that the 1D cuts (right panels) have Maxwellian shape. The temperatures of H$^+$ and He^{++} beams in the SW are $T_\parallel = 6.09$ eV; $T_\perp = 5.68$ eV (top two panels) and $T_\parallel = 7.33$ eV and $T_\perp = 14.38$ eV (bottom two panels).

Question 4.4 Discuss how the temperatures of H$^+$ and He^{++} are computed in the upstream SW and downstream magnetosheath. Compare and discuss the significance of the temperatures.

Note that while $T_\parallel$ of H$^+$ and He^{++} are not so different, the temperatures $T_\perp$ of the two ion species in the perpendicular direction are very different as He^{++}

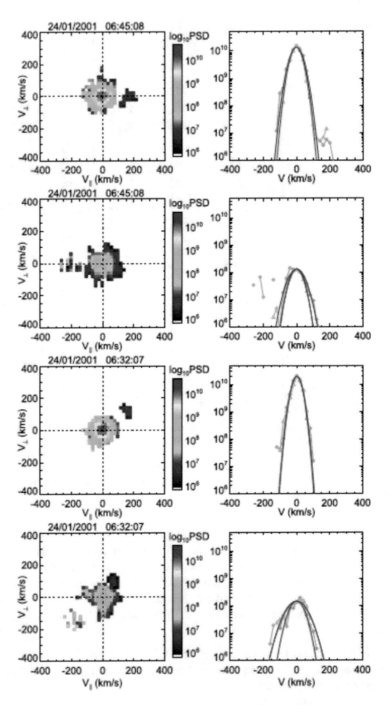

Fig. 4.4 Temperature of MS H^+ and He^{++} beams are estimated in the plasma frame of reference. The bow shock was crossed at 0645:06 UT. In the 1D distributions, the data points are circles and the solid lines are Maxwellian fits. Red is $T_\parallel$ and blue is $T_\perp$. Velocity scales are ± 400 km s^{-1} (from Parks et al., *Astrophys. J. Lett.*, **825**, 27, 2016)

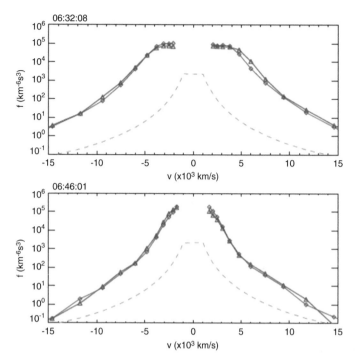

Fig. 4.5 The top panel shows 1D distributions of electrons in the MS and the bottom panel in the SW. The electron distributions in the MS have flat-top shape. Red and blue correspond to distributions in the parallel and perpendicular directions relative to the magnetic field. The dashed line is one count level. The time resolution of the data is 4 s. Electrons are measured by an experiment called PEACE (Johnstone et al. 1997). The data near zero velocity are due to photoelectrons and have been removed (from Parks et al., *Astrophys. J. Lett.*, **825**, 27, 2016)

is heated more than twice that of H^+ ions. These observations indicate that the interaction at the bow shock depends strongly on charge and mass.

4.2.5 SW Electrons Across the Bow Shock

The behavior of SW electrons across the bow shock was first observed by Montgomery (1973) and subsequently by Scudder et al. (1973) and Feldman (1982). They showed that the SW distribution after crossing the shock becomes flat-topped. Cluster validates these earlier observations (Fig. 4.5). The 1D electron distribution in the SW is nearly an isotropic Maxwellian (bottom, plasma frame of reference) but in the MS develops (top panel) a flat top shape. The edge of the flat top is around $4–6 \times 10^3$ km s^{-1} for electrons coming from the SW side and $4–7 \times 10^3$ km s^{-1} for electrons coming from the Earthward side. Flat-top electron distributions have also been observed in the geomagnetic tail current sheet regions (Chen et al. 2000;

Zhao et al. 2017) and in the polar cusps (Parks et al. 2008). Note that simulation of electrons crossing the shock indicates that the SW electrons encountering the bow shock can be accelerated by "surfing" mechanism (Hoshino and Shimada 2002).

Question 4.5 Discuss the different ways to produce the flat top distribution and where else in the magnetosphere they are observed.

4.3   Entropy Across Earth's Bow Shock

Fluid theories indicate that shocks are produced by irreversible processes. Entropy is a fundamental quantity in thermodynamics and the second law of thermodynamics states that in a closed system entropy (ΔS) increases in irreversible processes. The entropy change is driven by heat flow $\Delta S = (dQ/T)$ and the heat flow determines the sign of ΔS. For example, in an exothermic reaction with a constant temperature, heat flow to the surroundings causes ΔS to be positive. Collisions play a critical role in entropy generation. However, space plasmas are collisionless and it is not known how the process of thermalization is achieved without collisions. We will use Cluster data to determine how the entropy behaves across the bow shock.

Entropy in statistical physics is a measure of the number of possible microstates of a system in equilibrium. The observable properties of a system depend on how many phase points lie in each cell of phase space and the number of phase points in each cell defines a microstate of a system. In Ludwig Boltzmann's study of neutral gases, he noted that the dynamics of the individual particles are reversible but when all of these particles are put together into a macroscopic system it becomes irreversible. Boltzmann thus invented the concept of entropy to explain this mystery.

Boltzmann defined entropy as $S = -k_B H$, where the H-function is defined as $H = \int f \ln f d^3 v$. Here f is one particle distribution at a given point in space and dependent only on velocity $f(\mathbf{v})$ and k_B is Boltzmann's constant. The integration is performed over all velocities. Differentiation of H with respect to time leads to the famous H-theorem, $dH/dt = \int (1 + \ln f) \partial f / \partial t d^3 v \leq 0$. The right hand vanishes *only* if f is a Maxwellian. The H-function is always negative and for a system can be in many different configurations, H will decrease to a minimum as f evolves to the most probable distribution corresponding to a state of maximum entropy.

Question 4.6 Discuss how one would use the Cluster data to study the entropy change across the bow shock.

Our particle instruments measure the distribution function only at the spacecraft and not the total number of particles in a flux tube required to determine the total entropy. We have no information about the magnetic flux tube needed to compute the total number of particles. Instead of H, we can only compute a *normalized H function* defined by

$$h_i = \Sigma \, p_i \ln p_i \qquad (4.1)$$

where

$$p_i = f_i \Delta v_i^3 / N \tag{4.2}$$

N is particle number density, and i is the ith volume element of the measured phase space. This h is proportional to entropy density (entropy per particle) at the spacecraft. dh/dt is calculated from successive measurements using the relation

$$dh/dt = [h(t) - h(t - \Delta t)]/\Delta t, \tag{4.3}$$

where Δt is the spin period of the spacecraft.

Boltzmann analyzed a homogeneous gas at rest that changed in time because of collisions. The plasma we examine for h and dh/dt is collisionless and rapidly moving. We examined more than 20 relatively quiet shock crossings and here report the details of one event that occurred on 1 February 2002 during an outbound pass (Fig. 4.6). Cluster 1 was in the MS and was outbound. It crossed the bow shock at $\sim$1940:24 UT as indicated by the magnetic field data (top panel, Fig. 4.6). The plasma in the downstream MS covers a broad energy range, from $\sim$10 eV to several keV. The SW appears in the energy flux spectrogram plot as a narrow red line centered around $\sim$600 eV (panel 2). The SW flow speed V_x was about -320 km s^{-1} that slowed down to -75 km s^{-1} as the foot of the shock was approached and subsequently deviated in y- and z-directions just before crossing the shock. The flow speed in the MS settled to about -150 km s^{-1} (panel 3).

The Boltzmann h-function for ions (panel 4, black) and electrons (panel 5, black) has been measured and the results show that h in the SW was -2.4 (ions) and -5.5 (electrons), and these values decreased across the magnetic ramp to -4.5 and -7.2 in the MS. The corresponding values of $s = -kh$ in the SW are 3.3×10^{-16} ergs K^{-1} and 7.6×10^{-16} ergs K^{-1} and in the MS $\sim 6.2 \times 10^{-16}$ ergs K^{-1} and 9.9×10^{-16} ergs K^{-1}. The changes of entropy/density across the shock are Δs of $\sim 2.9 \times 10^{-16}$ ergs K^{-1} for ions and 2.3×10^{-16} ergs K^{-1} for electrons. (Note that because Cluster starts downstream of the shock and later moves upstream, it observes a reversed time history of a convected plasma volume. To correct for this artifact of reference frame, the dh/dt traces have been inverted.)

The time variation of dh/dt has also been computed across the shock. We find that dh/dt starts out around $\sim$0 in the SW consistent with the SW distributions being nearly Maxwellian. Then dh/dt turns negative in the magnetic ramp to -0.07 s^{-1} for ions (panel 4) and -0.13 s^{-1} for electrons (panel 5). The corresponding rate of entropy change across the shock is thus $\sim$0.1 $\times 10^{-16}$ ergs K s^{-1} for ions and $\sim$0.18 $\times 10^{-16}$ ergs K s^{-1} for electrons. The departure of dh/dt from 0 at the ramp is indicating f is not Maxwellian there, and that the distributions are not in equilibrium.

The readers should note that dh/dt after crossing the shock turns positive for electrons before fluctuating about 0 in the MS ($dh/dt > 0$ has also been seen in ions in many bow shock crossings). This behavior is interesting because positive

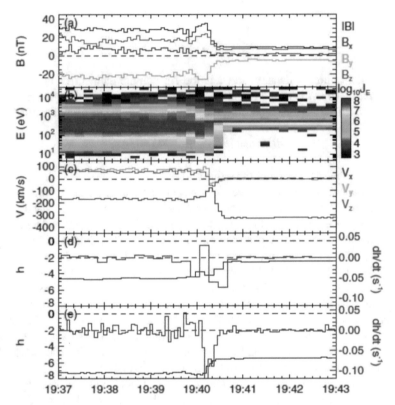

Fig. 4.6 Bow shock crossing on 1 February 2002. The ions are measured in the energy range D10–D35 keV. The SW appears in the energy flux spectrogram plot as a narrow red line centered around ∼600 eV (panel 2). The behavior of entropy density (h) across the bow shock has been calculated. The ion data are 3 spin (12 s) averages. Electrons are measured by an instrument called PEACE (Plasma Electron And Current Experiment) and the data are one spin averages. PEACE is also an electrostatic analyzer and measures electrons in the energy range D5–2.9 keV. The electron data shown are from SC2 (no 3D data on SC1) that was 600 km from SC1 (the bow shock crossing time has been shifted to coincide with SC1). From top to bottom: (**a**) B field and components in GSE coordinate system, (**b**) energy spectrogram of ions, (**c**) mean ion velocities computed from the 3D distribution functions, (**d**) h (black) and dh/dt (red) of ions, (**e**) h (black) and dh/dt (red) of electrons (from Parks et al., *Phys. Rev. Lett.*, **108**, 061102, 2012)

$dH/dt > 0$ is not predicted by the H-theorem. The full significance of this behavior is not understood but it could mean that the SW-bow shock-MS system is not closed and/or that the distribution function f depends on the position **r**.

To help us understand what could cause the entropy to change, Fig. 4.7 shows the distribution functions of the plasma in the vicinity of the shock. The top two panels show ions and the bottom panels, electrons. Panels (a) and (b) are measured at 19:40:09, which is at the top of the magnetic ramp, and panels (c) and (d) at 19:40:21 are from the foot of the shock. Multiple ion populations consisting of three

Fig. 4.7 Top two rows: 2D cuts of the 3D velocity ion distributions measured by SC1 on 1 February 2001 displayed in the SC coordinates. V_x, V_y and V_z are in the GSE coordinate system. Panels (**a**) and (**b**) show measurements made at the top of the magnetic ramp (1940:09 UT). Panels (**c**) and (**d**) were measured at the foot of the shock (1940:21 UT). Bottom row panels (**e**) and (**f**) are 1D cuts of the 3D electron distributions along velocities parallel and perpendicular to the direction of the magnetic field. Electron data come from SC2, which was 600 km from SC1. The different colors represent different times: solar wind (magenta; 1941:47 UT), magnetic ramp (blue 1941:43 UT and green 1941:39 UT), magnetosheath (red 1941:35 UT and black 1941:31 UT) (from Parks et al., *Phys. Rev. Lett.*, **108**, 061102, 2012)

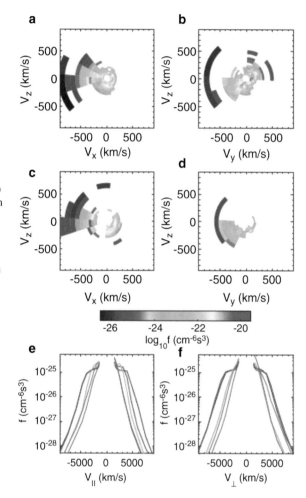

different populations are observed: Panel (a) shows the solar wind beam and diffuse beam moving away from the shock and panel (b) shows the gyrating population. In panel (a), the solar wind ion beam can still be seen after going through the magnetic ramp indicating that the SW distribution was not fully thermalized. The diffuse SW beam is moving away from the shock along V_x direction (panel a) and the gyrating particles (red, panel b) are along $V_y - V_z$ directions. The multiple distributions were first reported by ISEE observations.

The electron distribution in panel (e) $V_\parallel$ is along the magnetic field and in panel (f), $V_\perp$ is perpendicular to the magnetic field. The different colors represent different times, from the SW to MS (magenta, blue, green, red, and black). The changes of the electron distributions from the SW to MS are quite clearly seen in panel (e). The electron distributions normally show beam-like structure on the SW side and "flat" topped shaped distribution on the MS side. The green line shows a small beam-like enhancement along $V_{\parallel}$, but for this particular pass, the 1D cuts were not well optimized to show that the flow-associated beam as the B field was almost perpendicular to the SW.

Question 4.7 Discuss what one assumes to distinguish spatial from temporal variations in the data in Fig. 4.6?

Question 4.8 These entropy changes are for per particle and they are extremely small. What are possible explanations for the small values?

Since space plasmas are essentially collisionless, the behavior just reported should be consistent with the Vlasov theory, $\partial f/\partial t + \mathbf{v} \cdot \partial f/\partial \mathbf{r} + \mathbf{a} \cdot \partial f/\partial \mathbf{v} = 0$. Now multiply this equation with $\log f$ changing it to

$$\log f \partial f/\partial t + \log f \mathbf{v} \cdot \nabla f + \log f \mathbf{a} \cdot \nabla_v f = 0 \tag{4.4}$$

where the symbols have their usual meaning. Rewrite the above equation using the derivative of a product rule as

$$\partial(f \log f)/\partial t + \nabla \cdot (\mathbf{v} f \log f) + \mathbf{a} \cdot \nabla_v (f \log f) - (\partial f/\partial t + \mathbf{v} \cdot \nabla f + \mathbf{a} \cdot \nabla_v f) = 0 \tag{4.5}$$

The last term on the left side in the parenthesis vanishes because we are dealing with a Vlasov plasma. Integrate this equation over the velocity space and obtain,

$$\partial(ns)/\partial t + \nabla \cdot \int (-k_B \mathbf{v} f \log f) d\mathbf{v} = 0 \tag{4.6}$$

Here $n = \int f d\mathbf{v}$ is the density and $ns = \int f \log f d\mathbf{v}$ is the entropy flux. Note also that the integral of the term in the bracket vanishes for $\mathbf{a}$ equal to the Lorentz force. Equation (4.6) is the *entropy conservation equation* and the second term is the divergence of the entropy flux computed kinetically.

To proceed further, we define $\mathbf{v} = \mathbf{U} + \mathbf{c}$ where $\mathbf{U}$ is the velocity moment. Now change the variables $f'(\mathbf{c}) = f(\mathbf{U} + \mathbf{c})$ and Eq. (4.6) rewritten becomes

$$\partial(ns)/\partial t + \nabla \cdot (\mathbf{U} ns) = k_B \nabla \cdot \int \mathbf{c} f' \log f' d\mathbf{c} \tag{4.7}$$

The right side of this equation vanishes for Vlasov plasmas in equilibrium and corresponds to the adiabatic fluid case. The distribution function at the bow shock is however non-Maxwellian and the value of the integral is therefore finite.

Question 4.9 Use the above Eq. (4.7) and apply it to the Cluster data. Assume a steady state situation $(\partial/\partial t = 0)$ and a 1D bow shock with x-direction normal to the shock.

Question 4.10 Summarize what we have just learned about entropy change in collisionless shocks.

4.4 ICME Shock

The interplanetary space is often pervaded by transient shocks produced by solar disturbances such as the CME (Coronal Mass Ejection) events (Crooker et al. 1997). CMEs originate from the closed magnetic field regions of the corona not generally contributing to the SW (Crooker et al. 1997; Gosling 2000). Bremsstrahlung X-rays can also identify CME active area on the Sun.

The expanding speeds of the coronal CME flux tubes through the ambient SW speed are faster than the local Alfvén speed of the SW, and the edges can compress and steepen into shock waves. The CME driven shocks are called Interplanetary Coronal Mass Ejection Shocks (Fig. 4.8). While the ICME shocks and the bow shock share the same basic physics, there is a difference between the ICME shocks and the Earth's bow shock: the ICME shocks are moving through the SW and the bow shock standing in front of Earth. Hence, the flow structures in the sheaths and the sheath thicknesses (compared to the size of the obstacle) are expected to be very different (Liu et al. 2006, 2008; Odstrcil et al. 2008). Moreover, these shocks are in different physical regimes, and studying and comparing the results should enhance our understanding of the physics involved in shock dynamics.

Figure 4.8 shows an ICME shock event observed by Cluster on 22 January 2004. Data from both the g- (top three panels) and G- (next three panels) detectors are shown. The ICME shock was crossed around 0145 UT into the ICME sheath which was observed until around 0830 UT when Cluster exited the ICME sheath and began detecting the CME *ejecta*. Note that the region of the shocked SW plasma in the ICME sheath is compressed and heated. The ejecta includes He^{++} which is clearly seen in the figure. Note also that the G-detector shows there are particles above and below the shocked SW energies (recall that the G-detector does not measure the SW). These particles are reflected and gyrating particles regularly seen with the bow shock.

4.4.1 Distribution Function Across ICME Shocks

Well-defined beams are observed inside the ICME shock. We now determine their temperatures to learn how the SW was heated across the shock. Figure 4.9 shows the 2D distributions in velocities parallel and perpendicular to the direction of the magnetic field in the SW frame (S'-frame). The sheath plasma was fairly hot and overlapped into the He^{++} channels and consequently the He^{++} ions were not separated. The top two panels come from the SW and the bottom two from inside the ICME sheath. The temperatures for these are at 0133.12 UT, $T_\parallel = 11.8\,eV$ and $T_\perp = 11.4$. At 0133.12 UT $T_\parallel = 13.7\,eV$ and $T_\perp = 12.9\,eV$. Inside the ICME sheath, the temperatures were higher clearly showing that the SW has been heated. At 0137:30 UT, $T_\parallel = 33.0\,eV$ and $T_\perp = 34.3\,eV$ and at 0151:30 UT, $T_\parallel = 48.4\,eV$ and $T_\perp = 41.4\,eV$. These temperatures were obtained by fitting 1D distributions

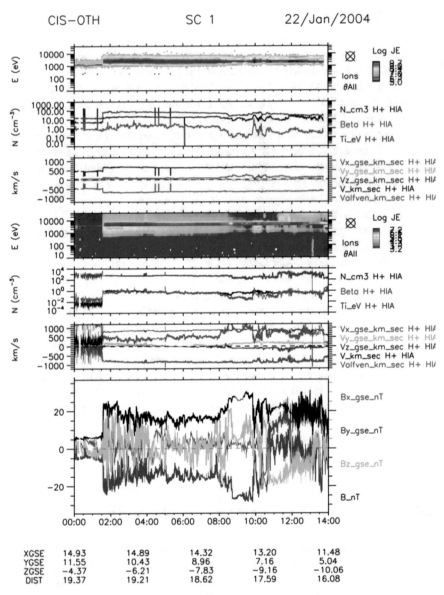

Fig. 4.8 SW crossing the ICME shock into the ICME sheath observed by Cluster. The top three panels show data from the SW g-detector which obtained data in 31 energy steps. On this day, the spin averaged 3D distributions here were transmitted every 84 s. Data of the next three panels come from the G-detector which typically measures above the SW energy, 1 keV. Hence there are no data before ~0130 UT. The bottom panel shows the B-field intensity and components

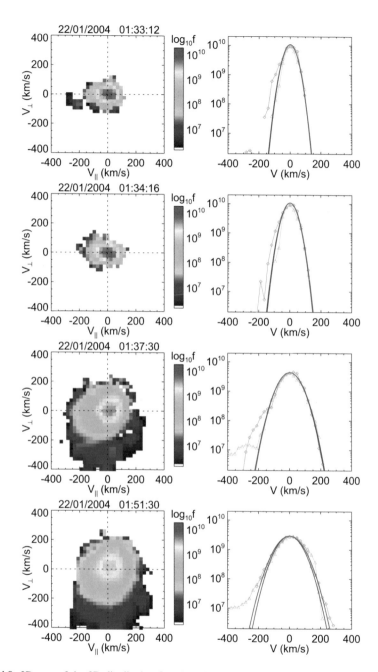

Fig. 4.9 2D cuts of the 3D distribution function of the plasma in the SW and inside the ICME event (left). The plots are made in the SW frame and the velocities are in directions parallel and perpendicular to the magnetic field. The right plots are 1D cuts in the directions parallel and perpendicular to magnetic field directions. To obtain estimates of the temperature, the 1D distributions were fitted with a Maxwellian function. The time resolution of the distributions is 4 s

with Maxwellian shape. Note that in the temperature determination, the integral did not include the more energetic particles that fall away from the Maxwellian approximation.

We find that the 2D distributions in the ICME sheath were not quite as isotropic as was the case inside the bow shock (Parks et al. 2016). Instruments flown in the future to study the ICME events must consider how the energy channels should be selected. Readers interested in the earlier observations of ICME shock behavior by Voyager 2 should consult the article by Richardson et al. (2008). A study that includes a large number of ICME shocks detected on Cluster to learn about the general behavior of the ICME shocks is required to show the differences and similarities with the bow shock. For low frequency waves observed in ICME shocks, see Wilson (2016).

Question 4.11 Computing the temperatures for the ICME shocked plasma using the detector designed for SW is not as accurate as determining temperature from the SW. Can the readers give an explanation why this is so?

4.4.2 Tidmann–Krall Entropy

Tidmann and Krall (1971) have suggested that entropy generation can be represented by fluctuating quantities derivable by expanding the Vlasov equation with $f = \langle f \rangle + \delta f$, $\mathbf{E} = \langle \mathbf{E} \rangle + \delta \mathbf{E}$, $\mathbf{B} = \langle \mathbf{B} \rangle + \delta \mathbf{B}$ where quantities in $\langle \ \rangle$ are mean quantities averaged over the ensemble and fluctuating quantities about the mean are given in δ. The fluctuating quantities obey the coupled Maxwell-Boltzmann equations,

$$\nabla \times \langle \mathbf{E} \rangle = -\frac{\partial \langle \mathbf{B} \rangle}{\partial t}$$

$$\nabla \cdot \langle \mathbf{D} \rangle = q \int \langle f \rangle d^3 \mathbf{v}$$

$$\nabla \times \langle \mathbf{B} \rangle = \frac{\partial \langle \mathbf{D} \rangle}{\partial t} + q \int \mathbf{v} \langle f \rangle d^3 v$$

$$\nabla \cdot \langle \mathbf{B} \rangle = 0, \tag{4.8}$$

where $\mathbf{D} = \epsilon_o \mathbf{E}$ and $\mathbf{B} = \mu_o \mathbf{H}$, and

$$\frac{\partial \delta f}{\partial t} + \mathbf{v} \cdot \frac{\partial \delta f}{\partial \mathbf{r}} + \frac{q}{m} \left[\langle \mathbf{E} \rangle + \mathbf{v} \times \langle \mathbf{B} \rangle \right] \cdot \frac{\partial \delta f}{\partial \mathbf{v}}$$

$$+ \frac{q}{m} \left[\delta \mathbf{E} + \mathbf{v} \times \delta \mathbf{B} \right] \cdot \frac{\partial \langle f \rangle}{\partial \mathbf{v}}$$

$$= -\frac{q}{m} \left[\delta \mathbf{E} + \mathbf{v} \times \delta \mathbf{B} \right] \frac{\partial \delta f}{\partial \mathbf{v}} - C \tag{4.9}$$

and the corresponding Maxwell equations for the fluctuating electric δE and magnetic δB fields. Here C "represents" the collision term expressed by

$$C = \frac{q}{m}(\delta E + v \times \delta B) \cdot \frac{\delta f}{\delta t}. \tag{4.10}$$

They then calculated the change of the entropy flux across the shock, showing that the entropy change is given by

$$\int v_x [\langle f \rangle_2 \ln\langle f \rangle_2 - \langle f \rangle_1 \ln\langle f \rangle_1] = \frac{q}{m} \int dx \int \frac{dv}{\langle f \rangle} \left\langle \frac{\delta \langle f \rangle}{\partial v} \cdot (\delta E + v \times \delta B)\delta f \right\rangle \tag{4.11}$$

where the subscripts 1 and 2 are quantities measured on the two sides. The fluctuating fields δE and δB are nonzero in the region of the shock and this region contributes to the integral of the right side of Eq. (4.11). Thus they suggest that the wave turbulence term mimics collision and is the source of entropy in the shock layer.

4.4.3 Entropy and Waves

The Cluster wave experiments Whisper (Décréau et al. 1997) and Staff (Cornilou-Wehrlin et al. 1997) have seen waves across the bow shock. We show the waves encountered on 1 February 2002 crossing shown above for particles (Fig. 4.6). Figure 4.10 shows the high temporal resolution magnetic field data sampled at 22.5 Hz of the total intensity and the components (top panel).

A considerable amount of complex variations is observed and they are roughly coincident with the occurrence of intense low frequency ES and EM waves. The waves have the largest amplitudes at the peak of the shock and subsequently become bursty inside the magnetosheath. The ES waves are broadband and occur above and below and above the electron cyclotron frequency and they are likely to be Doppler shifted ion acoustic waves. The EM waves are observed to just below the electron cyclotron frequency and are probably whistler mode waves. The narrow band EM waves are similar to lion roars commonly observed in the magnetosheath.

Although the general behavior of entropy across the bow shock for the above event is roughly correlated with the wave intensity. We see the largest changes when the wave amplitudes are the most intense. Although in principle one could examine the particle distributions and perform instability analysis to validate these observations, the 4 s time resolution particle data is not adequate for such studies. Unfortunately, we do not have faster time resolution data on the Cluster mission. Calculation of the turbulent term on the right side of Eq. (4.11) of electrostatic (ES) and electromagnetic (EM) waves requires quantification of the wave and particle measurements which is one of the tasks yet not performed.

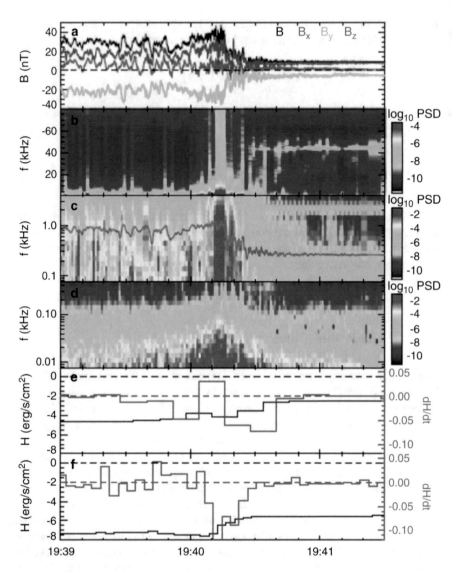

Fig. 4.10 The top panel (**a**) shows high-resolution magnetic field data, panel 2 (**b**) shows wave data from Whisper (frequency to 40 kHz), and panels 3 (**c**) and 4 (**d**) from ES component (64–1000 Hz) and EM component (0.01–180 Hz) from Staff. Electron cyclotron frequency is shown as the purple line in panel 3 (**c**). The bottom two panels (**e**) and (**f**) are h and dh/dt for ions and electrons, shown earlier in Fig. 4.6

Question 4.12 The formulation of Tidmann and Krall (1971) begins with the Vlasov equation, which is time symmetric and reversible. Is Eq. (4.11) a true entropy in the sense defined by Boltzmann?

4.5 Nonlinear Structures Upstream of Bow Shock

The shape of the bow shock is essentially a parabloid and the IMF interacting with the surface on the dayside forms an upstream foreshock region populated by the suprathermal particles whose distributions are distinct from the incident SW. The foreshock region is also permeated by a variety of large-amplitude waves (tens of nT). These waves interact with the SW (Eastwood et al. 2005) and sometimes, isolated and nearly monochromatic magnetic field structures could be triggered as in SLAMS (short large amplitude magnetic structures), embedded in the ultra low frequency ULF waves. We still do not know what processes could select a particular wave to grow while the neighboring waves remain unchanged.

The topic of upstream particle populations and waves in the upstream region has been extensively reviewed (Parks et al. 2017; Ipavich et al. 1981; Fuselier and Schmidt 1994; Fuselier and Thomsen 1992). Here we will only summarize briefly some of the salient features and discuss some of the issues that were prominent at the time when the upstream populations were discovered.

4.5.1 Beams, Intermediate, and Diffuse Distributions

Different types of distributions are observed depending strongly on the angle the IMF makes with the shock normal **n**. The dusk region defined by $\theta_{BN} \geq 50°$ is referred to as quasi-perpendicular shock region and on the morning side $\theta_{BN} \leq 50°$ is quasi-parallel region. and "*intermediate*" (Paschmann et al. 1981) distribution in the transition region. The reflected SW particles observed at the foreshock boundary are field aligned. The intermediate particles are found deeper in the foreshock region and their pitch-angles are wider than the distributions observed in the reflected SW beams. The particles in diffuse distributions are found in the far region of the foreshock region, in the quasi-parallel shock region, and their pitch-angles are nearly isotropic. The maximum energies of the reflected SW particles are about twice that of the incident SW beams, of the intermediate populations $\sim$10 keV, and of the diffuse particles $\sim$200 keV.

It was speculated originally that the three types of distributions may have been exemplifying the various stages of instabilities evolving from a beam to isotropic distributions. However, later studies indicated that while diffuse ion populations show a strong correlation in their He^{++}/H^{+} ratio with that of the SW (Ipavich et al. 1981), and because FABs have much smaller fraction of He^{++} relative to H^{+} by nearly two orders of magnitude, Fuselier and Thomsen (1992) suggested that FABs are not the seed population of the diffuse ions upstream of the quasi-parallel bow shock.

Initially, it was thought that the origin of the gyrating particles observed close to the bow shock boundary and the reflected SW are two distinct distributions produced by the dynamics of the quasi-perpendicular shock. Möebius et al. (2001) studied this problem using simultaneous Cluster data at various distances from the bow shock and found that the flux level of the FAB remained the same whether the beam population was close or farther away from the shock. They thus interpreted that the FABs most likely came from the gyrating ion population produced at the shock ramp by specular reflection of the SW. However, this interpretation assumed an efficient pitch-angle scattering of the gyrating particles by waves at the shock front. Another possible interpretation would be that the FAB ions have encountered the shock multiple times and gained energy.

An alternative interpretation of the upstream source is that particles are leaking out of the magnetosheath particles and propagating to the upstream region. This interpretation assumes that the energies of the upstream suprathermal particles were acquired in the magnetosheath. Kucharek et al. (2004) studied this problem using observations of FABs in low Mach number quasi-perpendicular shocks and showed that the phase space region of the beam was actually empty immediately downstream of the shock, while the angular distributions of the gyrating ions in the shock ramp provided the flux necessary to feed the ion beam found further upstream of the shock. These observations give strong support to the idea that for the events studied, the magnetosheath could not have been the source of the upstream suprathermal particles and favor the idea that wave-particle scattering is responsible for the reflected particles by low frequency waves.

The above results and conclusions are preliminary and need to be firmed up. More studies are needed with different Mach numbers and shock parameters to cover the different SW conditions to validate the conclusions. We also need to quantify better what types of wave-particle interaction could be efficient enough to scatter the particles in the region where the SW is reflected. There are also questions about the role of the electrons, whether electron data could be used to further help us understand the underlying physics.

4.6 Density Holes and Hot Flow Anomalies

We now discuss another class of particle distributions than those discussed above. These distributions are dynamic and the structures may evolve in space and time. Figure 4.11 shows a summary plot of data from Cluster 1 during an outbound pass on the duskside. Data in the top three panels come from the g-detector, the next two panels from the G-detector, and the bottom panel shows the magnetic field. Immediately after crossing the bow shock at ∼2036 UT, as shown in panel 2 are short intervals when the density (black) drops below the average SW values while the temperature (red) increases (for example, ∼2045 UT).

These structures include hot flow anomalies, HFAs (Schwartz et al. 1988; Vaisberg et al. 2016), also known as hot diamagnetic cavities, HDCs (Thomsen 1988), foreshock cavities, FCs (Sibeck et al. 2002) and density holes, and DHs (Parks et al. 2006). FCs result from the higher pressure region expanding into adjacent region. Although they have similar features as HFAs including plasma

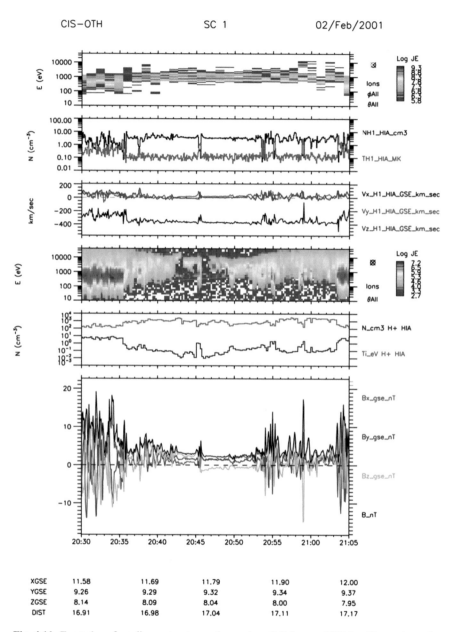

Fig. 4.11 Examples of nonlinear structures observed on 2 February 2001 by Cluster 1 on the outbound pass after crossing the bow shock at ~2036 UT. From top to bottom are shown energy flux spectrogram, density and temperature, and bulk velocities from the g-detector and the next two panels, energy flux spectrogram and density and temperature from the G-detector. The bottom panel shows the magnetic field intensity and components (GSE frame)

compression at the edges, their interior plasma does not show a temperature increase and there is no strong flow deviation. FCs will not be discussed further. The structures we discuss here do not have any connection to the solar wind (SW) magnetic holes, which are attributed to non-propagating current sheets within the SW (Turner et al. 1977) or ion Larmor scale magnetic holes observed in the plasma sheet during active geomagnetic times (Ge et al. 2011; Sundberg et al. 2015). However, HFA-like events are ubiquitous: they have also been observed in the environment of Venus, Mercury, Mars, and Saturn (Uritsky et al. 2014; Collinson et al. 2014; Slavin et al. 2009; Øieroset et al. 2001).

Some of these structures can also be recognized in the G-detector even though this detector does not measure in the X_{GSE} direction while the SC is in the SW indicating that the upstream structures are three-dimensional structures. To highlight some of the detailed features, we show in Fig. 4.12 four examples in higher time resolution (data are 4 s averages). These examples come from more than a hundred events randomly selected from four Cluster orbits. The density holes are accompanied by nearly identically shaped magnetic holes with reduced intensity in the hole and steepened magnetic field B at the edges.

Moreover, the density depletions are accompanied by temperature increase and reduced bulk velocities. We also see one of the magnetic field components changes sign in the density depleted region, indicating crossing of a current sheet. Note that some DHs last only one data point (4 s).

Question 4.13 Discuss the basic features of DHs and how they are correlated to other features shown in Fig. 4.12.

A sample of 147 randomly chosen DHs events during 6 orbits in 2001 have been analyzed (Fig. 4.13). The particle data used have 4 s time resolution. The panel on the left shows DHs with a mean duration of 17.9 ± 10.4 s, and the middle panel shows a mean $\Delta n/n$ of 0.69 ± 0.15. Here Δn represents the difference between the solar wind density and the minimum density in the hole. The right-hand panel shows that the mean rotation of the magnetic field across the neutral sheet was $\sim 36 \pm 24°$. DHs are associated with magnetic shears indicating that the IMF current sheet is intimately involved. What caused the double peak is not understood. We have examined high-resolution spacecraft potential data (a proxy for the electron density) and 22.5 Hz magnetic field for a few holes and find that there were no significant substructures, although fluctuations were present. Also, the four SC observations of a few DHs indicate that they can be as small as a few gyro-radii of keV ions.

Except for the duration and depth of the density depletions, the features of DHs are very similar to those seen in HFAs. The durations of HFAs are a few minutes and DHs are shorter and have smaller scale lengths (a few gyro-radii vs. several R_E) and they occur more frequently. Nevertheless, the similar features suggest that DHs may involve some of the same mechanisms active in HFAs.

Thomsen et al. (1986) observed that ions in HFAs can have two types of distributions: the first type consisted of two plasma populations that include the SW beam and the reflected SW particles from the bow shock and the second type included only a single hot population. Thomsen et al. (1986) called the first type

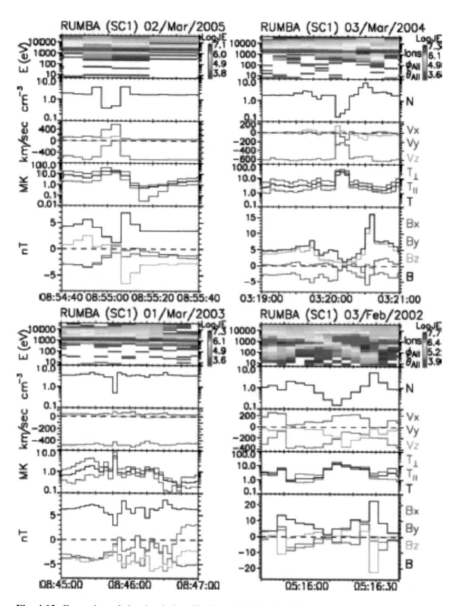

Fig. 4.12 Examples of density holes. Each panel includes, from top to bottom, energy flux spectrogram, density, bulk velocities and temperature, and magnetic field. The scale B is linear while for *n* it is logarithmic. The density holes on 2 March 2005 were detected in GSE coordinates (12.0, 3.2, 4.9) R_E, on 3 March 2004 at (9.2, −4.0 10.5) R_E, on 1 March 2003 at (12.6, 3.3, 6.3) R_E and 3 February 2002 at (9.6, 1.2, −8.5) R_E (from Parks et al., *Phys. Plasmas*, **13**, 050701, 2006)

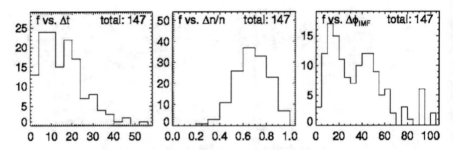

Fig. 4.13 Number of density holes as a function of duration (Δt), fractional density depletion ($\Delta n/n$), and magnetic field rotation from times before to after the hole for 147 DHs observed on six bow shock crossings (from Parks et al., *Phys. Plasmas*, **13**, 050701, 2006)

"*early phase* HFA" and the second type "*mature* HFA" events. Noting that HFAs are dynamic they suggested that the single hot population is produced by a streaming instability merging with the incoming SW and the reflected particles from the bow shock. Two stream instability has been theoretically predicted since the early days of space age (Pappadopoulos 1971). However, recently, Xiao et al. (2015) reported that some of the early stage HFAs could be stable and stationary while others are contracting and expanding because of pressure imbalance in HFAs.

4.6.1   *Density Holes and HFA Model*

There are several outstanding questions about DHs and HFAs: What relationship, if any, do DHs have to the well-studied HFAs? Could DHs be "early stage" HFAs that fail to fully develop for some reason?

There is strong observational evidence that HFAs are produced by the interaction of the IMF current sheet with the bow shock. A key feature of this model is that the SW electric field points inward along the normal of the current sheet and channels the reflected ions from the bow shock into the current sheet allowing the reflected particles to escape along the IMF. A test of all of the thirty HFA events detected by ISEE, AMPTE UKS, and IRM has shown that the SW electric (E) field is directed inward on at least one or both of the edges, consistent with the prediction of the model (Onsager et al. 1990; Thomsen et al. 1993; Schwartz et al. 2000).

To test if DHs are also produced by the interaction of the IMF current sheet, we hypothesized that DHs are "early phase" structures of HFAs and have tested if the HFA model could also explain DHs. For this study, we chose DHs without steepened edges (Fig. 4.14) and tested about 35 of such DHs in Cluster data obtained

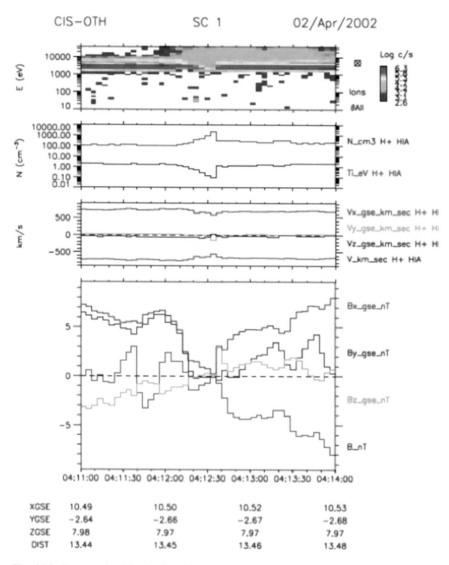

Fig. 4.14 An example of "early phase" DH. From top to down, energy flux spectrogram, density and temperature, bulk velocities, and magnetic field and components. Note that the increase of T and decrease of V_x are small. Note that B_y switches sign in the hole as in regular DHs (from Wilber et al., *Ann. Geophys.*, **26**, 3741, 2008)

during 2001. We then applied the same HFA test to DHs and the results displayed in Fig. 4.15 show the downstream angles (left plot) have mean values $84.0 \pm 7.9°$ and $83.8 \pm 8.4°$ for C1 (black) and C3 (green). The upstream angles are $84.9 \pm 8.5°$ and $88.5 \pm 10.5°$ (right plot).

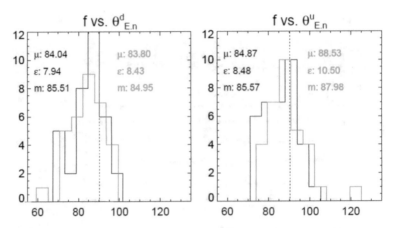

Fig. 4.15 Distributions of the angles the convection electric field E make with the inferred current sheet normals for the upstream (right) and downstream (left) sides. The angles <90° correspond to E-field pointing outward. Cluster 1 black, Cluster 3 green (from Wilber et al., *Ann. Geophys.*, **26**, 3741, 2008)

Question 4.14 What is incorrect about the plot shown in Fig. 4.14? Discuss the possible meaning of the results shown in Fig. 4.15.

4.6.2 Electrons in Nonlinear Structures

Thomsen et al. (1986) reported that electron distributions inside the density depleted region are Maxwellian and isotropic. Zhang et al. (2010a) reported from Themis data that electron distributions have flat-top shape near the steepened edges. Recent Cluster observations however found that while electron energy spectra in early phase HFAs can be fit with a single drifting κ function, the spectra inside mature HFAs can be fit by a combination of drifting Maxwellian below $\sim$10 eV and a heated distribution above 10 eV (Wang et al. 2013).

We have further studied the electron distributions in HFA-like events and an example from 2 February 2001 will be shown. Figure 4.16 summarizes the ion bulk parameters for this day using 4 s data. The top two panels show the energy flux spectrograms from g- and G-detectors. The next panels show density (black) and temperature (red), bulk velocities in the SC frame in GSE from the g-detector and the bottom panel shows the magnetic field intensity and components.

To characterize the behavior of electrons, we have studied a sequence of the electron distribution function from before to after the detection of the nonlinear structure detected $\sim$2045 UT immediately after the SC cross the bow shock into the SW.

The panels in Fig. 4.17 are 1D plots in directions parallel and perpendicular to the magnetic field. The many lines in each plot come from two sides of the distribution captured at 2 s intervals as the SC spins (spin rate is 4). The black lines represent

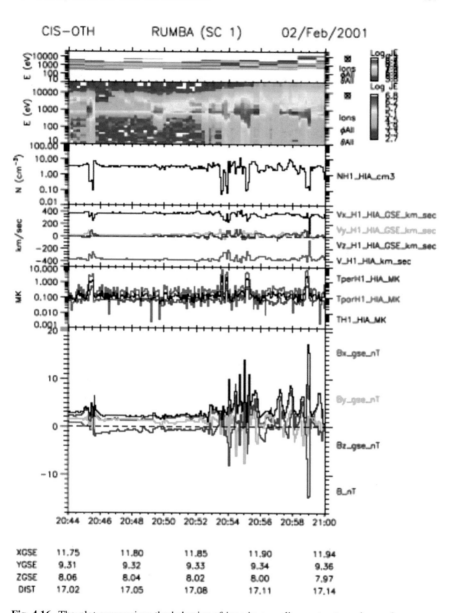

Fig. 4.16 The plot summarizes the behavior of ions in a nonlinear structure observed upstream of the bow shock at 2045–2046 UT on 2 February 2001. From top to bottom panels, energy flux, density (black) and temperature (red), bulk velocities, and magnetic field intensity and components. The data have 4 s time resolution and they are from SC1

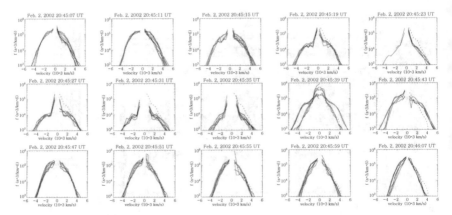

Fig. 4.17 This plot comes from just outside the bow shock when the nonlinear structure (2045:07 UT) was detected. A sequence of electron distributions from the MS to the SW is shown. The third row panels come from the density depleted region. The time resolution of each plot is 4s. The velocity scales are -6×10^6 to $+6 \times 10^6$ m/s. The electron data are obtained by the PEACE instrument on Cluster

the sectors between 0 and 67.5° and between 112.5° and 180° (that is, all non-perp directions). The red lines correspond to angles between 67.5° and 112.5° (perp direction). The closeness of these lines attests to the fact the different detectors have been cross calibrated. The dashed lines are one count levels. While Fig. 4.17 is complicated, it clearly shows that the electrons are changing in space and time. The phase space density can change by an order in magnitude and flat top shape is observed both inside the hole and near the edges. Wave-particle interaction is likely responsible for the complex behavior but details have not been worked out because of the limited particle time resolution. The flat top shape is another piece of evidence that the dynamics going on in these upstream nonlinear structures are similar to those seen at the bow shock.

Question 4.15 Discuss the behavior of electron distributions shown in Fig. 4.17. Note that the fourth panel (top row) shows the flattop is one sided, that is perpendicular on the left side and parallel on the right side. Discuss what could cause this feature?

4.6.3 Slowdown and Temperature Increase in HFAs

Two of the clearest bulk features associated with both HFAs and DHs are that the mean velocity slows down and the temperature increases in the density-depleted regions. A long standing question is what could cause the bulk velocity to slow down and the temperature to increase. To investigate the possible mechanisms, we have examined the data at the most detailed level provided by the Cluster ion experiment. An upstream structure that has been identified as a HFA event by Lucek et al. (2004)

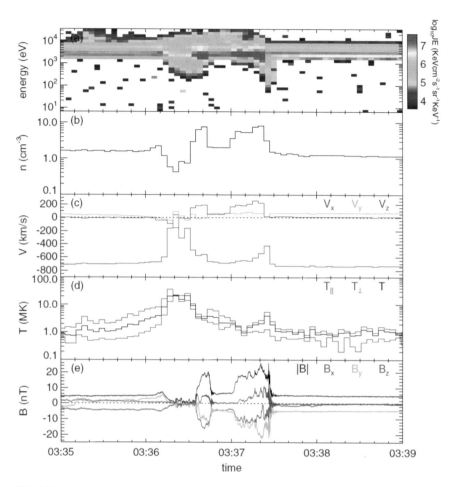

Fig. 4.18 A nonlinear plasma structure detected upstream of the bow shock previously studied by Lucek et al. (2004). From top to bottom shows energy flux spectrogram (**a**), density (**b**), mean velocity (GSE) (**c**), temperatures parallel and perpendicular to magnetic field (**d**), and magnetic field and components (**e**). The velocity scales are ±1800 km/s for both V_x and V_y (from Parks et al., *Astrophys. J. Lett.*, **77**, L39, 2013)

has been reexamined to study this problem further. Figure 4.18 shows the summary plot for this event which shows the G-detector data and includes the typical features showing slowing down of the mean speed and increase of the temperature in the density depleted region (∼0336:10–0336:20 UT).

This HFA in Fig. 4.18 was detected upstream of the bow shock, which was crossed at ∼0339:18 UT (not shown). Two large peaks were observed in both the density n (second panel) and magnetic field B (bottom panel) at ∼0336:40 UT and 0337:20 UT, another small peak at ∼0336:10 UT. The first peak was followed by a clear density depletion centered around 0336:20 UT whereas the density following the second peak returned to ∼SW level. The density and magnetic field depletion

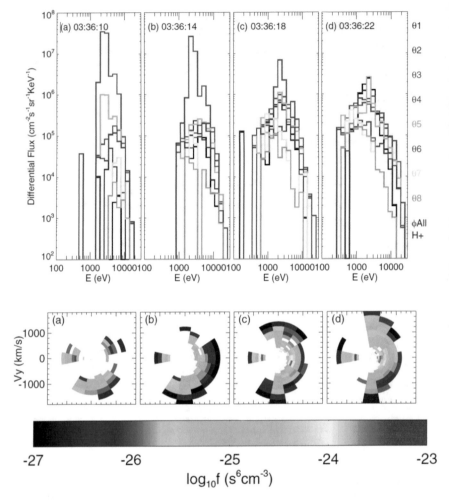

Fig. 4.19 Differential energy spectra from eight detectors in the polar direction covering polar angles +90 to −90° at 22.5° apart. The unit of Y-axis is number $(cm^2\text{-s-ster-keV})^{-1}$ and the X-axis is energy in eV. The times are start times. The corresponding plots in the velocity space are given below (from Parks et al., *Astrophys. J. Lett.*, **77**, L39, 2013)

for the first event went below the SW values attaining a minimum n $\sim$0.4 cc^{-1} and B $\sim$1.5 nT (0336:25 UT). The peak density at the leading edge had a value of $\sim$7.5 cc^{-1} and B of $\sim$18 nT.

The four top plots in Fig. 4.19 are differential energy spectra from the eight θ detectors from beginning to end of the first peak detected from 0336:10 UT to 0336:22 UT. Each plot is spin averaged 4 s apart. Here we see the fluxes of the detectors that detect the SW ($\theta = 4$ and 5) are decreasing, while particles from other directions are increasing. The four plots shown below in Fig. 4.19 are velocity space plots and correspond to the same time interval of the energy spectra shown above. The SW is the bright beam at $V_x = -650 \, \mathrm{km \, s^{-1}}$ (0336:10 UT) and it is always present and the speed remained constant. However, the SW beam intensity decreased steadily toward the minimum of the density depletion corroborating the energy flux spectrogram. The width of the SW beam remained nearly the same for the entire interval, indicating T of the SW beam was not changing. At the same time, other particles that included the back streaming and gyrating population evolved and increased, covering larger and larger velocity space. The increase of T in the density depleted region computed from the second moment thus comes from not only the SW but also other particles including the back streaming and gyrating populations that occupied a larger velocity space. Integration in the velocity space interpolates across the different distribution and the resulting temperature neither represents the SW beam or the reflected and gyrating population. Is the SW beam intensity decreasing because the particles are being redistributed? Are they scattered and reflected at the edges? Can the reduced intensity be accounted for by the increase of the reflected particles?

The last piece of information we can extract from our SW detector is to examine the fluxes of ions as a function of θ and ϕ (Fig. 4.20). As SC1 spins, the high voltage of each detector sweeps at a rate of 125 ms, and in one spin obtains ion distributions in 32 energy steps and 16 ϕ directions (for the operating mode at this time). The top plot of Fig. 4.20 shows the SW at $\phi = \pm 180°$, which is the direction of the detector pointing toward the Sun. The next plot shows the appearance of fluxes at $\phi = 0°$ which is the opposite of the SW direction. The subsequent two plots show buildup of the fluxes centered around $\phi = 0°$ at the same time as the SW intensity diminished.

We stress that the last few plots have clearly shown that the moments do not give correct information on the bulk quantities when the distributions in the velocity space include multiple populations. The examples on the bottom panels shown in Fig. 4.19 clearly demonstrate that the moments calculation interpolates between the two different distributions giving values that neither represents the SW beam or the diffuse particles of the reflected and gyrating population. As mentioned above, the SW beam bulk velocity is not deviated or slowed down nor the temperature changed. The behavior shown by this example describes HFAs including early and mature HFAs as well as the spontaneous hot flow anomaly (SHFA) events (Zhang et al. 2010b). The mechanism for HFA events need to be reexamined taking into account the results presented above.

The last piece of information we convey about this HFA event is that we have also used the four SC data to calculate the direction and speed of propagation of the structures assuming the front was uniform and constant speed. The results show that the phase velocities V_{ph} in the SW frame for the two cases were -302 and $-329 \, \mathrm{km \, s^{-1}}$ (not shown). The SW Alfvén velocity during these times was $\sim 65 \, \mathrm{km \, s^{-1}}$, indicating both structures were propagating toward the Sun against the SW at super-Alfvénic speeds, compressing and steepening the edges into shock-like waves.

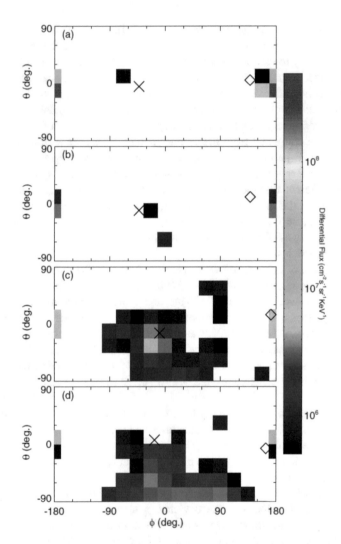

Fig. 4.20 Plots showing θ and ϕ behavior. Each θ and ϕ panel (**a** to **d**) contains the SW beam. There are 8 bins on the Y-axis (θ) and 16 on the X-axis (ϕ). Each panel is separated by 4 s (from Parks et al., *Astrophys. J. Lett.*, **77**, L39, 2013)

Question 4.16 Discuss the two magnetic peaks (Fig. 4.18) focusing on similar and different particle features observed.

Question 4.17 Summarize what you have learned about the change of bulk velocity and temperature in HFAs.

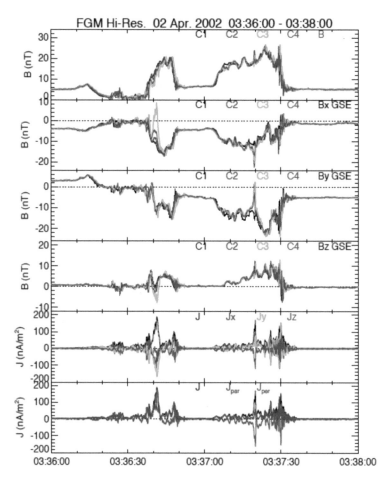

Fig. 4.21 High resolution magnetic field data of the HFA event shown in Fig. 4.18. The magnetic field data are color coded, with C1 black, C2 red, C3 green, and C4 blue. The data here also show that shorter time magnetic structures can *spontaneously* grow at the edge of the gradient (~0336:40 UT and 0337:40 UT). For the first pulse, B_x was most intense on SC3, then C2 and C4. The bottom two panels show currents flowing in GSE and magnetic directions

4.6.4 Currents Associated with HFA

Currents associated with the HFAs have been calculated by solving Ampere's equation $\nabla \times \mathbf{B} = \mathbf{J}$ using the four Cluster magnetic field measurements (ignore displacement current) and the "curlometer" technique (Dunlop et al. 2002).

Figure 4.21 shows the high resolution magnetic field sampled at a rate of 22.5 Hz for the same event studied in Fig. 4.18. The top four panels show the magnetic field from the four SC and the bottom two panels currents in GSE and in the directions parallel and perpendicular to the magnetic field. The total magnetic field intensity black (top) and the three components (subsequent panels) have been color-coded with B_x red, B_y green, and B_z blue. At the leading edges are whistler mode waves similar to those seen in front of the bow shock. We also see the magnetic field $|\mathbf{B}|$ and the profiles of B_x, B_y, and B_z are modulated by low amplitude ultra low-frequency (ULF) waves ($\sim$5–10s period) that permeated the upstream region.

Question 4.18 Discuss the behavior of the magnetic field and current shown in Fig. 4.21.

Another feature is that the field-aligned currents (FAC, $\mathbf{J}_\parallel$) were larger than the current perpendicular to B ($\mathbf{J}_\perp$). In the first case, the magnitude of $|\mathbf{J}|_\parallel \sim 150\,\mathrm{nAm}^{-2}$ and $|\mathbf{J}|_\perp \sim 25\,\mathrm{nAm}^{-2}$ and in the second $|\mathbf{J}|_\parallel \sim 100\,\mathrm{nA\,m}^{-2}$ and $|\mathbf{J}|_\perp \sim 10\,\mathrm{nAm}^{-2}$. While both currents had similar magnitudes, they were flowing in opposite directions, along B in the first case and perpendicular to $\mathbf{B}$ in the second. The magnetic field on SC 3 was predominantly in the $-y$-direction, hence the FAC was propagating against the $\mathbf{B}$-direction from dawn to dusk for the first pulse and in the opposite direction for the second pulse. The increase of $|\mathbf{J}|_\perp$ is due to the front moving against the SW that compressed the edge. However, $\mathbf{J}_\parallel$ comes from field-aligned beams.

We also note that the current responsible for the spontaneous B-field growth had a large field-aligned component. Note also that the total magnetic field intensity during 0336:39–0336:41 UT remained nearly constant indicating that this pulse behaved like a transverse Alfvén wave. The behavior of the pulse with its rotation is characteristic of a flux rope, seen in other larger SW structures, for example, in ICMEs. More work is needed to validate if the field-aligned pulses are flux ropes, but if they are, our observations indicate that the flux rope can have small dimensions, of the order of an ion cyclotron radius (a few hundred km). The second pulse had a duration of less than 1 s. It was observed on SC 2, 3, and 4 with the pulse largest on SC2 and 3. The magnitude was nearly the same as the first. B_y increased from -15 to $+3\,\mathrm{nT}$, B_z component changed from -10 to $-20\,\mathrm{nT}$, and B_z changed only a few nT. The interplanetary current sheet was nearby but the SC did not cross it during this event.

4.6.5   Waves Associated with HFA

Wave data from Whisper and Staff are shown to demonstrate what types of waves were observed on 2 April 2002 HFA event (see Fig. 4.18). The top panel shows Whisper data (up to 40 kHz). The second panel shows the electric component from Staff (64–4000 Hz), the third and fourth panels are magnetic fluctuations from Staff (64–4000 Hz) and (0.06–180 Hz). The red lines refer to cyclotron frequency, f_{ce}. The units for the power spectra from Whisper waves are $\mathrm{V/(Hz)}^{1/2}$ and from Staff ES $(\mathrm{mV}^{22}/\mathrm{m}^2/\mathrm{Hz})$. Power spectrum of EM waves is in units of $(\mathrm{nT}^2/\mathrm{Hz})$. Waves in nonlinear events behave very similar to those in the vicinity of the bow shock (Gurnett 1985) (Fig. 4.22).

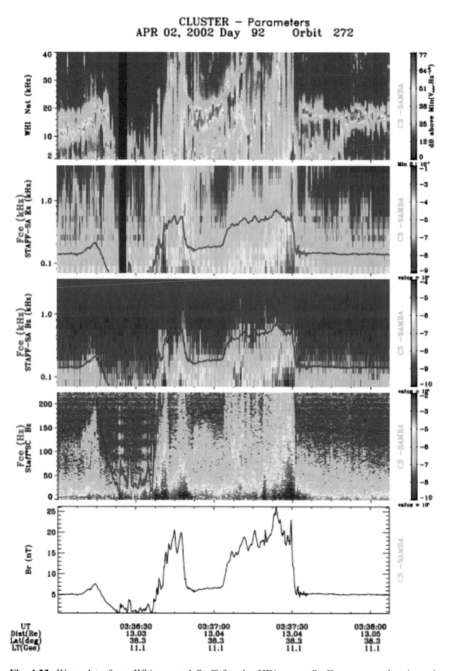

Fig. 4.22 Wave data from Whisper and Staff for the HFA event Staff measures electric and magnetic field fluctuations from ~8 Hz to 4 kHz and Whisper measures a few kHz to ~50 kHz. The red line is the calculated cyclotron frequency of electrons. The magnetic field and components come from FGM

We have studied wave and particle behavior for a few nonlinear events (Parks et al. 2007; Lin et al. 2008) and briefly discuss here what we know. The fluctuating electromagnetic (EM) waves (panels three and four) show strong whistler mode emissions at the magnetic ramps (~03:36:35 UT) and also when exiting the density depleted region (~03:36:40 UT). Strong emissions were also observed near the time of the FAC at 0336:40 UT, and at ~03:36:47 UT and 03:36:30 UT. Other weaker and more diffuse emissions observed between the 2 magnetic and density ramps 03:36:50 and 03:37:00 in the frequency range ~50 to ~100 Hz are also whistler mode waves. The narrowband EM emissions observed are similar to the lion roars in the magnetosheath.

Wave theory predicts that the ES emissions detected above and below the local electron cyclotron frequency (F_{ce}) are triggered by instability of currents. Balikhin et al. (2005) and Hull et al. (2006) have identified them as possibly Doppler shifted ion acoustic waves. Electron Solitary Waves (ESW) have been seen in substructures of the shock measurements (Matsumoto et al. 1997; Hull et al. 2006; Bale and Mozer 2007) and in SLAMS (Behlke et al. 2004). Strong electric spikes were seen in the EFW data for the first event (033:40 UT), but the available time resolution (2.2 ms) was not adequate to unambiguously identify the bipolar characteristics of ESWs (not shown).

Question 4.19 Discuss the waves associated with DHs and HFAs and compare them to waves observed at the bow shock.

4.7   Growth of Nonlinear Structures

The Cluster mission has provided an opportunity for greatly improved in situ measurements of the spatial and temporal development. During the apogee passes in the solar wind in 2003, the Cluster spacecraft were aligned predominantly along the Sun-Earth line (*string of pearl*) with separation distances as large as ~ 1.6R_E (Earth radius). This configuration has enabled observations of the spatial and temporal development of structures moving with the solar wind as they pass by the different spacecraft. Figure 4.23 shows an example to demonstrate the temporal development of a compressional pulse formed at the upstream edge of the nonlinear structure (Lee et al. 2009). At ~1050 UT Cluster spacecraft 1 (SC1) was in the solar wind at $(9.8, -1.5, -9.7)R_E$ and separated from SC4 in the X-, Y-, and Z-directions by ~1.6, ~0.75, and ~0.58 R_E, respectively. Cluster was moving earthward, and SC1 encountered the bow shock later at 1210 UT at ~ $(8.4, -1.8, -9.7)R_E$ (not shown).

Question 4.20 The magnetic (B) fields from the four Cluster spacecraft are shown from 16 February 2003. Describe what could be going on for this upstream nonlinear structure.

On SC3 $\delta B/B_o$ increased slightly to ~ 0.95, indicating a small growth of the pulse without change in shape. On SC2, however, the amplitude has more than

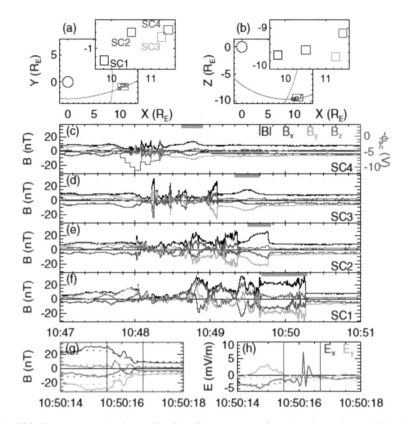

Fig. 4.23 Upper two panels show orbit of the Cluster spacecraft projected onto the (**a**) XY and (**b**) XZ planes in the geocentric solar ecliptic coordinates on 16 February 2003 (blue line) including the configuration of the Cluster spacecraft at 1050 UT. The gray curve represents the model bow shock location. Magnetic field measurements are shown from (**c**) to (**f**). Full resolution (22.5 Hz) data were used. The compressional pulse and shock-like structure are marked by the orange bar. The spacecraft potential, sc, is also plotted in (**c**). Bottom panels show (**g**) magnetic and (**h**) electric fields measured by SC1 (solid lines) and SC2 (dotted lines). The time on SC2 was shifted by 30 s to match the edges (from Lee et al., *Phys. Rev. Lett.*, **103**, 031101, 2009)

doubled, $\delta B/B_o \sim 2.4$. From time delays between spacecraft we have deduced that the pulse grew impulsively between SC3 and SC2. The estimated average growth rate $(\delta B/B_o)/\delta t$ between SC4 and SC3 is $\sim 0.0024\,\mathrm{s}^{-1}$, while between SC3 and SC2, $(\delta B/B_o)/\delta t \sim 0.097\,\mathrm{s}^{-1}$, 40 times larger. Moreover, the edge of the pulse facing the solar wind (1049:45 UT) rapidly steepened. Given the larger separation along the plasma flow direction, we interpret the variation between SC2 and SC3 as temporal rather than spatial. The distance between SC2 and SC3 was ~ 0.75, ~ 0.11, and ~ 0.16 R_E in the X-, Y-, and Z-directions, respectively, and since the structure was moving in the X-direction embedded in the solar wind, it is reasonable to assume that a similar plasma region was sampled by SC2 and SC3. Thus, the

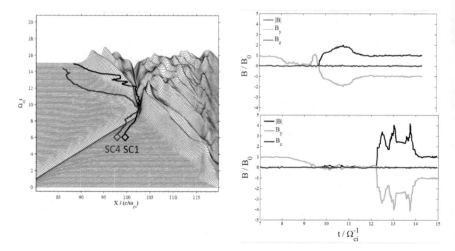

Fig. 4.24 Simulation of an IMF current sheet interaction with the bow shock, originally suggested by Burgess and Schwartz (1988). The left panel shows the simulation results and the right predicted behavior of magnetic field measured by SC 1 and 4. For this simulation, we assumed that the leading edge of DH develops into a perpendicular shock. The shock parameters used are ($\theta_{Bn} = 90°$, $M_A = 4.5$, $\beta_i = 1$, $\beta_e = 0.5$) (Courtesy of Z.W. Yang)

variations observed from SC3 to SC2 can be interpreted consistently as growing and breaking of a nonlinear pulse.

On SC1 it was possible to see that the pulse had developed into a shock-like structure with a ramp, overshoot, and magnetosheath-like downstream region. The amplitude was comparable to that on SC2 indicating that after steepening the amplitude ceased to grow. An interesting activity was observed in the steepened edge on SC1. Panels g and h below show an expanded view of **B** and **E** fields measured on SC1 (solid lines) and SC2 (dotted lines). On SC1, a few Hz frequency oscillations occurred across the edge (within vertical bars) in both **B** and **E** fields, while the transition on SC2 was smooth. These embedded oscillations may be an early phase of whistler mode waves or electromagnetic oscillations that could propagate to the upstream region of the shock.

PIC Simulation We have simulated the above observations using 1D PIC simulation code and here we show an example to illustrate how the simulation helps us interpret Cluster observations. The upstream structures are 3D phenomena but we have begun with 1D code as done previously. In particular, we simulate the event shown in Fig. 4.23 using the model of the current sheet interaction with the bow shock (Burgess and Schwartz 1988). For this simulation, we assumed that the leading edge of DH develops into a perpendicular shock. Figure 4.24 shows the results from 1D PIC electromagnetic full-particle code that simulate the shock. As in previous simulations, a reflecting wall method was used to generate the shock.

Particles are injected on the right-hand side of the simulation box with an inflow/upstream drift speed V_{inj} and are reflected at the left-hand side. The distribution functions of the ions and electrons are Maxwellian in the velocity space centered at V_{inj}. The shock builds up and moves with a speed V_{ref} from the left-hand side toward the right (in the downstream rest frame). The shock normal angle used is $90°$ and the velocities are given in units of the upstream Aflvén velocity V_A. The upstream Alfvén Mach number of the shock is $M_A = 5.5$. The plasma grid size length $dx = dz = 0.058c/\omega_{pi}$, where $c/\omega_{pi} = 1$ (ion inertial length), $c/\omega_{pe} = 0.1$ (electron inertial length), $\omega_{pe}/\omega_{ce} = 2$ (ω_{pe} is electron plasma frequency, ω_{ce} is electron cyclotron frequency), mass ratio $m_i = m_e = 100$, ion beta $\beta_i = 0.01$, and electron $\beta_e = 0.1$, time resolution $dt = 0.001\omega_{ci}$ where ω_{ci} is ion gyro-frequency.

The magnetic field is given in units of the upstream magnetic field magnitude B_o. Initially, there are 16 ions and 16 electrons per grid cell and the simulation box length is 2048 grids. The black and magenta traces (left panel) show the magnetic field sampled by SC1 and SC4. The right panels show the components of magnetic field traces recorded by SC4 (top) and SC1 (bottom) and demonstrate that the details and duration of the magnetic field measured depend on the spacecraft location relative to the bow shock and how long the IMF current sheet stays connected.

4.8 Acceleration of Reflected Particles

The reflected SW particles in the upstream region have energies higher than the incident solar wind particles. A typical reflected particle has an energy a few keV, several times that of the SW. Sonnerup (1969) developed a GC particle theory to show how the reflected SW particles could be accelerated at the shock. Subsequently, Paschmann et al. (1980) improved Sonnerup's model using a more general geometry. Paschmann's formulation will be used here.

Although shocks are formed by the collective interaction of many particles, studying the acceleration of particles using single particle formulation is useful as one can clearly see the dynamics. The theory is essentially formulated in the de Hoffman-Teller frame assuming the electric and magnetic fields are steady. Figure 4.25 shows the simple geometry of the shock region adapted from Paschmann et al. (1980). Here, θ is the angle between the IMF $\mathbf{B}$ and the normal $\mathbf{n}$ of the shock plane, $\mathbf{b}$ is along the IMF, $\mathbf{v}$ is along the direction of the incident solar wind and the angle ψ is between $\mathbf{B}$ and $\mathbf{v}_i$. Since the vectors $\mathbf{n}$, $\mathbf{b}$, and $\mathbf{v}$ are not in general coplanar, $\psi \neq \theta + \phi$. In this frame, the total energy remains constant. For $\psi \neq 0$, there will be an electric field $\mathbf{E} = \mathbf{v}_i \times \mathbf{B}$ in the shock frame (S-frame) which is essentially the SC-frame of reference.

We seek a reference frame (S'-frame) in which the SW flows parallel to $\mathbf{B}_{IMF}$. Now let $\mathbf{v}_i$ be the guiding center of the SW particle incident on the bow shock. It is now decomposed into

$$\mathbf{v}_i = \mathbf{v}_{\|i} + \mathbf{v}_t, \tag{4.12}$$

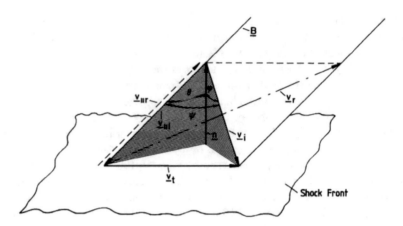

Fig. 4.25 The velocities are shown in a three-dimensional space to treat the problem of a SW incident on a shock surface. In the HT frame which is obtained by sliding along the shock surface with a velocity $\mathbf{v}_{HT}$, the solar wind electric field $\mathbf{E} = -\mathbf{v}_i \times \mathbf{B}$ vanishes (from Paschmann et al., *J. Geophys. Res.*, **85**, 4689, 1980)

where $\mathbf{v}_{\|i}$ is parallel to $\mathbf{B}$ and $\mathbf{v}_t$ is parallel to the shock surface. If $\mathbf{v}_t$ slides along the shock surface, it will transform $\mathbf{v}_i$ into the S'- frame. In the diagram, $\mathbf{v}_r$ represents the velocity of the reflected particles and we decompose $\mathbf{v}_r$ into

$$\mathbf{v}_r = \mathbf{v}_{\|r} + \mathbf{v}_t, \tag{4.13}$$

where $\mathbf{v}_{\|r}$ is the component of the reflected velocity along $\mathbf{B}_{\text{IMF}}$.

Independent of any reflection mechanisms of which there are many and they will not be specified here, the first invariant may or may not be conserved. If the reflection process conserves the magnetic moment, we find

$$\mathbf{v}_{\|r} = -\mathbf{v}_{\|i}. \tag{4.14}$$

If the magnetic moment is not conserved, then Eq. (4.14) must be modified to,

$$\mathbf{v}_{\|r} = -\delta \mathbf{v}_{\|i} \tag{4.15}$$

where δ is a positive constant and ≤ 1.

Since energy is conserved, we can write

$$v_{\|i}^2 + v_{\perp i}^2 = v_{\|r}^2 + v_{\perp r}^2, \tag{4.16}$$

where $v_{\perp}$ is the velocity component perpendicular to the magnetic field. If the incident SW is assumed " cold," the thermal energy can be neglected and $v_{\perp i} \approx 0$. Then,

$$v_{\perp r}^2 \approx v_{\|i}^2 - v_{\|r}^2$$

$$= (1 - \delta^2) v_{\|i}^2. \tag{4.17}$$

where we have used Eq. (4.15). This equation shows that $\delta \leq 1$ for it to be physically significant. To evaluate (4.16), use (4.12), and (4.13) and obtain

$$v_i^2 = v_{\|i}^2 + v_t^2 + 2\mathbf{v}_{\|i} \cdot \mathbf{v}_t$$
$$v_r^2 = v_{\|r}^2 + v_t^2 + 2\mathbf{v}_{\|r} \cdot \mathbf{v}_t. \qquad (4.18)$$

Subtract the bottom equation from the top equation and using (4.15) yields

$$v_i^2 - v_r^2 = v_{\|i}^2 - \delta v_{\|i}^2 + 2\mathbf{v}_{\|i} \cdot \mathbf{v}_t + 2\delta \mathbf{v}_{\|i} \cdot \mathbf{v}_t$$
$$= (1 - \delta^2)\mathbf{v}_{\|i}^2 + 2(1 + \delta)\mathbf{v}_{\|i} \cdot \mathbf{v}_t. \qquad (4.19)$$

Now note that

$$\mathbf{v}_{\|i} \cdot \mathbf{v}_t = \mathbf{v}_{\|i} \cdot (\mathbf{v}_i - \mathbf{v}_{\|i}). \qquad (4.20)$$

The scalar product can be evaluated noting from Fig. 4.25 that

$$v_{\|i} \cos \theta_{Bn} = v_i \cos \theta_{vn}. \qquad (4.21)$$

Hence (4.20) becomes

$$\mathbf{v}_{\|i} \cdot \mathbf{v}_{HT} = v_{\|i} v_i \cos \theta_{Bv} - v_{\|i}^2 \qquad (4.22)$$

and inserting (4.22) into (4.19) and solving for v_r^2 yields

$$v_r^2 = v_i^2 + (1 + \delta^2)v_{\|i}^2 - 2(1 + \delta)v_{\|i} v_i \cos \theta_{Bv}. \qquad (4.23)$$

Now use (4.15) in (4.16) and obtain

$$v_{\perp r}^2 = v_{\perp i}^2 + (1 - \delta^2)v_{\|i}^2$$
$$\approx (1 - \delta^2)v_{\|i}^2. \qquad (4.24)$$

If $\delta = 1$, we note that the thermal velocities do not change upon reflection. If $\delta \neq 1$, a fraction $(1 - \delta^2)$ of the incident energy will be converted to the reflected thermal energy. The second line follows from the assumption that the SW is cold. The ratio of the reflected and incident energy of the particle is shown to be

$$\frac{\epsilon_r}{\epsilon_i} = 1 + 2(1 + \delta) \left(\frac{\cos^2 \theta_{vn} - \cos \theta_{vn} \cos \theta_{Bn} \cos \theta_{Bv}}{\cos^2 \theta_{Bn}} \right). \qquad (4.25)$$

This equation gives the energy of the reflected particles as a function of the incident energy dependent only on the shock geometry. As is evident, a large range of energies can be obtained for the reflected particles.

Question 4.21 The incident and reflected energies of the solar wind in the shock frame (rest frame) are, respectively,

$$\epsilon_i = \frac{m}{2}(v_{\parallel i}^2 + v_{\perp i}^2) \approx \frac{m}{2}v_{\parallel i}^2$$

$$\epsilon_r = \frac{m}{2}(v_{\parallel r}^2 + v_{\perp r}^2).$$

Use the various relations derived above and derive Eq. (4.25).

The formulation presented above requires finding a frame of reference in which the electric field vanishes. This is possible only because the fields are assumed homogeneous and particles behave adiabatically. If the fields are inhomogeneous, a *HT* frame cannot be found and one must resort to full analysis of the Lorentz equation of motion. In this case, particles are still accelerated because they drift in the direction of the electric field. In the limit of the guiding center approximation, this analysis will yield the same result as derived above. This analysis is however much more general and can yield new results, one of which is that particle acceleration occurs even for transmitted particles.

Note that there are other particle acceleration mechanisms. For instance, particles can be accelerated by stochastic processes because shocks in nature are turbulent and intense large-amplitude waves permeate shock regions. One stochastic process that has been studied and is applicable for high-energy particles (i.e., cosmic ray particles), invokes the well-known Fermi process, which accelerates particles by repeated encounters with the shock.

4.9   Concluding Remarks

We have presented a brief overview of the bow shock physics, emphasizing new observations made recently by the Cluster mission. We have shown many features that are not predicted by existing theories and models. The Earth's bow shock involves nonlinear dynamics of collisionless plasmas and fundamental issues about the bow shock physics still remain not well understood. A key question is how collisionless shocks can thermalize the plasma behind the shock. Also important is to understand how the upstream beam and diffuse populations are produced, what creates distributions of non-gyrotropic ions and electrons (Anderson et al. 1985; Gurgiolo et al. 1981) that are observed in the upstream region, and how the shock accelerates ions to MeV energies (Skoug et al. 1996; Meziane et al. 1999). Two competing models explain the superthermal ions observed in the upstream region: escape of magnetospheric ring current particles across the magnetopause (Anagnostopoulos et al. 1986; Sarris et al. 1987; Sibeck et al. 1988) and Fermi acceleration (Terasawa 1979; Freeman and Parks 2000). Some models have predicted that the ring current particles can escape into interplanetary space under certain magnetic field geometry.

We summarize briefly what we have learned and also ask some questions.

1. The various upstream nonlinear structures are part of the bow shock dynamics. However, how the various structures are related to each other remains unknown. How are the reflected, intermediate, and diffuse distributions related to the more dynamic HFAs and DHs?

2. No studies have as yet identified the specific criteria for when these nonlinear structures are observed, or a mechanism for their production. Do all shorter temporal variations like DHs evolve to the longer HFAs as they are convected toward the bow shock? Are the longer duration HFA events produced by merging many shorter DHs? How do the upstream nonlinear structures affect the dynamics or the reformation of the bow shock?

3. Upstream nonlinear structures are observed only when the back streaming particles are present but not all back streaming particles produce the nonlinear structures. Cluster results show that the beams and gyrating ions come from the same source. The current sheet interaction model with the bow shock is crucial for the upstream structures. Do the upstream particles play a role in the production of the current sheet? The upstream particles interact with Alfvén waves. For a discussion of nonlinear dynamics of Alfvén waves and ions upstream of the bow shock, see Teresawa (1988).

4. Sometimes the solar wind can penetrate the bow shock with little change in the temperature indicating little or no interaction occurring at the bow shock. This behavior is not explained by any of the theoretical or simulation models and requires new innovations. We showed one example of ICME shock and find here that the SW is heated considerably. More studies are needed to identify the similarities and differences of ICME shock and the bow shock.

5. Behlke et al. (2004) have reported multi-point measurements of electric fields associated with short large-amplitude magnetic structures (SLAMS) in the vicinity of the parallel shock. Horbury et al. (2002) have made four-point measurements in the quasi-perpendicular region to determine the orientation and motion of the shock. How can we use the wave and electric field data to further study the nonlinear dynamics of ESWs?

6. Our observations indicate that the behavior of entropy density across the bow shock is consistent with the Vlasov description of collisionless plasmas. Our analysis included only the distribution function of charged particles. The spacecraft measurements are local and cannot measure the total particles in a flux tube. How important is it to know about the behavior of the total entropy?

7. Particle In Cell (PIC) models have simulated the behavior of the bow shock showing that the SW thermalizes across the boundary. But the PIC codes are formulated using the Vlasov equation and the theory does not include heating particles. What artificial processes are going on in the PIC code that show SW is heated across the shock? Are PIC codes showing the results of phase mixing

artificially produced because there the waves are interacting with fluctuations? Although collisions rarely occur, should the PIC code be modified to include the full Boltzmann equation to study space plasma shocks?

The coupled Lorentz-Maxwell equations do not distinguish a shock boundary from any other electromagnetic boundary. However, shock models based on MHD fluid concepts have called the first boundary formed by the solar wind interaction with the geomagnetic field a collisionless bow shock. The Earth's bow shock behaves very differently from ordinary fluid shocks. Not only is a portion of the SW reflected into the upstream region where it interacts with the oncoming SW, the penetrating component of the SW is not immediately thermalized across the bow shock.

The magnetosheath distributions are very complex with multiple populations that include the magnetospheric and ionospheric population. For example, O^+ ions have been observed in the magnetosheath (Marcucci et al. 2004). A question still not answered is where all of the particles in the magnetosheath come from and what fraction of the solar wind beam contributes to magnetosheath population. Can experiments be designed to identify all of the particle distributions in the magnetosheath?

There are still no accepted mechanisms for producing collisionless shocks. In view of this, we now speculate about the bow shock and magnetosheath. Suppose collisionless plasmas do not form shock waves. What that would mean is that the boundary we call the bow shock is simply an electromagnetic boundary formed by the SW interacting with the geomagnetic field as in the original model of Sydney Chapman (Chapter 8, Parks 2004). As the SW approaches the geomagnetic field, electromagnetic induction produces an $\mathcal{EMF}$ that drives a current which then steepens the magnetic field (first boundary). A considerable amount of complex dynamics is going on at this boundary, including reflection and transmission of incident SW particles. The reflected particles interact with the incident SW and produce instabilities, for example, counter streaming instability. The currents formed at the induced boundary could also be unstable generating waves and wave-particle interaction could heat and accelerate the SW particles. The heated and accelerated plasma expand to the downstream of the boundary and a second boundary will form as this newly formed hot plasma balances the geomagnetic field. This boundary is the magnetopause. The readers are urged to criticize and discuss the strengths and weakness of this speculative idea.

4.10   Solutions to Questions in Chap. 4

Solution 4.1 First note that the energy flux spectrograms have units of counts/s. The unit counts/s is not absolute since different instruments give different counts/s detecting the same plasma. The correct unit is ergs/cm^2 $-$ s^{-1} which is independent of the instrument. This unit is obtained by dividing the counts/s by the geometrical factor of the instrument. The density determined in the SW from the g-detector is $4.6\,cc^{-1}$ and total density in the downstream MS determined from the G-data was $\sim 20\,cc^{-1}$. The temperature of the SW determined from the second moment is $20\,eV$ and in the MS $kT \sim 80$–$100\,eV$ (panel 6). The mean plasma flow of the SW was $\sim 410\,km\,s^{-1}$ and in the MS $\sim 200\,km\,s^{-1}$ (panel 7). Heat flux computed in the plasma frame (bottom panel) showed that heat was dissipated in the MS at the rate of ~ 0.05–$0.1\,ergs\,cm^{-2}\,s^{-1}$. The g-detector shows the SW deviated going across the shock into the MS, increasing $\langle V_y \rangle$ and $\langle V_z \rangle$ to $\sim 50\,km\,s^{-1}$. Both detectors show that the SW speed (V_x, black) in the magnetosheath was $\sim 200\,km\,s^{-1}$, larger than the Alfvén speed (magenta) which was $\sim 50\,km\,s^{-1}$. While the SW bulk speed in the MS decreased, the bulk speed remained super-Alfénic in the MS. The SW-like beams in the magnetosheath are probably the directly transmitted SW beams discussed by Sckopke (1995). However, the beams have slightly lower energies than the original SW. The usual explanation of this feature is explained by the presence of a potential across the bow shock. Because the g-detector is looking at limited field of view, it does not detect particles coming from other directions. The G-detector is fully 3D except it does not count particles measured by the g-detector. The energy dispersion is due to the gyrating particles at the bow shock. The higher energy particles extend further into space and hence are detected first.

Solution 4.2 Distribution functions from both the MS and SW are shown. The distributions in the MS show complex structures, sometimes showing multiple beams indicating considerable variations of the penetrated SW beams in the MS, with some showing features of H^+ and He^{++} nearly the same as in the SW (0632:06 UT) while others have been scattered and heated. In addition to the penetrated SW, other particles such as particles leaking out of the magnetosphere and ionosphere (Marcucci et al. 2004) could be present including locally accelerated particles. The density computed from the moment includes all of these particles in the MS. The total density varied from ~ 1.7 to $9.7\,cc^{-1}$, the bulk velocity from 247 to 396 km/s, and temperature ~ 13–$132\,eV$. The foot of the shock was crossed at ~ 0645:07 UT and the distribution at 0644:07 UT was near the peak of the ramp. To learn about the other particles present, more detailed velocity space distributions need to be studied.

Solution 4.3 Each panel shows phase space density in the velocity space from each of the eight detectors. The two columns on the left show data from the SW and the two on the right from the MS. In the upstream region, SW beam was observed near the equatorial plane fifth panel. The SW beam appears between $-177°$ and $171°$ where $\phi = 180°$ is toward the Sun. The intense beam (red) at $\sim 414\,km\,s^{-1}$ is H^+ and the less intense beam (blue) at $589\,km\,s^{-1}$ is He^{++}. The particles around the SW include reflected and leakage population from the magnetosheath. In the magnetosheath, SW was observed in the seventh panel which was looking at the particles coming from the south of the equator indicating the particles have been scattered. The deviation probably resulted from the fluctuations of the magnetic field at the shock and magnetosheath.

Solution 4.4 When temperature is calculated in the SW frame of reference (S'-frame), the measured velocity must be shifted by the amount of shift from the origin, in this case approximately $-297\,\mathrm{km\,s^{-1}}$ for H^+ and $-501\,\mathrm{km\,s^{-1}}$ for He^{++}. These distributions are then transformed to directions parallel (X-axis) and perpendicular (Y-axis) to the magnetic field and the 1D cuts were then taken to obtain parallel (red) and perpendicular (black) temperatures assuming that the shapes of the distributions are Maxwellian. We note that the ratio of bulk speeds for He^{++} to H^+ in the MS is ~ 1.7 (instead of 1.4 as in the SW) because He^{++} with greater M/Q slows down less than H^+. The distributions were anisotropic and the parallel temperatures were slightly higher than the perpendicular temperatures (as in the SW). The surprising new result is that the temperatures of H^+ and He^{++} remained essentially the same crossing the shock into the downstream magnetosheath region for both ion species. The readers should consider what the results imply. How can the SW beam remain the same crossing the bow shock? Note that the temperatures estimated here for the beams are a factor of two lower than those shown in Fig. 4.1 that included the other particles ($T \sim 12\,\mathrm{eV}$). The SW temperatures measured here are more "pristine" because they included only the beams.

Solution 4.5 The accepted interpretation about flat top shape is that the lower energy electrons cannot overcome the electric potential, resulting in loss of these particles (Feldman 1985). The potential deduced on the SC frame is $\sim 50\,\mathrm{V}$ (S-frame) slightly lower than previous observations by Formisano et al. (1982) and Wygant et al. (1987). However, the value of the potential is frame dependent and must be computed in the deHoffman-Teller frame (Goodrich and Scudder 1984). Goodrich and Scudder have argued that reference frame is important. The readers are recommended to read and discuss how the frame dependence can physically affect the value of the potential. Flat top shape distributions are ubiquitous in space and they have been detected in the polar cusp (Parks et al. 2008) and in the central plasma regions (Chen et al. 2000; Zhao et al. 2017).

Solution 4.6 We can measure the entropy from observations because three-dimensional (3D) distributions $f(\mathbf{r}, \mathbf{v}, t)$ of the SW plasma are measured by Cluster satellites near the bow shock. With Cluster g-detector, we can measure the distribution function upstream, downstream, and across Earth's bow shock. However, it is not possible to compute H and dH/dt using these data. As mentioned in the text, we can measure the distribution function only at the spacecraft and therefore cannot obtain the total number of particles in a flux tube required to compute H and dH/dt. Therefore, we compute instead a *normalized H function* by modifying the original definition (see Eqs. (4.1)–(4.3)).

Solution 4.7 Space plasmas are essentially collisionless. We can thus assume that Boltzmann's situation is analogous to an instrument moving with the magnetic flux tube of steady SW across the Earth's bow shock. For our case, the SW speed was moving much faster than the spacecraft by nearly two orders of magnitude, and moreover, the spacecraft is also moving relative to the bow shock. We can interpret the observed time variations as due to spacecraft motion through spatial structures.

It is also necessary to assume that the measurements along the spacecraft track represent the same plasma volume that traveled the spacecraft track.

Solution 4.8 The measured Δs across the bow shock is of the same order as entropy change observed in isolated free expansion of ideal gas when the volume changes by a factor of 2, $\Delta s = 0.95 \times 10^{-16}$ ergs K^{-1}, and in ice at 0 K melting to water at the same temperature, $\Delta s = 3.3 \times 10^{-16}$ergs K^{-1}. In these calculations, the entropy change per mole has been divided by Avagadro's number so they can be compared to space plasma per particle values. We speculate that these numbers are all small because the original energy per particle has an order of magnitude value of kT and when the state change involves an amount of energy corresponding roughly to the original amount of energy, the associated entropy change will be of the order of Boltzmann's constant. The bow shock results may simply be telling us that the compression ratio at Earth's bow shock is not some huge number ($\sim$3 for this crossing; typical compression for Earth's bow shock is $\sim$2–4). If the compression ratio was really large as it might be for a big astrophysical shock, then the entropy per particle would be expected to be much larger than k_B. Can the readers come up with a different reason?

Solution 4.9 The entropy conservation equation (4.7) will simplify to

$$\frac{d(U_x ns - F_x)}{dx} = 0.$$

This equation states that $(U_x ns - F_x) = $ constant indicating that the entropy flux in the X-direction is continuous. We can rewrite this equation as $(U_{x1} n_1 s_1 - F_{x1}) = (U_{x2} n_2 s_2 - F_{x2})$ where the subindices 1 and 2 are quantities measured in the upstream and downstream regions. The left side $(s_2 - s_1)$ has already been computed. The new terms $(F_{2x} - F_{1x})$ have been calculated for ions, shown in Fig. 4.26 (no 3D electron data on SC1). Since the mass is conserved, we use the mass conservation equation $U_1 n_1 = U_2 n_2$, and rewrite the equation as

$$s_2 - s_1 = (F_{2x} - F_{1x})/U_1 n_1,$$

where U_1, the flow in the normal direction, is determined from the minimum variance analysis. For processes that produce non-Maxwellian distribution functions, the right side gives the amount of per particle entropy change in this steady state $1D$ Vlasov model.

The data used here were obtained when the SW flow was not varying significantly during the time it took to measure both sides of the shock, and we assume that we are measuring particles in the same flux tube of plasma but at earlier and later times. The fact that the behavior of dh/dt (red) and $(F_2 - F_1)/U_1 n_1$ (black) is "similar" is not a proof of the Vlasov theory; rather, it indicates that our results generally support the idea that the Vlasov model applies to the data.

Solution 4.10 We measured entropy density that increased across Earth's bow shock. Our observations are consistent with the Vlasov model of entropy that predicts entropy density can be locally generated when the distribution function

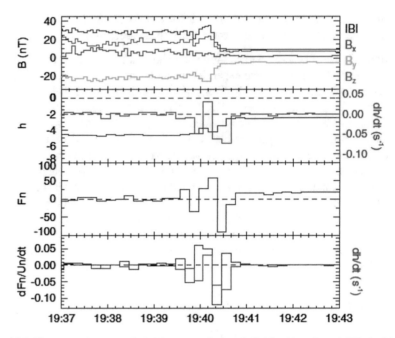

Fig. 4.26 Shown are the magnetic field (top panel), h and dh/dt of ions [panel (2)], the kinetic function F_n [panel (3)], and the local change of per particle kinetic entropy flux, Eq. (3) [panel (4), black] superposed on dh/dt (red) (from Parks et al., *Phys. Rev. Lett.*, **108**, 061102, 2012)

is non-Maxwellian. The reader should note that the analysis we performed only included the distribution function of charged particles. However, complex electromagnetic waves permeate the shock region and they should be included (see Sect. 4.4.2 that discusses the suggestion of entropy change by Tidmann and Krall).

Solution 4.11 The computation is more difficult because during ICME shocks the speed of the SW is very high relative to the measurement platform (SC). Hence the distribution functions are shifted to much higher energies. What this means for an instrument is important. In particular, the energy width of our SW instrument becomes broader at high energies. Since the energy width of an instrument is normally chosen with the constraint to obtain about an equal number of counts for all energy channels for a given accumulation time and since the energy spectrum of any phenomena generally falls with energy, one selects a wider energy width at higher energies. Hence caution must be exercised to obtain accurate 2D distributions in the $V_\parallel - V_\perp$ space for ESAs designed to measure the SW.

Solution 4.12 Classical definition of entropy generation requires collisions to diffuse particles in velocity space. Consequently, the above definition of entropy is not consistent with the original definition. However waves can scatter particles from one region of velocity space to another region. Studying wave-particle interaction problem will take a full treatment of the Boltzmann equation which is beyond the scope of this book. In spite of the many unanswered questions about the Vlasov approach, we nevertheless encourage continuing data analysis combined with computer modeling with measured shock parameters. Hopefully they will reveal new clues about energy dissipation and entropy generation in collisionless plasmas observed throughout the Universe.

Solution 4.13 These examples indicate that some DHs are deep with only a few measurable particles inside them while others are shallow, only a factor of 2 smaller in density n. The edges are steepened and bulk velocity in the X-direction is considerably reduced inside DHs, sometimes approaching ~ 0 km/s. The temperature T inside the hole is very hot, more than ten times the solar wind T. A component of the magnetic field is observed to change sign across the hole showing the IMF current sheet is active. Note that DHs have many similarities to HFAs but no studies have thus far identified the specific criteria for when the DHs are observed or a mechanism for their production.

Solution 4.14 The units for the energy flux plot is in counts/s which is incorrect. Counts/second is instrument dependent because the instruments have different geometrical factors. The correct unit is ergs cm^{-2}-s^{-1}. Statistical analysis shows the E-field tends to point away from the IMF current sheets. The fact that there are many cases with angles $<90^o$ is significant and contrasts with the determinations made of HFAs, which have normal components of the field pointing inward. This analysis indicates that DHs may be different from HFAs and HDCs. How the mechanisms are different is not answered at this time.

Solution 4.15 The g-detector data show that the electron distributions in the neighborhood of upstream structures are dynamic both inside and outside of the structure. The presence of flattop shape distributions is clearly evident and the distributions are evolving in a complicated way. The sequence from 2045:07 UT to 2046:07 UT includes the time when the SC was in the SW (2045:07 UT) covering the vicinity of the structure, and finally back into the SW (2046:07 UT) as indicated by the third distribution on the third row obtained at 2045:55 UT. The distributions in the density depleted region, 1045:20–1035:34 UT (panels 4 and 5, second row) show that the distributions are the widest indicating that the temperature is the highest. The one-sided flattop distributions observed for electrons moving toward the Sun suggest that these electrons could have been back scattered by the steepened edge. Also similar to the bow shock, there could be a potential drop across the steepened edges. The large variations of the distributions observed is probably due to the interaction of the varying SW and the waves albeit the mechanisms have not yet been identified. Clearly further work is needed to learn more about the complex dynamics going on in these nonlinear structures.

Solution 4.16 The density of the event that began at 0336:21 UT recovered and returned to the SW value at 0336:00 UT. A small enhanced edge was encountered on the trailing edge side with n increasing to 2.5 cc^{-1} and B to 7 nT. Coincident with the density depleted region (0336:00–0336:45 UT), V_x slowed down to ~ 0 km s^{-1}

and at minimum n, V_x even went slightly positive (one data point, 4s). We also see that V_z, which was ~ 0 in the SW, attained ~ 200 km s^{-1} at the edges and V_y increased from ~ 50 to 75 km s^{-1}. The temperature in the density depleted region was $\sim 3 \times 10^7$ K (0336:00–0336:45 UT), which is thirty times higher than the SW $T \sim 10^6$ K. The magnetic field B_y component switched sign in the density depleted region from positive to negative (0336:30 UT).

The density n for the peak at 0337:10–0337:30 UT increased from ~ 1.4 to ~ 8.3 cc^{-1} and B from ~ 5 to 22 nT. Plasma during this interval behaved differently from the first one discussed above as it did not include a density-depleted region and the magnetic field did not change sign in the minimum. The peak on the upstream side lasted until $\sim 0337:00$ UT after which n decreased to ~ 2 cc^{-1} and the magnetic field to ~ 7 nT. These values (during 0336:40–0337:10 UT) are close to the SW values but slightly higher indicating that Cluster had not completely returned to SW. Following the enhanced edge, V_x did not decrease in the "low-density" region (0336:41–0337:33 UT). At the compressed edge, however, V_x did decrease from 725 to ~ 500 km s^{-1}. A temperature increase was *not* observed during the density dip region although a slight increase was observed at the compressed edge (0337:29 UT).

Depending on the flux of the sunward and anti-sunward going particles, V_x could even vanish or become positive. The temperature in the density-depleted region increases because the new population occupied a larger volume in velocity space and the second velocity moment increased. Note also the appearance of ions with energies lower than the SW energy. These particles may be accelerated at the shock or at the boundary of the HFA structure. These particles have different distributions from the back streaming SW population coming from the bow shock (not shown), which is usually present when upstream DHs and HFAs are observed.

Solution 4.17 We used the distribution function information and in addition output from each of the eight detectors in the θ direction (Fig. 4.19). Velocity space plots, bottom row in Fig. 4.19, show that the beam intensity was reduced considerably, corroborating the energy flux plot. Further examination of the beam feature has been made using the spin resolution 3D distribution functions inside the density-depleted region, 0336:10–0336:38 UT. Moreover, the behavior of the particles has been examined using the differential energy spectra of the ions observed by the 8 detectors that look from $-90°$ to $+90°$ in θ-direction (Fig. 4.19). Each detector is $22.5°$ apart in the θ-direction and the data are averaged over all azimuth angles, ϕ-direction. The detectors that measure the SW beam are generally on the ecliptic plane ($\theta 4$ and $\theta 5$) looking toward the Sun. Fluxes of the SW peak (red and green) steadily became lower inside the HFA corroborating the 2D velocity space plots. Coincidentally, we see that fluxes from other directions are increasing. In particular, note the substantial increase of particle flux away from the ecliptic plane. For example, we see fluxes of $\theta 1$ (black, $\theta = +90°$) creeping up in intensity opposite to the SW beam. At 0336:22 UT, the fluxes in the $\theta 1$ detector have become nearly equal to the SW flux ($\theta 5$). These particles all contribute to the velocity and temperature moments. The mean velocity and V_x were reduced because the SW beam was counteracted by the particles traveling in the opposite direction. The temperature increased because of the back streaming particles, which occupied a large velocity space.

Cluster data have shown that the apparent decrease of the SW mean velocity V_x observed with density depleted regions of HFAs as shown in the moment data is not due to slowing down of the SW but an *artifact* of the moment calculations. The SW beam persisted in the HFA event for the entire duration with little change in velocity. Our interpretation of the decrease of the beam intensity and the appearance of particles from other directions is that the SW is interacting with the magnetic edges, being reflected as in the bow shock. The lesson learned here is that studying velocity moments alone does not give the full information about the dynamics and could even lead to incorrect conclusions.

Solution 4.18 A striking feature is the growth of spontaneous magnetic pulses on the inside edges of both steepened magnetic fields (between 0336:38–0336:42 UT and 0337:12–0337:13 UT). The first pulse was observed on SC 1, 2, 3, and 4 and the magnitude was the largest on SC3 and smallest on SC1. On SC3, B_x component decreased from 0 nT reaching a minimum -15 nT, then maximum to $+7$ nT and turning over. B_y increased from -6 to 0 nT subsequently decreasing to -15 nT, and B_z increased from 0 nT $+ 5$ nT then crossing zero and decreasing to -7 nT and turned over. These observations are consistent with the suggestion that a current sheet must be involved in producing a nonlinear structure upstream of the bow shock (Burgess and Schwartz 1988).

Solution 4.19 Generally, ES emissions are observed up to the plasma frequency. Waves are most intense at the edges where the gradients are large. The broadband ES waves near the whistler range intensify at both steepened edges. They are also bursty below and above f_{ce}. Waves associated with the two spontaneous events show intense emissions of electrostatic (ES) waves up to the electron plasma frequency for the first event (Fig. 4.22; $\sim$0336:40 UT) and a much weaker emission for the later event at $\sim$03:37:20 UT. ES emissions are Langmuir waves ($\sim$20 kHz) very similar to ones that permeate the electron foreshock region upstream of the bow shock (Gurnett 1985; Matsumoto et al. 1997). The wave amplitude variations are correlated to the particle density variations indicating that particles and field are closely coupled. For example, at $\sim$03:36:30 when SC3 was crossing the density depleted region, the ES waves diminished. On the other hand, the intensity of the waves rapidly increased (to $\sim$40 kHz) during density enhancements associated with the steepened magnetic edges.

Solution 4.20 We follow the description given by Lee et al. (2009). Consider a nonlinear structure observed on SC4 (furthest from the bow shock) between 1047:50 and 1048:20 UT. This event can be identified by a depression in $|B|$ and electron density (we use spacecraft potential, Φ_{sc}, which is a proxy for electron density). A compressional pulse is observed at the upstream edge of the DH ($\sim$1048:45 UT, marked by an orange bar on top of Fig. 4.23). The amplitude of the pulse was $\delta B / B_o \sim 0.84$, where B_o is the magnitude of the upstream B-field and δB is the enhancement at the pulse. Subsequent observations from the other spacecraft indicate the DH developed in a complicated way. Here let us focus only on the development of the compressional pulse observed at the upstream edge of the DH (marked by orange bars in Fig. 4.23).

Solution 4.21 We use Eqs. (4.21)–(4.24) in the energy equations given above. Then we note that

$$\frac{\epsilon_r}{\epsilon_i} = \frac{v_r^2 + v_{\perp_r^2}}{v_{\|_i^2}}.$$

But we note that

$$v_{\perp_r^2} = (1 - \delta^2) v_{\|_i}^2.$$

Hence, we can thus write

$$\frac{\epsilon_r^2}{\epsilon_i^2} = 1 + (1 + \delta)^2 \frac{\cos^2 \theta_{vn}}{\cos^2 \theta_{bn}} - 2(1 + \delta) \frac{\cos \theta_{Bv} \cos \theta_{vn}}{\cos^2 \theta_{Bn}} + (1 - \delta^2) \frac{\cos \theta^2 \theta_{vn}}{\cos^2 \theta_{Bn}}.$$

Collecting similar terms, this equation will lead to the desired expression shown in Eq. (4.25),

$$\frac{\epsilon_r}{\epsilon_i} = 1 + 2(1 + \delta) \left(\frac{\cos^2 \theta_{vn} - \cos \theta_{vn} \cos \theta_{Bn} \cos \theta_{Bv}}{\cos^2 \theta_{Bn}} \right).$$

This theory has been tested by Paschmann et al. (1980) using ISEE data. The ISEE detector is a 2D detector. The reflection of the SW from the bow shock is a 3D problem. Sonnerup's reflection theory should be tested using data from a 3D detector on Cluster.

References

Anagnostopoulos, A., et al.: Magnetospheric origin of energetic (at least 50 keV) ions upstream of the bow shock - The October 31, 1977, Event. J. Geophys. Res. **59**, 2859 (1986)

Anderson, K.A., et al.: A component of nongyrotropic (phase-bunched) electrons upstream from the earth's bow shock. J. Geophys. Res. **90**, 10809 (1985)

Bale, S., Mozer, F.: Measurement of large parallel and perpendicular electric fields on electron spatial scales in the terrestrial bow shock. Phys. Rev. Lett. **98**, 205001 (2007)

Bale, S.D., et al.: Quasi-perpendicular shock structure and processes. Space Sci. Rev. **118**, 161 (2005)

Balikhin, M., et al.: Ion sound wave packets at the quasi-perpendicular shock front. Geophys. Res. Lett. **32**, L24106 (2005)

Balogh, A., Treumann, R.: Physics of Collisionless Shocks: Space Plasma Shock Waves. Springer, New York (2013)

Behlke, R., et al.: Solitary structures associated with short large-amplitude magnetic structures (SLAMS) upstream of the earth's quasi-parallel bow shock. Geophys. Res. Lett. **31**, L16805 (2004)

Burgess, D., Schwartz, S.: Colliding plasma structures current sheet and perpendicular shock. J. Geophys. Res. **93**, 11327 (1988)

Burgess, D., et al.: Ion acceleration at the earth's bow shock. Space Sci. Rev. **175**, 5 (2012)

Burgess, D., Scholer, M.: Collisionless Shocks in Space Plasmas. Cambridge Press, London (2015)

Chen, L.J., et al.: Multicomponent plasma distributions in the tail current sheet associated with substorms. Geophys. Res. Lett. **27**, 843 (2000)

Collinson, G., et al.: A survey of hot flow anomalies at Venus. J. Geophys. Res. **119**, 978 (2014)

Cornilou-Wehrlin, N., et al.: The Cluster Spatio-Temporal Analysis of Field Fluctuations (STAFF) Experiment. Space Sci. Rev. **79**, 107 (1997)

Crooker, N., et al.: Transients associated with recurrent storms. J. Geophys. Res. **102**, 14041 (1997)

Décréau, P.M., et al.: A resonance sounder and wave analyzer: performance and perspectives for the cluster mission. Space Sci. Rev. **79**, 157 (1997)

Dunlop, M., et al.: Four-point cluster application of magnetic field analysis tools: the curlometer. J. Geophys. Res. **107**, 1384 (2002)

Eastwood, J.P., et al.: The foreshock. Space Sci. Rev. **118**, 41 (2005)

Feldman, W.C.: Quantitative tests of a steady state theory of solar wind electrons. J. Geophys. Res. **87**, 7355 (1982)

Formisano, V., et al.: Measurement of the potential drop across the earth's collisionless bow shock. Geophys. Res. Lett. **9**, 1033 (1982)

Formisano, V., Palmiotto, F., Moreno, G.: α-particle observations in the solar wind. Sol. Phys. **15**, 479 (1970)

Freeman, T., Parks, G.K.: Fermi acceleration of suprathermal solar wind oxygen ions. J. Geophys. Res. **105**,15715 (2000)

Fuselier, S., Schmidt, W.: H+ and He2+ heating at the Earth's bow shock. J. Geophys. Res. **99**, 11539 (1994)

Fuselier, S., Thomsen, M.: He(2+) in field-aligned beams - ISEE results. Geophys. Res. Lett. **19**, 437 (1992)

Fuselier, S.A., et al.: Ion distributions in the earth's foreshock upstream from the bow shock. Adv. Space Res. **15**, 43 (1995)

Ge, Y.S., et al., Case studies of mirror mode structures observed by THEMIS in the near-earth tail during substorms. J. Geophys. Res. **116**, A01209 (2011)

Gold, T.: Gas Dynamics of Cosmic Clouds, vol. 103. North Holland, Amsterdam (1955)

Goldstein, M., et al.: Multipoint observations of plasma phenomena made in space by cluster. J. Plasma Phys. **81**, 325810301 (2015)

Goodrich, C., Scudder, J.: The adiabatic energy change of plasma electrons and the frame dependence of the cross-shock potential at collisionless magnetosonic shock waves. J. Geophys. Res. **89**, 6654 (1984)

Gosling, J.: Coronal mass ejection. AIP Conf. Proc. **516**, 59 (2000)

Gurgiolo, C., et al.: Non-E × **B** ordered ion beams upstream of the earth's bow shock. J. Geophys. Res. **86**, 4415 (1981)

Gurnett, D.A.: In: Tsurutani, B., Stone, R.G. (eds.) Plasma Waves and Instabilities in Collisionless Shocks in the Heliosphere: Review of Current Research. Geophysical Monograph Series, vol. 35, pp. 207–224. American Geophysical Union, Washington, DC (1985)

Horbury, T., et al.: Four spacecraft measurements of the quasiperpendicular terrestrial bow shock: orientation and motion. J. Geophys. Res. **107** (2002). https://doi.org/10.1029/2001JA000273

Hoshino, M., Shimada, N.: Nonthermal electrons at high mach number shocks: electron shock surfing acceleration. Astrophys. J. **572**, 880 (2002)

Hull, A., et al.: Large-amplitude electrostatic waves associated with magnetic ramp substructure at earth's bow shock. Geophys. Res. Lett. **33**, L15104 (2006)

Ipavich, F., et al.: A statistical survey of ions observed upstream of earth's bow shock: energy spectra, composition and spatial variations. J. Geophys. Res. **86**, 4337 (1981)

Johnstone, A., et al.: Peace: a plasma electron and current experiment. Space Sci. Rev. **79**, 351 (1997)

Kennel, C.F., et al.: A quarter century of collisionless shock research. In: Stone, R.G., Tsurutani, B.T. (eds.) Collisionless Shocks in the Heliosphere: A Tutorial Review. Geophysical Monograph, vol. 34. American Geophysical Union, Washington, DC (1985)

Krall, N.A.: What do we really know about collisionless shocks? Adv. Space Res. **20**, 715 (1997)

Lee, E., et al.: Nonlinear development of shock like structure in the solar wind. Phys. Rev. Lett. **103**, 031101 (2009)

Lembège, B., et al.: Selected problems in collisionless-shock physics. Space Sci. Rev. **110**, 161 (2004)

Leroy, M., et al.: The structure of perpendicular bow shock. J. Geophys. Res. **87**, 5081 (1982)

Liu, Y., et al.: Thermodynamic structure of collision-dominated expanding plasma: heating of interplanetary coronal mass ejections. J. Geophys. Res. **111**, A01102 (2006)

Liu, Y., et al.: A comprehensive view of the 2006 December 13 CME: from the sun to interplanetary space. Astrophys. J. **689**, 563 (2008)

Lucek, E., et al.: Cluster observations of hot flow anomalies. J. Geophys. Res. **109** (2004)

Marcucci, M.F., et al.: Energetic magnetospheric oxygen in the magnetosheath and its response to IMF orientation: cluster observations. J. Geophys. Res. **109**, A07203 (2004)

Matsumoto, M., et al.: Plasma waves in the upstream and bow shock regions observed by geotail in 1997. Adv. Space Res. **20**, 683 (1997)

Meziane, K., et al.: Evidence for acceleration of ions to $\sim$1 Mev by adiabatic-like reflection at the quasi-perpendicular earth's bow shock. Geophys. Res. Lett. **26**, 2925 (1999)

Möebius, E., et al.: Observations of the spatial and temporal structure of field-aligned and gyrating ring distributions at the quasi-perpendicular bow shock with cluster CIS. Ann. Geophys. **19**, 1411 (2001)

Montgomery, M.: The solar wind in the outer solar system. Rev. Space Sci. **14**, 559 (1973)

Ness., N., et al.: Initial results of the IMP 1 magnetic field experiment. J. Geophys. Res. **69**, 3531 (1964)

Odstrcil, D., et al.: Numerical simulations of solar wind disturbances by coupled models. ASP Conf. Ser. **385**, 167 (2008)

Øieroset, M., Mitchell, D.L., Phan, T.D., et al.: Hot diamagnetic cavities upstream of the Martian bow shock. Geophys. Res. Lett. **28**, 887 (2001)

Onsager, T., et al.: Survey of coherent ion reflection at the quasi-parallel bow shock. J. Geophys. Res. **95**, 2261 (1990)

Papadopoulos, K.: Ion thermalization in the earth's bow shock. J. Geophys. Res. **76**, 3806 (1971)

Parks, G.K.: Physics of Space Plasmas, An Introduction. Westview Press, Boulder, CO (2004)

Parks, G.K., et al.: Larmor radius size density holes discovered in the solar wind upstream of earth's bow shock. Phys. Plasmas **13**, 050701 (2006)

Parks, G., et al.: Density holes in the upstream solar wind. AIP Conf. Proc. **932**, 9 (2007)

Parks, G.K., et al.: Transport of transient solar wind particles in earth's cusps. Phys. Plas. **15**, 080702 (2008)

Parks, G.K., et al.: Entropy generation across earth's collisionless bow shock. Phys. Rev. Lett. **108**, 061102 (2012)

Parks, G.K., et al.: Reinterpretation of slowdown of solar wind mean velocity in nonlinear structures observed upstream of earth's bow shock. Astrophys. J. Lett. **77**, L39 (2013)

Parks, GK.: Physics of Space Plasmas. An Introduction, Westview Press. Boulder, Colorado, (2004)

Parks, G.K., et al.: Transport of solar wind H^+ and He^{++} ions across earth's bow shock. Astrophys. J. Lett. **825**, 27 (2016)

Parks, G., Lee, E.S., Fu, S.Y., et al.: Shocks in collisionless plasmas. Rev. Modern Plasma Phys. **1**, 1 (2017)

Paschmann, G., et al.: Energization of solar wind ions by reflection from earth's bow shock. J. Geophys. Res. **85**, 4689 (1980)

Paschmann, et al.: Characteristics if reflected and diffuse ions upstream from the earth's bow shock. J. Geophys. Res. **86**, 4355 (1981)

Richardson, J., et al.: Cool heliosheath plasma and deceleration of the upstream solar wind at the termination shock. Nature **454**, 63 (2008)

Sarris, E., et al.: Simultaneous measurements of energetic ion (50 keV and above) and electron (220 keV and above) activity upstream of earth's bow shock and inside the plasma sheet - magnetospheric source for the November 3 and December 3, 1977 upstream events. J. Geophys. Res. **92**, 12083 (1987)

Schwartz, S., et al.: Active current sheets near the earth's bow shock. J. Geophys. Res. **93**, 11295 (1988)

Schwartz, S., et al.: Conditions for the formation of hot flow anomalies at earth's bow shock. J. Geophys. Res. **105**, 12639 (2000)

Sckopke, N.: Ion heating at the Earth's quasi-perpendicular bow shock. Adv. Space Res. **15**, 261 (1995)

Scudder, J., et al.: Electron observations in the solar wind and magnetosheath. J. Geophys. Res. **78**, 6535 (1973)

Sibeck, D., et al.: Wind observations of foreshock cavities. J. Geophys. Res. **107**, 1271 (2002)

Skoug, R., et al.: Upstream and magnetosheath energetic ions with energies to 2 MeV. Geophys. Res. Lett. **23**, 1223 (1996)

Slavin, J.A., et al.: Messenger and Venus express observations of solar wind interaction with Venus. Geophys. Res. Lett. **36**, L09106 (2009)

Sonnerup, B.U.Ö.: Acceleration of particles reflected at a shock front. J. Geophys. Res. **74**, 1301 (1969)

Sonnet, C., et al.: The distant geomagnetic field, 3. Disorder and shocks in the magnetopause. J. Geophys. Res. **68**, 1233 (1963)

Sundberg, T., et al.: Properties and origin of subproton-scale magnetic holes in the terrestrial plasma sheet. J. Geophys. Res. **120**, 2600 (2015)

Terasawa, T.: Origin of 30–100 keV protons observed in the upstream region of the earth's bow shock. Plan. Space Sci. **27**, 365 (1979)

Teresawa, T.: Nonlinear dynamics of Alfvén waves: interactions between ions and shock upstream waves. Comput. Phys. Comm. **49**, 193 (1988)

Thomsen, M.F.: Multi-spacecraft observations of collisionless shocks. Adv. Space Sci. **8**, 157 (1988)

Thomsen, M., et al.: Hot, diamagnetic cavities upstream from the earth's bow shock. J. Geophys. Res. **91**, 2961 (1986)

Thomsen, M., et al.: Observational test of hot flow anomaly formation by the interaction of a magnetic discontinuity with the bow shock. J. Geophys. Res. **98**, 15319 (1993)

Tidmann, D., Krall, N.: Shock Waves in Collisionless Plasmas. Wiley, New York (1971)

Treuman, R.: Fundamentals of collisionless shocks for astrophysical application, 1. Non-relativistic shocks. Astron. Astrophys. Rev. **17**, 409535 (2009)

Turner, J., et al.: Magnetic holes in the solar wind. J. Geophys. Res. **82**, 1921 (1977)

Uritsky, V., et al.: Active current sheets and candidate: hot flow anomalies upstream of Mercury's bow shock. J. Geophys. Res. **119**, 853 (2014)

Vaisberg, O., et al.: Origin of the backstreaming ions in a young hot flow anomaly. Plan. Spac. Sci. **131**, 102 (2016)

Wang, L., et al.: Quiet time solar wind super halo electrons at solar minimum. AIP Conf. Proc. **1539**, 299 (2013b)

Wilber, M., et al.: Foreshock density holes in the context of known upstream plasma structures. Ann. Geophys. **26**, 3741 (2008)

Wilson III, L.B.: Low frequency waves at and upstream of collisionless shocks, In: Keiling, A., Lee, D.-H., Nakariakov, V. (eds.) Low-Frequency Waves in Space Plasmas. Geophysical Monograph, vol. 216, p. 269. American Geophysical Union, Washington, DC (2016)

Wu, C.S.: Physical mechanisms for turbulent dissipation mechanisms in collisionless shock waves. Space Sci. Rev. **32**, 83 (1982)

Wygant, J., et al.: Electric field measurements at subcritical, oblique bow shock crossings. J. Geophys. Res. **92**, 11109 (1987)

Xiao, T., et al.: Propagation characteristics of young hot flow anomalies near the bow shock: cluster observations. J. Geophys. Res. **120**, 4142 (2015)

Zhang, H., et al.: Time history of events and macroscale interactions during substorm observations of a series of hot flow anomaly events. J. Geophys. Res. **115**, A 12235 (2010a)

Zhang, H., et al.: Spontaneous hot flow anomalies at quasi-parallel shocks: 1. Observations. J. Geophys. Res. **118**, 3357 (2010b)

Zhao, D., et al.: Electron flat-top distributions and cross-scale wave modulations observed in the current sheet of geomagnetic tail. Phys. Plasma **24**, 082903 (2017)

Additional Reading

Advances in Space Science: Proceedings of the D2.1 Symposium of COSPAR Scientific Commission D, Hamburg, 11–21 July 1994, vol. 15(8–9), pp. 1–544 (1995)

Anderson, K.A.: Measurements of bow shock particles far upstream from the earth. J. Geophys. Res. **86**, 4445 (1981)

Anderson, K.A., et al.: Thin sheets of energetic electrons upstream from the earth's bow shock. Geophys. Res. Lett. **6**, 401 (1979)

Auer, P.L., Kilb, R.W., Crevier, W.F.: Thermalization in the earth's bow shock. J. Geophys. Res. **76**, 2927 (1971)

Balikhin, M., et al.: Experimental determination of the dispersion of waves observed upstream of a quasi-perpendicular shock. Geophys. Res. Lett. **24**, 787 (1997)

Balikhin, M.A., et al.: Determination of the dispersion of low frequency waves downstream of a quasi-perpendicular collisionless shock. Ann. Geophys. **15**, 143 (1997)

Balogh, A.: Cluster at the earth's bow shock: introduction. Space Sci. Rev. **118**, 1 (2005)

Balogh, A., et al.: The cluster magnetic field investigation. Space Sci. Rev. **79**, 65 (1997)

Barnes, A., Hung, R.: On the kinetic temperature of He++ in the solar wind. Cosmic Electrodyn. **3**, 416 (1973)

Bennett, L., Kivelson, M.G., Khurana, K., et al.: A model of the earth's distant bow shock. J. Geophys. Res. **102**, 26927 (1997)

Burgess, D.: Cyclic behavior at quasi-parallel collisionless shocks. Geophys. Res. Lett. **16**, 345 (1989)

Burgess, D.: Foreshock-shock interaction at collisionless quasi-parallel shocks. Adv. Space Res. **15**, 159 (1995)

Burlaga, L., Ogilvie, K.: Heating of the solar wind. Astro. J. **159**, 659 (1970)

Chapman, J.F., Cairns, I,: Modeling of earth's bow shock: applications. J. Geophys. Res. **109**, A11201 (2004)

Chapman, S., et al.: Perpendicular shock reformation and acceleration. Space Sci. Rev. **121**, 5 (2006)

Coates, A.J., et al.: AMPTE-UKS three-dimensional ion experiment. IEEE Trans. GeoSci. Remote Sens. **GE 23**, 287 (1985)

Coroniti, F.: Dissipation discontinuities in hydromagnetic shock waves. J. Plasma Phys. **4**, 265 (1970)

Dubouloz, N., Scholer, M.: On the origin of short large-amplitude magnetic structures upstream of quasi-parallel collisionless shocks. Geophys. Res. Lett. **20**, 547 (1993)

Ellacot, S.W., Wilkinson, W.P.: Heating of directly transmitted ions at low mach number perpendicular shocks: new insights from a statistical physics formulation. J. Geophys. Res. **108**, 1409 (2003)

Evans, D.: Precipitating electron fluxes formed by a magnetic field aligned. J. Geophys. Res. **79**, 2853 (1974)

Eviatar, A., Schulz, M.: Ion-temperature anisotropies and the structure of the solar wind. Plan. Space Sci. **18**, 321 (1970)

Facskó, G., et al.: Studies of hot flow anomalies using cluster multi-spacecraft measurements. Adv. Space Res. **45**, 541 (2010)

Fazakerley, A., et al.: AMPTE-UKS observations of velocity distributions associated with magnetosheath waves. Adv. Space. Res. **15**, 349 (1995)

Fazakerley, A.N., et al.: Observations of upstream ions, solar wind ions and electromagnetic waves in the earth's foreshock. Adv. Space Res. **15**, 103 (1995)

Feldman, W., et al.: Plasma and magnetic fields from the sun. In: White, O.R. (ed.) The Solar Output and its Variations, vol. 351. Colorado Associated University Press, Boulder (1977)

Feldman, W.C.: Electron velocity distributions near collisionless shocks. In: Tsurutani, B., Stone, R.G. (eds.) Collisionless Shocks in Heliosphere: Review of Current Research. Geophysical Monograph. American Geophysical Union, Washington, DC (1985)

Filbert, P., Kellogg, P.J.: Electrostatic noise at the plasma frequency beyond the earth's bow shock. J. Geophys. Res. **84**, 1369 (1979)

Fisk, L., Gloeckler, G.: The global configuration of the heliosheath inferred from recent voyager 1 observations, Astrophys. J. **776**, 79 (2013)

Fitzenreiter, R.J., et al.: Detection of bump-on-tail reduced electron velocity distributions at the electron foreshock boundary. Geophys. Res. Lett. **11**, 496 (1984)

Fitzenreiter, R.J., et al.: The electron foreshock. Adv. Space Res. **15**, 27 (1995)

Forsland, D., Shock, C.R.: Numerical simulation of electrostatic counterstreaming instabilities of ion beams. Phys. Rev. Lett. **25**, 281 (1970)

Gedalin, M.: Ion dynamics and distribution at the quasi-perpendicular collisionless shock front. Surv. Geophys. **18**, 541 (1997)

Gosling, J., Robson, A.: Ion reflection, gyration and dissipation at supercritical shocks. In: Stone, R.G., Tsurutani, B.T. (eds.) Collisionless Shocks in the Heliosphere: Reviews of Current Research. Geophysical Monograph, vol. 35. American Geophysical Union, Washington, DC (1985)

Gosling, J.T., Hildner, E., Mac Queen, R.M., Munro, R.H., et al.: Direct observations of a flare related coronal and solar wind disturbance. Solar Phys. **40**, 439 (1975)

Gosling, J., et al.: Ion reflection and downstream thermalization at the quasi-parallel bow shock. J. Geophys. Res. **94**, 10027 (1989)

Gosling, J.T., et al.: Counterstreaming suprathermal electron events upstream of corotating shocks in the solar wind beyond approximately 2 AU: ULYSSES. Geophys. Res. Let. **20**, 2335 (1993)

Greenstadt, E., et al.: Dual satellite observations of earth's bow shock, III: field determined shock structure. Cosmic Elect. **1**, 316 (1970)

Greenstadt, E., Fredricks, R.: Shock systems in collisionless plasmas. In: Lanzerotti, L., Kennel, C., Parker, E.N. (eds.) Solar System Plasma Physics, vol. III. North Holland, Amsterdam (1979)

Gringaus, K., et al.: Some results of experiments in interplanetary space by means of charged particle traps on soviet space probes. Space Res. **2**, 539 (1961)

Hellinger, P., et al.: Whistler waves in 3D hybrid simulations of quasi-perpendicular shocks. Geophys. Res. Lett. **22**, 2091 (1995)

Hellinger, P., Mangeney, A.: Electromagnetic ion beam instabilities: oblique pulsations. J. Geophs. Res. **104**, 4669 (1999)

Hirshberg. J, et al.: The helium component of the solar wind streams. J. Geophys. Res. **79**, 934 (1974)

Hong, J., et al.: Effect of ion-to-electron mass ratio on the evolution of ion beam driven instability in particle-in-cell simulations. Phys. Plasmas **19**, 092111 (2012)

Hoppe, M., et al.: Upstream hydromagnetic waves and their association with backstreaming ion populations - ISEE 1 and 2 observations. J. Geophys. Res. **86**, 4471 (1981)

Hull, A., et al.: Electron heating and phase space signatures at strong and weak quasi-perpendicular shocks. J. Geophys. Res. **103**, 2041 (1998)

Hundhausen, A., et al.: Vela satellite observations of solar wind ions. J. Geophys. Res. **72**, 1979 (1967)

Kaufmann, R., et al.: Shock observations with the Explorer 12 magnetometer. J. Geophys. Res. **72**, 2323 (1967)

Kucharek, H., Scholer, M.: Quasi-perpendicular to quasi-parallel shock transitions. Adv. Space Sci. **15**(8/9), 171 (1995)

Kucharek, H., et al.: On the origin of field-aligned beams at the quasi-perpendicular bow shock: multi-spacecraft observations by cluster. Ann. Geophys. **22**, 2301 (2004)

Lembège, B., Savioni, P.: Formation of reflected electron bursts by the nonstationarity and nonuniformity of a collisionless shock front. J. Geophys. Res. **107**, 1037 (2002)

Lembège, B., et al.: Nonstationarity of a two-dimensional perpendicular shock: competing mechanisms. J. Geophys. Res. **114** (2009) https://doi.org/10.1029/2008JA013618

Leroy, M., et al.: Simulation of perpendicular bow shock. Geophys. Res. Lett. **8**, 1269 (1981)

Lin, N., et al.: Nonlinear low frequency wave aspect of foreshock density holes. Ann. Geophys. **26**(12), 3707 (2008)

Lin, R.P., et al.: A three dimensional plasma and energetic particle investigation for the WIND spacecraft. Space Sci. Rev. **71**, 125 (1995)

Lin, Y.: Global hybrid simulation of hot flow anomalies near the bow shock and in the magnetosheath. Planet. Space Sci. **50**, 577 (2002)

Longmire, C.: Elementary Plasma Physics. Interscience Publishers. A Division of Wiley, New York (1963)

Marsch, E., Zhao, L., Tu, C.Y.: Limits on the core temperature anisotropy of solar wind protons. Ann. Geophys. **24**, 2057 (2006)

Masters, A., McAndrews, H., Steinberg, J., et al.: Hot flow anomalies at Saturn's bow shock. J. Geophys. Res. **114**, A08217 (2009)

Matsukio, S., Scholer, M.: On microinstabilities in the foot of high mach number perpendicular shocks. J. Geophys. Res. **111**, A06104 (2006)

Mazelle, C., et al.: Production of gyrating ions from nonlinear wave-particle interaction upstream from the earth's bow shock: a case study from cluster-CIS. Planet. Space Sci. **51**, 785 (2003)

Meziane, K., et al.: Three-dimensional observations of gyrating ion distributions far upstream from the earth's bow shock and their association with low-frequency waves. J. Geophys. Res. **106** 5731 (2001)

Meziane, K., et al.: Simultaneous observations of field-aligned beams and gyrating ions in the terrestrial foreshock. J. Geophys. Res. **119**, A05107 (2004)

Ogilvie, K.: Differences Between the Bulk Speeds of Hydrogen and Helium in the Solar Wind, J. Geophys. Res. **80**, 1335 (1975)

Ogilvie K.W., Zwally, H.J.: Hydrogen and helium velocities in the solar wind. Solar Phys. **2**(4), 236 (1972)

Omidi, N., Sibeck, D.: Formation of hot flow anomalies and solitary shocks. J. Geophys. Res. **112**, A01203 (2007)

Omidi, N., et al.: Spontaneous hot flow anomalies at quasi-parallel shocks: 2. Hybrid simulations. J. Geophys. Res. **118**, 173 (2013)

Onsager, T., et al.: High frequency electrostatic waves near earth's bow shock. J. Geophys. Res. **13**, 397 (1989)

Park, J., et al.: Particle-in-cell simulations of particle energization from low Mach number fast mode shocks. Phys. Plasmas **19**, 062904 (2012)

Pappadopoulos, K.: Ion thermalization in the earth's bow shock. J. Geophys. Res. **76**, 3806 (1971)

Paschmann, G., et al.: Observations of gyrating ions in the foot of the nearly perpendicular bow shock. Geophys. Res. Lett. **9**, 881 (1982)

Quest, K.: Simulations of high-Mach-number collisionless perpendicular shocks in astrophysical plasmas. Phys. Rev. Lett. **54**, 1872 (1985)

Report of the workshop on opportunities in plasma astrophysics: Princeton, NJ, January 18–21 (2010)

Robbins, D., et al.: Helium in the solar wind. J. Geophys. Res. **75**, 1178 (1970)

Scholer, M., Matsukiyo, S.: Nonstationarity of quasi-perpendicular shocks: a comparison of full particle simulations with different ion to electron mass ratio. Ann. Geophys. **22**, 2345 (2004)

Sckopke, N., et al.: Evolution of ion distributions across the nearly perpendicular bow shock: specularly and non-specularly reflected-gyrating ions. J. Geophys. Res. **88**, 6121 (1983)

Sckopke, N., et al.: Ion thermalization in quasi-perpendicular shocks involving reflected ions. J. Geophys. Res. **95**, 6337 (1990)

Scholer, M., et al.: Quasi-perpendicular shocks: length scale of the cross potential, shock reformation and implications for shock surfing. J. Geophys. Res. **108**, 1014 (2003)

Scholer, M., et al.: Cluster at the bow shock: status and outlook. Space Sci. Rev. **118**, 223 (2005)

Schwartz, S.J., Burgess, D.: Quasi-parallel shocks: a patchwork of three-dimensional structures. Geophys. Res. Lett. **18**, 373 (1991)

Scudder, J.D., et al.: The resolved layer of a collisionless, high beta, supercritical quasi-perpendicular shock wave, I, II, III. J. Geophys. Res. **91**, 11019 (1986)

Sibeck, D., et al.: The magnetosphere as a sufficient source for upstream ions on November 1, 1984. J. Geophys. Res. **93**, 14328 (1988)

Shestakov, A., Vaisberg, O.L.: Study and comparison of the parameters of five hot flow anomalies at a bow shock front. Cosmic Res. **54**, 77 (2016)

Strong, I.B., et al.: Measurements of proton temperatures in the solar wind. Phys. Rev. Lett. **16**, 632 (1966)

Teste, A., Parks, G.K.: Counter streaming beams and flat-top electron distributions observed with Langmuir, whistler, and compressional Alfvén waves in earth's magnetic tail. Phys. Rev. Lett. **102**(7), id. 075003 (2009)

Thomsen, M.F.: Upstream suprathermal ions. In: Tsurutani, B.T., Stone, R.G. (eds.) Collisionless Shocks in Heliosphere: Reviews of Current Research. Geophysical Monograph, vol. 35, p. 253. American Geophysical Union, Washington, DC (1985)

Thomsen, M., et al.: Observational evidence on the origin of ions upstream of the earth's bow shock. J. Geophys. Res. **88**, 7843 (1983)

Thomsen, M., et al.: Magnetic pulsations at the quasi-parallel shock. J. Geophys. Res. **95**, 957 (1990)

Thomas, V.A., Brecht, S.H.: Evolution of diamagnetic cavities in the solar wind. J. Geophys. Res. **93**, 11341 (1988)

Thomas, V.A., et al.: Hybrid simulation of the formation of a hot flow anomaly. J. Geophys. Res. **96**, 11625 (1991)

Tsurutani, B., Stone, R.G. (eds.): Collisionless Shocks in the Heliosphere: Review of Current Research. Monographical Series, vol. 35. American Geophysical Union, Washington, DC (1985)

Umeda, T., et al.: Modified two-stream instability at perpendicular collisionless shocks: full particle simulations. J. Geophys. Res. **117**, A03206 (2012)

Wang, S., Zong, Q.-G., Zhang, H.: Hot flow anomaly formation and evolution: cluster observations. J. Geophys. Res. **118**, 957 (2013)

Wilkinson, W.P., et al.: Nonthermal ions and associated magnetic-field behavior at a quasi-parallel earth's bow shock. J. Geophys, Res. **98**, 3889 (1993)

Wilkinson, W.P.: The earth's quasi-parallel bow shock: review of observations and perspectives for cluster. Planet. Space Sci. **51**, 629 (2003)

Wilkinson, W., Schwartz, S.: Parametric dependence of the density of specularly reflected ions at quasi-perpendicular collisionless shocks. Planet. Space Sci. **38**, 419 (1990)

Wilson III, L.B., et al., Low-frequency whistler waves and shocklets observed at quasi-perpendicular interplanetary shocks. J. Geophys. Res. **114**, A10106 (2009)

Wilson III, L.B., et al.: Quantified energy dissipation rates: electromagnetic wave observations in the terrestrial bow shock: 2. Waves and Dissipation. J. Geophys. Res. **119**, 6475 (2014)

Wilson III, L.B., et al.: Shocklets, SLAMS, and field-aligned ion beams in the terrestrial foreshock. J. Geophys. Res. **118**, 957 (2013)

Wu, C.S., A fast Fermi process-energetic electrons accelerated by a nearly perpendicular bow shock. J. Geophys. Res. **89**, 8857 (1984)

Wu, C.S., Yoon, P.: Kinetic Hydromagnetic instabilities due to a spherical shell distribution of pickup ions. J. Geophys. Res. **95**, 10273 (1990)

Chapter 5
Boundaries and Current Sheets

5.1 Introduction

Magnetohydrodynamic (MHD) theories have defined space plasma boundaries of large-scale structures through discontinuous macroscopic parameters such as density and temperature in the same way ordinary fluid boundaries are defined. However, space plasmas are coupled to electric $\mathbf{E}$ and magnetic $\mathbf{B}$ fields, and therefore the boundary requirements must also include the behavior of discontinuous electromagnetic fields and current sheets separating the boundaries.

The boundaries in solar system plasma near Earth include the bow shock, magnetopause and cusps on the dayside, and the high and low latitude plasma sheet boundaries on the night side. Inside the magnetosphere, boundaries include those of trapped radiation belt and plasmapause. The boundaries consist of the usual perpendicular diamagnetic currents as well as field-aligned currents running along the magnetic field direction. Moreover, transient ring current encircles the inner radiation belts during disturbed solar wind (SW). The questions still not well understood are how the SW interaction with the geomagnetic field drives currents and transports plasmas across the boundaries.

When the thickness of the boundary currents is much larger than an ion Larmor radius, MHD theory classifies the boundaries as tangential, contact, rotational, and shock. These boundaries define how mass, momentum, and energy are transported across them (Willis 1971). Contact discontinuity separates two regions of space at rest with different densities and temperatures.

Open boundaries are formed by "merging" two different magnetic field lines. Visualize two ideal MHD fluid elements with frozen-in-magnetic field moving toward each other. If the magnetic fields have opposite polarity, a *magnetic neutral point or line* is produced and fluids from one side can flow into the other side. MHD theory allows field lines from different sources to *"connect, disconnect and reconnect."*

© Springer Nature Switzerland AG 2018   191
G. K. Parks, *Characterizing Space Plasmas*, Astronomy and Astrophysics Library,
https://doi.org/10.1007/978-3-319-90041-4_5

Space plasma researchers have had great interest in connecting magnetic fields from different sources because superposition of two magnetic fields from different sources can change the overall topology of the large-scale magnetic field. Space physicists look at this as dynamical, asserting that magnetic field lines are *annihilated* and the field energy goes into accelerating the particles. The concept of magnetic field connection and annihilation is considered fundamentally important for solar, astrophysical, and laboratory fusion plasmas. However, space plasmas are collisionless and models based on MHD theory and concepts do not always give a correct picture because field lines are *not consistent* with Maxwell's equations.

Below we begin with a short history of how the "open" boundary concept was first introduced to space plasma physics. We then examine how MHD theory looks at this concept and discuss some of the issues. Then we show observations of the SW particles entering the magnetosphere through the cusp regions. Particle motions in the vicinity of magnetic neutral points are reviewed. We conclude the chapter by showing how magnetic boundaries can be treated self-consistently using Vlasov theory for boundaries with the same and opposite polarities of the magnetic field.

5.2 Magnetic Reconnection on Earth

The concept of *magnetic reconnection* grew out of the work of Ronald Giovanelli (1949) who was interested to learn how electrons are accelerated in solar flares. He noted that oppositely directed magnetic fields of sunspots include magnetic *neutral points* and when sunspot fields evolved in space and time, the induced electric fields near the neutral points could accelerate particles.

F. Hoyle thought that auroral particles could also be accelerated in the same way and asked J. Dungey, his PhD student, to develop Giovanelli's ideas about particle acceleration in magnetic neutral points and apply the theory to auroras. Soon after the interplanetary magnetic field (IMF) was discovered, Dungey (1961) realized the presence of IMF would affect the dynamics of solar wind (SW) interaction with the geomagnetic field. He noted the static model proposed by Sydney Chapman would not work because it did not include the IMF. He published a 2D diagram to illustrate how the IMF would modify the geomagnetic field in the distant regions (Fig. 5.1). The diagram shows the IMF and Earth's geomagnetic field lines being "cut" and merged on the dayside and subsequently convected with the SW to the nightside where the original field lines "reconnect." Presumably, field lines on the nightside *"find their mate"* to satisfy $\nabla \cdot \mathbf{B} = 0$ (Fig. 5.1). But in drawing this diagram, he cautioned the readers,

> ... the use of lines of force is a mathematical device and that they are not physical objects;
> the motion of lines of force is a further device...
>
> **James Dungey**

Dungey suggested the superposition of southward IMF with the geomagnetic field creates two magnetic *neutral points* and the SW flowing in the vicinity of a neutral point could drive the ionospheric current system. He also noted that plasmas

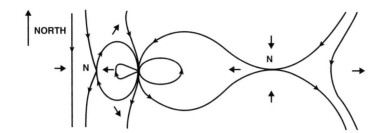

Fig. 5.1 A sketch of 2D magnetic field combining IMF and the outer geomagnetic field. The point N represents a magnetic null point, called an X-point (from Dungey, *Phys. Rev. Lett.*, **6**, 47, 1961)

flowing near the neutral point are controlled by a strong *current density* induced by time varying magnetic field there. Dungey is often credited with the concept of magnetic reconnection, but we did not find the term "magnetic reconnection" in any of his early papers (Dungey 1953, 1961, 1975). His *sketch* however has become a major paradigm in space, laboratory, solar, and astrophysical plasma physics.

Dungey's picture shows an X-type neutral point on the dayside equatorial plane when the IMF has a southward component. The common interpretation is that the southward IMF connects with Earth's dipole field (which is northward on the equator) and the connected field line is subsequently convected frozen in the SW in the anti-sunward direction. This process "stretches" Earth's dipole field and forms a geomagnetic tail which includes another X-point that separates the field lines in the northern and southern hemispheres. Since Dungey's original picture, researchers have added numerous modifications. For example, some models produce a neutral line instead of the X-point. Some have reconnecting tail field occurring $\sim$30 R_E but others have this occurring very far in the tail, $\sim$200 R_E. Recently, Lui et al. (2005) have discussed what one is expected to observe in 3D magnetic reconnection geometry.

Because the reconnected field lines have been "stretched," there is excess stress energy in the magnetic field (energy comes from the solar wind) and this energy is released by convecting the reconnected stretched field lines back toward the Sun. In this process, the footprints of the convecting magnetic field in the ionosphere will trace out a "D-shaped" pattern, which resembles the ionospheric disturbed storm current distribution, which was an explanation originally sought by Dungey.

In Fig. 5.1, one can see there is a $\nabla \times \mathbf{B}$ near the magnetic null point, which is the strong current density $\mathbf{J}$ that Giovanelli and Dungey referred to. In the 2D case, $\mathbf{J}$ is coming out of the page (dayside and nightside). Physically what this means is that the reconnection process is inducing a large-scale current at the X-point. The induced current must produce magnetic fields strong enough to change the pre-existing magnetic field topology. The key question still not understood is what mechanism could produce such strong currents.

Reconnection of IMF and dipole field produces an "open" magnetosphere which is an attempt to answer an important question of how the solar wind particles enter the magnetosphere. An open magnetosphere allows the solar wind and magnetosheath particles to directly enter the magnetosphere. In "closed" magnetospheric models, particles are turned around by the $\mathbf{v} \times \mathbf{B}$ force. Solar wind and magnetosheath particles however can cross the magnetopause boundary if in the frame of the particles there is an $\mathbf{E} \times \mathbf{B}$ velocity. A diffusive transport process across a closed boundary involving waves interacting with particles has also been examined when plasma is treated as an ensemble of many particles. However, transport by diffusion is too slow to explain some dynamical processes and the space physics community chose to adopt the open model.

5.2.1 Issues with Magnetic Reconnection

Magnetic field reconnection models are based on MHD equations and concepts which allow magnetic field lines to exist. MHD theory treats magnetic field as a physical entity. A magnetic field line is however an abstraction and no identity can be attached to field lines especially if the fields are time varying. Field lines are mathematical constructs for "visualizing" continuous fields. Like longitude and latitude lines on a globe, *field lines do not actually exist!* However, in everyday life, we draw magnetic and electric field lines for convenience.

Magnetic charges from which lines of $\mathbf{B}$ might emerge (like electric fields) do not exist. Hence, $\nabla \cdot \mathbf{B} = 0$ and $\mathbf{B}$ lines do not begin or end and they must close back on themselves. This statement is true not only for static fields but also for dynamic fields. Do Maxwell's equations allow us to define magnetic field lines without ambiguity? As shown below, one can define magnetic field lines only if the fluid is ideal.

5.2.2 Frozen-In Concept

To discuss this problem, we examine Ampere's equation, $\nabla \times \mathbf{E} = -\partial \mathbf{B}/\partial t$ which we have already done in Chap. 1. We remind the readers that this equation relates the curl of the electric field to time variations of the magnetic field and buried in this equation are important and subtle physics not immediately apparent. In particular, Faraday showed that

$$\mathcal{EMF} = \frac{d\phi}{dt} \tag{5.1}$$

Question 5.1 Readers remind yourselves how Faraday derived Eq. (5.1) and discuss briefly the meaning of this equation.

In deriving Eq. (5.1), one notes that there is an electric field $\mathbf{E}'$ in the moving frame. However, suppose the electric field in the moving plasma frame $\mathbf{E}' = 0$

(Often assumed in MHD theory because plasma has high electrical conductivity). Then the Lorentz transformation equation reduces to $\mathbf{E} = -\mathbf{V} \times \mathbf{B}$ and Faraday's equation $\partial \mathbf{B}/\partial t = -\nabla \times \mathbf{E}$ becomes

$$\frac{\partial \mathbf{B}}{\partial t} = \nabla \times (\mathbf{V} \times \mathbf{B}) \tag{5.2}$$

Now if we set up a contour C as before and perform the integration of the modified equation as we did with the original equation, we obtain

$$\frac{d\phi}{dt} = 0 \tag{5.3}$$

In the absence of $\mathbf{E}'$, $\mathcal{EMF}$ is not generated and the total magnetic flux is conserved. Equation (5.3) leads to the well-known *frozen-in-field* theorem, invented by Alfvén and Fälthammar (1963) to illustrate the behavior of ideal fluids with infinite conductivity ($\sigma = \infty$).

5.2.3   *Magnetic Field Line in Ideal Plasmas*

Plasmas with infinite (∞) conductivity are also known as *perfect fluids*. In perfect fluids, it turns out we can define a *magnetic field line*. Imagine a material line defined by intersecting two material surfaces. We can choose these surfaces everywhere tangent to the magnetic field at $t = 0$ (Fig. 5.2). Consequently the magnetic flux through both of these surfaces is zero and their intersection defines a field line at this particular time. Because of the flux conservation theorem, the surfaces will remain flux free and in the course of motion the intersection remains a field line. The field line can then be defined without ambiguity.

We reiterate once more, a magnetic field line can only be defined if $\mathbf{E}' = 0$ because then the total magnetic flux is conserved and no $\mathcal{EMF}$ is induced. Since the $\mathcal{EMF}$ represents work done on the charge, no work can be done on any charges when the $\mathcal{EMF}$ vanishes. The particles experience no force, and the particles are "stuck" on the magnetic field line. Particles cannot participate in any dynamics when the particles are stuck because there is no force to move them.

Fig. 5.2   A sketch of two magnetic flux tubes

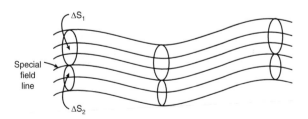

However, no matter how high the value of conductivity is for plasmas, the value is finite. Ideal fluids do not exist in nature and the total magnetic flux is therefore *never* conserved if the plasma is not ideal. In general, a field line cannot be defined and a *unique* identity cannot be given to magnetic field. The general behavior of plasmas is described by Faraday's law (Eq. (5.1)) in which the $\mathcal{E}MF$ induced by the changing magnetic field drives a current.

5.2.4   MHD Point of View

The MHD equation that describes how the magnetic field changes in time is

$$\frac{\partial \mathbf{B}}{\partial t} = \frac{1}{\mu_o \sigma} \nabla^2 \mathbf{B} + \nabla \times (\mathbf{V} \times \mathbf{B}) \tag{5.4}$$

If the conductor were at rest, $\mathbf{V} = 0$, and the differential equation simplifies to

$$\frac{\partial \mathbf{B}}{\partial t} = \frac{1}{\mu_o \sigma} \nabla^2 \mathbf{B} \tag{5.5}$$

which has exactly the same form as the diffusion equation in gas and fluid dynamics. In MHD theory, one describes how $\mathbf{B}$ changes in time relative to spatial changes by *diffusing* the magnetic field $\mathbf{B}$. With σ finite, there are Ohmic losses and the currents responsible for the magnetic field will decay away and the strength of a magnetic field in a magnetized conductor will steadily decrease. The general solution of this partial differential equation can be obtained using Green's function. However, we can estimate the diffusion rate by letting L be the characteristic scale length of $\mathbf{B}$. Then substitute L^{-2} for ∇^2 and solve the equation $\partial \mathbf{B}/\partial t \approx \pm (1/\mu_o \sigma L^2)\mathbf{B}$ where $\pm$ allows for gain or loss of $\mathbf{B}$ with time. The solution of this differential equation is $\mathbf{B} = \mathbf{B}_o \exp \pm t - t/t_D$ where $\mathbf{B}_o$ is the initial value of the magnetic field and $t_D = \mu_o \sigma L^2$ is the characteristic time for the magnetic field to increase or decay to $1/e$ of its initial value.

For space plasma applications, although we do not know exactly what value σ has or even know how to define this term when collisions are absent, we know the value will be very large because space plasmas are good conductors. Depending on what we choose for L^2 (L = characteristic length), the diffusion time t_D for magnetic fields can be very long. In magnetic field merging models, L could represent electron or ion inertial lengths, c/ω_e or c/ω_p assuming ions are protons. If we choose $\sigma = 10^8$ mhos/m (conductivity of copper) and solar wind density of $5\,\mathrm{cc}^{-1}$, we obtain $\omega_e \approx 1.26 \times 10^5$ and $\omega_p \approx 2.9 \times 10^3$. The diffusion times for electron inertial scale are $\sim 7 \times 10^8$ s and ion inertial scale 1.3×10^{12} s. These are extremely long times, much longer than the dynamic time scales of space plasma phenomena and are thus not useful for reconnection models of space plasmas.

If the fluid is ideal, $\sigma = \infty$, $\mathbf{B}$ fields are frozen and there is no diffusion. In this case, the first equation on the right vanishes and Eq. (5.4) reduces to

$$\frac{\partial \mathbf{B}}{\partial t} = \nabla \times (\mathbf{V} \times \mathbf{B}) \tag{5.6}$$

which leads to flux conservation equation $d\Phi/dt = 0$ or the frozen-in-field theorem. MHD fluid picture of open field lines has serious problems and is not consistent with observations. A flux transfer event (FTE) is thought produced by connecting the interplanetary magnetic field line with the geomagnetic field line. Russell and Elphic (1979) have given us an example of FTE as an open flux tube. However, FTEs are populated by keV electrons. The readers are asked to examine significance of keV electrons on open field lines in Question 5.3.

Question 5.2 Derive Eq. (5.4) from one of Maxwell equations and the simple form of Ohm's law.

Question 5.3 What is the speed of 1 keV electrons along the magnetic field direction? Based on your calculation, estimate the time it would take to empty a flux tube. Are there electron transport mechanisms in the magnetosphere fast enough to replenish the lost electrons on time scales for a spacecraft experiment to detect the electrons?

5.2.5 Recent MMS Results

The magnetopause is a region where one could in principle test to see if magnetic field merging is active when the interplanetary magnetic field has a southward component. NASA's Magnetospheric Multiscale (MMS) was designed to study the microphysics of magnetic reconnection. Recently a special issue of Geophysical Research Letters (October 2016) was devoted to the first results from the MMS observations. The many interesting papers included several that showed the boundary is very dynamic consisting of many thin current sheets and the existence of multiple micro-scale current sheets (Chen et al. 2016; Lavraud et al. 2016; Phan et al. 2016; Hasegawa et al. 2017). The authors have interpreted their observations using a reconnection model and concepts. The readers are challenged to see if they can interpret the MMS observations self-consistently using Maxwell and Boltzmann concepts.

The MMS results also indicated that in these current sheets, the frozen-in-field law is violated. The violation of the frozen-in-field condition was originally predicted by Mozer et al. (2002) from the extremely high time resolution data from Polar. Interpreting their observations using the generalized Ohm's law, they deduced that electrons are decoupled from ions and thus concluded that the ideal MHD condition was violated in their observation. This violation is a necessary requirement if merging were taking place (but not a sufficient condition). The results of MMS have been interpreted in terms of merging, but here we suggest that the results are consistent with inductive effects produced by the changing and turbulent magnetic field that are persistent at the magnetopause.

Other observations that are relevant for understanding magnetopause dynamics include Cluster observations that have shown that the magnetopause region frequently includes an extremely *cold* plasma population whose energies are lower than most of the thresholds of the detectors and are normally not detected until accelerated by electric field induced by the motion of the boundary (Sauvaud et al. 2001). Moreover, past observations have also shown the low energy particles (a few eV) can behave differently from the higher energy keV particles. For example, the observations by Hapgood and Bryant (1990) showed that the temperature and density of low energy electrons are inversely correlated across the boundary, which is a feature that can be produced by a diffusive-like transport mechanism. Moreover, Hall et al. (1991) showed that the electrons in the outer part of the transition region originate in the magnetosheath while those closer to earth come from the magnetosphere and energized magnetosheath population. They also reported counter-streaming electrons parallel and antiparallel to the magnetic field direction, suggesting that the magnetic field topology is closed. Similar features have also been observed by an experiment on the short-lived Equator-S spacecraft (Parks et al. 1999).

5.3   SW Entry into Magnetosphere Through Cusps

Are there other ways to transport the SW particles into the magnetosphere? Observations of particles in the environment of cusps say yes. The geomagnetic field of Earth at high altitudes has funnel shaped regions called "cusps." Cusp magnetic fields received much attention in early fusion research when researchers sought how to trap and confine plasmas. Theoretical analysis has shown that particle trajectories depend on initial phase angles of injected particles and can pass through the "throat" without mirroring or interact with trapped particles exciting instabilities and heat plasma. Magnetic cusp geometry is observed with many magnetic structures, including pulsar magnetospheres. It is important to understand how particles move in cusp geometry.

Space plasma observations first showed the importance of cusps when magnetosheath particles were detected in the polar cusp regions. Particles in the polar cusp regions have since been systematically studied, recently by the four Cluster spacecraft. Cluster has apogee and perigee at 19 R_E and 4 R_E ($R_E \approx 6400$ km) and the orbit plane is perpendicular to the ecliptic plane. The four spacecraft frequently cut through the cusp regions at mid-altitudes, $\sim$4 R_E, and high altitudes, $\sim$10 R_E.

Figure 5.3 shows two examples of the SW transient disturbances observed at L1. These transient solar wind events were observed by ACE on 21 June 2007 (top left) and 21 October 2001 (top right). We use these transient SW particles as *markers* and show they are later detected in the high altitude and mid-altitude cusp regions

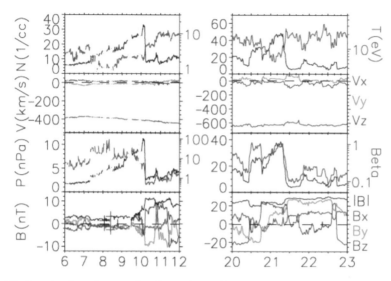

Fig. 5.3 Transient Interplanetary disturbance detected in the magnetosheath. The top panels show density and temperature, the second panel the bulk velocities, the third panel pressure and beta of the plasma, and the bottom panel shows the interplanetary magnetic field (from Parks et al., *Phys. Plasmas*, **15**, 080702, 2008)

depending on whether Cluster was near apogee or perigee. The SW transients are easy to recognize because they have higher densities (black) and lower temperatures (red) than the unperturbed SW (top panels, Fig. 5.3).

5.3.1 High Latitude Cusp Region

During the time intervals of the two SW transients, Cluster 1 was in the magnetosheath and perigee allowing us to see how the SW transients were modified crossing the bow shock. Figure 5.4 shows the data from the magnetosheath.

We discuss here only the behavior of density changes because temperature changes require knowledge about heating mechanisms that are still not clearly understood. Note that the SW transients increased plasma β to very high values, ~100–1000 and $V_\perp$ (black) in the magnetosheath exceeded the Alfvén speed, $V_A = B_o/(\mu_o \, nm)^{1/2}$. For this event, $V_\perp$ was $200\,\mathrm{km\,s}^{-1}$, $V_A \sim 30\,\mathrm{km\,s}^{-1}$, and $V_\perp$ exceeded V_A in the entire magnetosheath. Inside the magnetosphere, $V_\perp$ was $\sim 0\,\mathrm{km\,s}^{-1}$ indicating there was *no transport* across the magnetopause. This behavior has been seen for all magnetopause crossings studied, including times when V_{sw} was as low as $300\,\mathrm{km\,s}^{-1}$, when a lower hybrid soliton was detected at the magnetopause (Trines et al. 2007) and during flux transfer events (Dunlop et al. 2005) (not shown).

Fig. 5.4 Transient
Interplanetary disturbance
detected in the
magnetosheath. The particles
comprising this event were
subsequently detected at the
low altitude cusp with nearly
equal density (from Parks
et al., *Phys. Plasmas*, **15**,
080702, 2008)

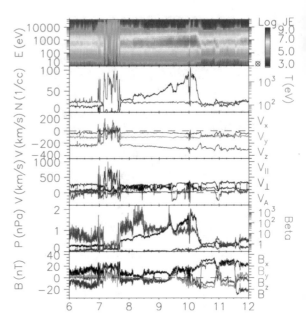

Question 5.4 Compare the features observed at L1 (Fig. 5.3) relative to what was observed in the magnetosheath shown in Fig. 5.4.

5.3.2   Mid-Latitude Cusp Region

On 21 October 2001, Cluster was on the dayside inbound from the southern hemisphere perigee when it encountered an abrupt increase of ion density at 2110 UT with flow along B (Fig. 5.5). The SC was at a distance of ~ 4 R$_E$ and detected a large deviation of the magnetic field. Note the profile of the density here is very similar to what was observed at L$_1$. We do not have direct measurements of magnetosheath density n_{msh} by Cluster for this event. However, the transient solar wind compressed the magnetopause and the Geotail spacecraft crossed into the magnetosheath in the dawn sector (3.4, -20.3, -0.3) R$_E$ measuring a density of 30 cc^{-1}. This lower density measured by Geotail was due to Geotail measuring a more restricted range of particle energies, 100 eV–45 keV. (Not shown. See Parks et al. 2008.)

Question 5.5 Discuss the salient features of Fig. 5.5 relative to what was observed at L1.

Fig. 5.5 Transient
Interplanetary disturbance
detected in the low-altitude
cusp region. The transient
particles from the SW are
detected by Cluster at $\sim$4 R$_E$
altitude (from Parks et al.,
Phys. Plasmas, **15**, 080702,
2008)

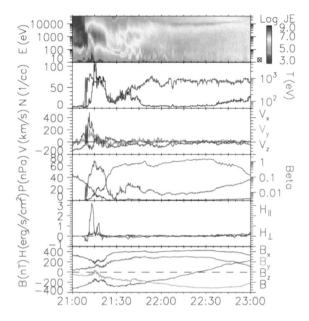

5.3.3   Velocity Space Distributions in Cusps

Two-dimensional cuts of 3D ion phase-space distributions (three spin averages,
12 s) have been computed after subtracting the perpendicular flow from 2107:13
to 2114:15 UT. Near the origin of the middle panel in Fig. 5.6 is a low-energy
ionospheric beam. The composition instrument has clearly identified these ions
as O$^+$ ions (not shown). The beam appears in the ESA spectrogram as the
monoenergetic $\sim$100 eV ions. The sequence shows a low energy O$^+$ field-aligned
beam propagating outward along B from the ionosphere, with a few keV H$^+$
beam propagating toward the ionosphere, surrounded by an isotropic high energy
magnetospheric population. These ions are seen to mirror below the spacecraft, and
initially exhibit a loss cone for the returning ions. Later, the distributions evolved to
a bi-Maxwellian with $T_\parallel \sim 2.5 \times 10^7$ K $\geq T_\perp \sim 1.5 \times 10^7$ K.

A sequence of one-dimensional (1D) cuts of spin averaged (4 s) electron phase
space distributions is shown in Fig. 5.6 (top panels). In the plot, $v_\parallel$ (black) and $v_\perp$
(red) are directions relative to **B**. Before entry into the high-density plasma region,
the distributions had both cold and hot components. The hot electrons with $v_\parallel$ are
electrons coming out from the ionosphere. The $-v_\parallel$ electrons going toward the
ionosphere are initially anisotropic and become nearly isotropic. The distribution
turns over at $-v_\parallel \sim 10^6$ m s^{-1}, where the "flattop" shape begins.

Question 5.6   Give a possible interpretation of the flattop shape electron distributions in Fig. 5.6. What do they say about the possible connection of the spacecraft
to the magnetosheath?

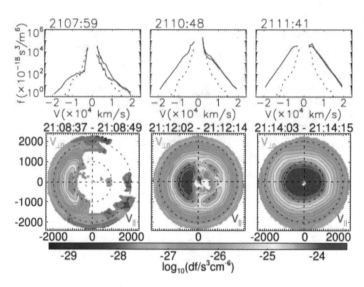

Fig. 5.6 Phase-space distributions of electrons are shown on the top panels and ions bottom panels. The electrons are for energies $\geq 15\,\text{eV}$. Here the subindices in $\pm v_{\parallel}$ correspond to directions parallel and antiparallel to B. The electron distributions have "flattop" shape and are similar to magnetosheath distributions. The ions show mixed magnetosheath, magnetospheric, and ionospheric distributions. Times when the distributions are samples are indicated on the top (from Parks et al., *Phys. Plasmas*, **15**, 080702, 2008)

5.3.4   Waves in Cusps

Previous observations have shown that the high-altitude cusps adjacent to the magnetopause are often permeated with large-amplitude waves (Cargill et al. 2005). Cluster also has observed energetic particles up to several hundred keV in the cusps, corroborating earlier observations. During disturbed SW, even energetic 20 keV O^+ ions have been detected coming out of the ionosphere (Fu et al. 2005). The waves accompanying the mid-altitude cusp particles include electromagnetic EM and broadband electrostatic ES emissions and electron Bernstein modes at multiples of $(n + 1/2)\, f_{\text{ce}}$, where f_{ce} is the electron cyclotron frequency (Fig. 5.7).

During 21 October 2001 pass, the EM emissions measured by Cluster included whistler mode waves. The broadband bursty ES waves are due to unresolved solitary electron holes (Canu et al. 2002; Pickett et al. 2004). The broadband ES and Bernstein mode waves are always observed with the transient particles in mid-altitude cusps. These observations raise many questions about the interaction of the transient solar wind with the magnetosphere. The IMF during the events studied fluctuated considerably and included both northward and southward directions. Our studies indicate solar wind plasma is indeed entering the magnetosphere via the cusps as previously suggested (Frank 1971; Hekkila and Winningham 1971).

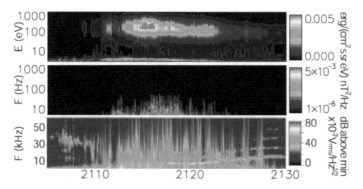

Fig. 5.7 A wave spectrogram 21 October 2001. EM waves are electron and ion cyclotron emissions. The broadband ES emissions are produced by deBye-scale electron holes. The narrow emissions at harmonics $(n + 1/2) f_{ce}$ are Bernstein mode waves. Density $\sim$22 cc^{-1} for electrons and protons and $B = 240$ nT at 2112 UT yield the following frequencies: cyclotron $f_{ce} = 6.73$ kHz and $f_{ci} = 3.67$ Hz, plasma $f_{pe} = 44.3$ kHz and $f_{pi} = 972$ Hz, lower and upper hybrid f_{lh} 155 Hz, f_{uh} 44.8 kHz (from Parks et al., *Phys. Plasmas*, **15**, 080702, 2008)

Fig. 5.8 Dispersion characteristics of Bernstein waves (from Krall and Trivelpiece, *McGraw Hill Book Company*, New York, NY, 1973)

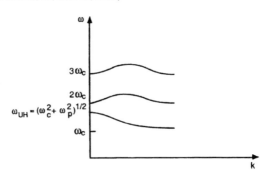

5.3.5   *Discussion of Particles and Waves in Cusps*

Electron holes could be produced locally by a two stream instability or they could have come from the high-altitude cusp (Pickett et al. 2004). The Bernstein mode waves are similar to those seen in the plasmasphere and the magnetosphere (Canu et al. 2006a,b). They can be excited by a mixture of cold and anisotropic hot electrons with background ions (Ashour-Abdalla and Kennel 1978; Willes and Robinson 1997), ion shell distributions (Janhunen et al. 2003) and ion beams propagating perpendicular to B (Brinca et al. 2003). No models have incorporated ion beams parallel to B.

Figure 5.8 shows the frequency of the Bernstein modes plotted as a function of k. The modes appear as narrow bands of frequency. When the plasma density is low ($\omega_p \ll \omega_c$), the Bernstein mode occurs at close to the harmonics of the cyclotron frequency,

$$\omega_n^2 = n\omega_c^2(1 + \alpha_n) \quad n = 1, 2, \ldots \tag{5.7}$$

where n is the harmonic number and

$$\alpha_n = \frac{2\omega_p^2 m}{k^2 k_B T} I_n \left(\frac{k^2 k_B T}{m\omega_c^2} \right) e^{-k^2 k_B T / m\omega_c^2} \tag{5.8}$$

Here $I_n(k^2 k_B T / m\omega_c^2)$ is the modified Bessel function of the first kind of order n. This result is very different from the results of the fluid theory, which predicts a cutoff between the lower and the upper hybrid frequencies for waves propagating perpendicular to $\mathbf{B}_0$ (see Fig. 9.13, Parks 2004). However, the Vlasov theory shows the Bernstein waves propagating in this gap at harmonics of the cyclotron frequency.

Transient particles in the magnetosheath create a considerable amount of disturbance. On 21 October 2001, the transient particles were not only injected into the cusp, but they could have triggered magnetospheric substorm instability that produced an intense aurora. Cluster by coincidence detected high fluxes of energetic 20 keV O^+ ions precipitating into the ionosphere, while cold ionospheric O^+ ions were flowing out (not shown). This solar wind transient initiated a moderate size magnetic storm.

Sixty SW transient events have been studied covering periods of fast and slow solar wind and northward and southward interplanetary magnetic fields (Parks et al. 2008). The same events measured in two different regions have shown that the transient particles inside the cusp can have densities nearly equal to magnetosheath densities. We still need to investigate what processes can transport particles in the cusps nearly unimpeded from the magnetosheath into the magnetosphere. We also need to understand the significance of ionospheric particles that extend out into the magnetosheath and the high super-Alfvénic bulk flows that accompany solitary waves and FTEs. Some of the cusp events we studied also included tens of keV particles in the transient solar wind. We need to evaluate how these particles are related to the energetic particles observed in high-altitude cusps. Future work should combine data analysis and modeling to enhance our understanding of the physics of solar wind interaction with the magnetosphere.

Question 5.7 Discuss how the Bernstein mode waves differ from ordinary X- mode EM waves.

Question 5.8 Estimate the number of particles entering the magnetosphere through the cusp.

5.4 Particle Motions in Magnetic Neutral Regions

The first-order orbit theory is not applicable when the magnetic field changes significantly in distances of the order of a gyroradius and the requirement $\mathbf{B}_0 \gg |(\mathbf{r} \cdot \nabla_0)\mathbf{B}|$ is violated. This can happen when magnetic fields become very weak, for example, at the X-point or the current sheet of the magnetopause and geomagnetic tail where the field intensity vanishes and the fields reverse directions. The cyclotron radius of a particle approaching the null magnetic field region becomes larger and larger and the GC approximation breaks down. The magnetic moment μ of a particle going through the neutral point becomes ∞, and it clearly is not physical to talk about adiabatic invariants. The description of the particle motion then requires the full Lorentz equation which must be solved analytically or numerically.

Particle orbits in current sheet geometries have been calculated for specific neutral line models since 1965 (Speiser 1965a,b). Later studies showed that the orbits in current sheets could become "chaotic" (we will not discuss chaotic orbits, interested readers are referred to Chen and Palmedaso 1986). Numerous studies have since expanded particle orbit calculations that include various types of neutral point and sheet models using computer simulation tools (Zhu and Winglee 1996; Eastwood 1972; Dungey 1975). The simple particle orbits shed light on the physics. Below orbit calculations of some simple models that previously appeared in Parks (2004) will be reviewed.

5.4.1   Particles in Magnetic Neutral Regions

We will show the behavior of particle motion in several different geometries appropriate for the current sheet. We begin with a fairly simple magnetic field geometry given by

$$
\begin{aligned}
B_x &= B_0 & z &\geq L \\
B_x &= \frac{B_0 z}{L} & L &\geq z \geq -L \\
B_x &= -B_0 & z &\leq -L
\end{aligned}
\tag{5.9}
$$

which supports a neutral line at $z = 0$. This geometry has been used in geomagnetic tail studies. Here, the thickness of the neutral sheet is $\pm L$ where the plasma sheet boundary is located and the lobe magnetic field intensity is B_0. The component equations of the Lorentz equation in the nonrelativistic limit are

$$
\ddot{x} = 0
$$

$$
\ddot{y} = \left(\frac{q B_0}{m L} \right) z \dot{z}
$$

$$
\ddot{z} = -\left(\frac{q B_0}{m L} \right) z \dot{y}
\tag{5.10}
$$

where the single and double dots are first and second time derivatives. These equations are solved by first multiplying the second equation by $\dot{y}$ and the third by $\dot{z}$. The addition of the equations then will yield

$$\frac{d}{dt}\left(\dot{y}^2 + \dot{z}^2\right) = 0 \tag{5.11}$$

This equation is just the energy conservation equation and it describes the motion of particles in the plane perpendicular to the direction of the magnetic field (yz-plane). Now note that $d\left(z^2/2\right)/dt = z\dot{z}$. Integration of the second equation of Eq. (5.10) then yields

$$\dot{y} = \dot{y}_0 + \left(\frac{qB_0}{2mL}\right)\left(z^2 - z_0^2\right) \tag{5.12}$$

Insert this into the third equation of Eq. (5.10) and after multiplying it by $2\dot{z}$ and integrating, this equation becomes

$$\dot{z}^2 = \dot{z}_0^2 + \left(\frac{qB_0}{mL}\right)\left[\left(\frac{qB_0}{2mL}\right)z_0^2 - \dot{y}_0\right]\left(z^2 - z_0^2\right) - \left(\frac{qB_0}{2mL}\right)^2\left(z^4 - z_0^4\right) \tag{5.13}$$

In the above equations, z_0, $\dot{z}_0$, y_0, and $\dot{y}_0$ are initial values. Equation (5.13) is rather complicated but it can be put into a simpler form,

$$\dot{z}^2 = \left(1 - k^2 + k^2 z^2\right)\left(1 - z^2\right) \tag{5.14}$$

where $k^2 = qB_0 z_m^2/4Ly_0$ and z_m represents the point on the z-axis where the particles turn around.

The solutions of Eq. (5.14) involve elliptic integrals and Jacobi elliptic functions. Basically, the solutions include three different classes of orbits depending on the values of k^2. For example, if $k^2 = 1$, the solutions correspond to orbits originating in the plasma sheet that asymptotically approach the neutral line. The orbits corresponding to $k^2 > 1$ are confined to one side of the neutral line and the orbits for $k^2 < 1$ correspond to trajectories that cross the neutral line. Examples of these orbits for protons in Earth's plasma sheet are found in Rothwell et al. (1984).

5.4.2 Neutral Sheet Including a Constant Electric Field

An improved magnetic field model includes a constant dawn-dusk electric field $\mathbf{E} = E_0\hat{y}$. The components of the Lorentz equation in this case are

$$\ddot{x} = 0$$
$$\ddot{y} = c_1 z\dot{z} + c_2$$
$$\ddot{z} = -c_1 z\dot{y} \tag{5.15}$$

where $c_1 = q B_0/mL$ and $c_2 = q E_o/m$. These equations are different from the previous case only in the y-equation where the constant electric field acts on the particles. The first integrals of the last two equations are obtained in the same way as above, yielding

$$\dot{y} = \dot{y}_o + \left(\frac{c_1}{2}\right)\left(z^2 - z_o^2\right) + c_2 t \tag{5.16}$$

$$\frac{1}{2}\left(\dot{z}^2 + \dot{y}^2\right) - c_2 y = \frac{1}{2}\left(\dot{z}_o^2 + \dot{y}_o^2\right) - c_2 y_o \tag{5.17}$$

Equation (5.17) is the energy conservation equation. As before, the subscript o refers to initial values. Substitution of Eq. (5.16) into the third equation of Eq. (5.15) yields

$$\ddot{z} = -c_1 z\left[\dot{y}_o + \left(\frac{c_1}{2}\right)\left(z^2 - z_o^2\right) + c_2 t\right] \tag{5.18}$$

Equation (5.18) is a nonlinear equation and the solutions include chaotic motions. While it is not our purpose to discuss chaotic dynamics here, it is instructive to compare Eq. (5.18) with the general form of nonlinear equation

$$\ddot{z} + 2\gamma\dot{z} + \alpha z + \beta z^3 = f(t) \tag{5.19}$$

We see that Eq. (5.18) has no damping term, hence the corresponding terms are $\gamma = 0$, $\alpha = (c_1 y_0 - c_1^2 z_0^2/2 + c_1 c_2 t)$, $\beta = c_1^2/2$, and $f(t) = 0$. Periodic solutions exist in this case where the natural frequency increases with amplitude if $\beta > 0$. (For nonlinear equations involving chaotic dynamics, see Chernikov et al. 1988, and Lichtenberg and Lieberman 1983.)

Physical insight is gained by considering the solution of Eq. (5.18) for large times. For large times, $c_2 t$ is positive and will monotonically increase, indicating the solution is oscillatory and bounded in $z(t)$. Hence, Eq. (5.18) for large times can be approximated by

$$\ddot{z} \approx -c_1 c_2 z t = -\left(\frac{q}{m}\right)^2 \left(\frac{E_0 B_0}{L}\right) z t \tag{5.20}$$

The solution of this differential equation is given by

$$z = \sqrt{t'} Z_{1/3}\left(\frac{2}{3}t'^{3/2}\right) \tag{5.21}$$

where $t' = (B_o E_0/L)^{1/3}(q/m)^{2/3} t$ and $Z_{1/3}$ is a linear combination of the Bessel function of the first and second kinds, of order one-third. For large times, (5.21) is approximately

$$z \approx -\frac{t^{-1/4}}{(E_o B_o/L)^{1/12}(q/m)^{1/6}} \left\{ A \cos \left[\frac{2}{3} \left(\frac{q}{m} \right) \left(\frac{B_o E_o}{L} \right)^{1/2} t^{3/2} \right] \right.$$

$$\left. + B \sin \left[\frac{2}{3} \left(\frac{q}{m} \right) \left(\frac{B_o E_o}{L} \right)^{1/2} t^{3/2} \right] \right\} \tag{5.22}$$

where A and B are constants that depend on the initial conditions. Equation (5.22) shows that for large times, the amplitude of oscillation decays as $1/t^{1/4}$. Taking this z-decay into account, Eq. (5.16) can now be integrated to yield

$$y \approx y_0 + \left[\dot{y}_0 - \left(\frac{c_1}{2} \right) z_0^2 \right] t + \frac{c_2 t^2}{2} \tag{5.23}$$

Equations (5.23) and (5.17) show that the kinetic energy increases as t^2. Thus, a particle executes a damped oscillation about $z = 0$ while accelerating positive ions in the y-direction and electrons in the $-y$-direction. That this indeed can happen is seen as follows. The electric field in Eq. (5.15) accelerates a proton in the y-direction. Then, ignoring the term $c_1 z \dot{z}$, we see that $\dot{y}$ is proportional to t. The third equation of Eq. (5.15) then becomes

$$\ddot{z} = -kz \tag{5.24}$$

where $k = c_1 \dot{y}$. This equation describes the motion of a harmonic oscillator with a spring constant k. Since k becomes larger with time, because of the constant E-field, the spring becomes stiffer, and the oscillation amplitude decays with time. The oscillation of $z(t)$ is due to the $\mathbf{v} \times \mathbf{B}$ force, which is always directed toward $z = 0$ for z positive or negative because the magnetic field reverses its direction across the neutral sheet. The trajectories of the charged particles in this simple electromagnetic field model remain always in the neutral sheet and their energy increases without bound (Zhu and Parks 1993).

5.4.3 Magnetic Tail with a Small Perpendicular Field

Observations show the geomagnetic field does not exactly have a Harris type field geometry but includes a small magnetic field component perpendicular to the neutral sheet. The existence of this small field is important because it brings in another degree of freedom and moreover the tearing mode instability cannot be excited (see below). To examine the effect of the perpendicular field, we use the magnetic field model

$$\mathbf{B} = B_0 \left(\frac{z}{L} \hat{\mathbf{x}} + \delta \hat{\mathbf{z}} \right) \tag{5.25}$$

where $\delta \ll 1$. Note that this magnetic field geometry strictly speaking is not a neutral sheet. We use the same procedure as was done above. We find that the component equations of the Lorentz equation are

$$\ddot{x} = c_3\dot{y} \tag{5.26}$$

$$\ddot{y} = -c_3\dot{x} + c_1 z\dot{z} + c_2 \tag{5.27}$$

$$\ddot{z} = -c_1 z\dot{y} \tag{5.28}$$

where c_1 and c_2 have already been defined and $c_3 = qB_o\delta/m$. The first integrals of these equations are

$$(\dot{x} - \dot{x}_o) = c_3(y - y_o) \tag{5.29}$$

$$\frac{1}{2}\left[\left(\dot{y}^2 - \dot{y}_o^2\right) + \left(\dot{z}^2 - \dot{z}_o^2\right)\right] = c_2(y - y_o) - c_3\int \dot{x}\dot{y}dt \tag{5.30}$$

$$(\dot{y} - \dot{y}_o) = \frac{c_1}{2}\left(z^2 - z_o^2\right) + c_2 t - c_3(x - x_o) \tag{5.31}$$

Integration of Eq. (5.31) yields

$$(y - y_o) = \frac{c_1}{2}\int z^2 dt + \frac{c_2 t^2}{2} - c_3\int xdt + c_4 t \tag{5.32}$$

where $c_4 = \dot{y}_o - c_1 z_o^2/2 + c_3 x_0$. Now, substitute Eq. (5.32) into Eq. (5.29) and obtain

$$\dot{x} = \dot{x}_o + c_5 t^2 + c_6 t + c_7\int z^2 dt - c_3^2\int xdt \tag{5.33}$$

where $c_5 = c_2 c_3/2$, $c_6 = c_3 c_4$, and $c_7 = c_1 c_3/2$. Now, from Eq. (5.29), obtain

$$\frac{1}{2}(\dot{x}^2 - \dot{x}_o^2) = c_3\int \dot{x}\dot{y}dt \tag{5.34}$$

and substituting this into (5.30) yields an energy integral

$$\frac{m}{2}(\dot{x}^2 + \dot{y}^2 + \dot{z}^2) + q\Phi = \frac{m}{2}(\dot{x}_o^2 + \dot{y}_o^2 + \dot{z}_o^2) + q\Phi_o \tag{5.35}$$

where $q\Phi$ is the electrostatic potential energy and the subindices o are initial values. We now consider a proton incident on the neutral sheet with a small velocity and let us discuss qualitatively some of the behaviors of this particle.

1. The particle will be initially accelerated in the $+y$ direction by the electric field E_o.
2. The particle will gain a velocity $\dot{y}$ proportional to t (Eq. (5.31)). It will also be accelerated in the x-direction proportional to $\dot{y}$ or t (Eq. (5.26)). Hence, $\dot{x}$ is proportional to t^2. This motion is new and not present in the previous models.

3. If $\dot{y} > 0$, motion in the z-direction is oscillatory (Eq. (5.28)). The oscillations will grow if $\dot{y}$ increases in time or damp if $\dot{y}$ decreases with time. As long as $\dot{y} > 0$, the term $c_1 z \dot{z}$ in (5.27) can be considered small.
4. The first term in (5.27) is proportional to t^2 (Eq. (5.33)) so that $\dot{y}$ grows negatively until y goes to zero, or even becomes negative. The z-motion will execute a damped motion until $\dot{y} = 0$ (Eq. (5.28)).
5. After $\ddot{y}$ becomes negative, and until $\dot{y} = 0$, z will execute decreasing oscillatory motion. After $\dot{y} < 0$, z no longer oscillates but increases exponentially (Eq. (5.28)) and the particle is ejected from the neutral sheet. Note the interesting feature that unlike the previous example, the particles do not remain in the neutral sheet but are ejected.

To see the time when the particles are ejected from the neutral sheet, note that we can write Eq. (5.28) with the help of Eq. (5.31) as

$$\ddot{z} = -kz \tag{5.36}$$

where $k = c_1^2 z^2/2 + c_1 c_2 t - c_1 c_3 x + c_8$, and $c_8 = c_1 c_3 x_0 + c_1 \dot{y}_0 - c_1 z_0^2/2$. Particle ejection time is obtained by letting $k \to 0$. This approximation shows

$$t = \frac{c_3 x}{c_2} - \frac{c_1 z^2}{2c_2} - \frac{c_8}{c_1 c_2} \tag{5.37}$$

This is the time when k becomes negative and z grows exponentially. The ejection time depends on initial conditions as well as on x and z^2. Now we integrate Eq. (5.33) and obtain

$$x = x_o + \dot{x}_o t + \frac{c_6 t^2}{2} + \frac{c_5 t^3}{3} \tag{5.38}$$

where we have assumed that the integral over z^2 is small since z is oscillating, and the integral over x is also small because it is multiplied by $c_3^2 = B_o^2 \delta^2$. We will choose the initial conditions as $x = \dot{x}_o = 0$, $c_4 = c_8 = 0$, which indicate that $\dot{y}_o = c_1 z_o^2/2$. Note also that since $c_6 = c_3 c_4$, we find that $c_6 = 0$.

The ejection time is obtained by combining Eqs. (5.38) and (5.37) with these initial conditions. The end result is ejection from the neutral sheet ($z = 0$) occurs at

$$t = \frac{\sqrt{6}}{q/m} \frac{1}{B_o \delta} \tag{5.39}$$

Equation (5.38) shows that at ejection,

$$x = \frac{c_5 t^3}{3}$$

$$= \frac{\sqrt{6E_o^2}}{q/m} \frac{1}{(B_o \delta)^2} \tag{5.40}$$

and the velocity of the particle is

$$\dot{x} = c_5 t^2$$

$$= \frac{3E_o}{B_o \delta} \tag{5.41}$$

It is interesting to note that the velocity of the particles is independent of q/m within the approximation made. Thus, protons and electrons are both ejected. Note, however, that electrons are ejected much sooner than protons because of the mass factor (Eq. (5.39)).

These results are important because they imply that particles in the neutral sheet are accelerated by the electric field and then the particles interacting with the neutral sheet are ejected toward the planet (x-direction). These ejected plasma sheet particles could be an important source for auroras (see Speiser 1965a,b; Sonnerup 1971; Zhu and Parks 1993, for examples of possible particle trajectories). The readers are requested to discuss whether the above results could be verified by observations. For observations that searched for neutral line in the plasma sheet, see Lui et al. (1997).

5.5 Kinetic Models of Current Sheets

We now discuss how the plasma boundary problem can be studied in the framework of the Vlasov equation. The general procedure for obtaining the solution is to use the distribution function constructed with the constants of motion as variables and use them in the Vlasov equation. Then, charge and current densities can be computed in terms of the distribution function, and these source terms are used in the Maxwell equations to obtain information on electric and magnetic fields. However, in practice, the self-consistent kinetic equations are difficult to solve and analytic solutions have been obtained only for 1D geometries. Nevertheless, these solutions have provided important insights about the behavior of collisionless plasmas at boundaries. For completeness, we begin with a discussion of conditions for electric and magnetic fields and currents consistent with Maxwell's equations for stationary boundaries. We then examine moving boundaries.

5.5.1 Stationary Boundaries

Suppose **E** is different in two regions separated by a stationary boundary. Choose a coordinate frame in which the boundary is at rest and consider only stationary phenomena. The boundary condition for **E** is obtained by integrating the equation $\nabla \times \mathbf{E} = 0$ across the interface and using Stoke's law. Let the segments AB and CD be represented by $\Delta \mathbf{l}$ and $-\Delta \mathbf{l}$ and let the segments AD and BC be negligibly small.

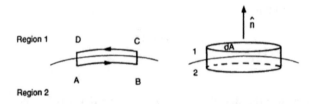

Fig. 5.9 A sketch showing boundary surfaces across which the electric and magnetic fields are different. The rectangular contour is partly on side 1 and side 2. The volume of the pill box also covers sides 1 and 2. The normal direction is indicated by **n**

Then $\mathbf{E}_2 \cdot \Delta \mathbf{l} + \mathbf{E}_1 \cdot (-\Delta \mathbf{l}) = 0$ and $(\mathbf{E}_2 - \mathbf{E}_1) \cdot \Delta \mathbf{l} = 0$ where the subscripts 1 and 2 refer to the two sides of the boundary. This equation states that the component of the electric field parallel (tangent) to the interface is continuous. In terms of the unit normal vector **n**, we see that $(\mathbf{E}_2 - \mathbf{E}_1) \times \mathbf{n} = 0$ (Fig. 5.9).

Let us denote the difference between the values of any electrodynamic quantity on the two sides of the discontinuity by enclosing the quantity in the square brackets [ ]. Thus, the continuity of the tangential component of the electric field $\mathbf{E}_t$ is written as

$$[\mathbf{E}_t] = 0 \qquad (5.42)$$

The boundary condition for **B** is obtained by the method similar to the boundary conditions for **E**. We construct a small pill box at the interface and shrink the side to a negligibly small height. Let **n** be the unit normal vector and, using Gauss's law, we obtain $(\mathbf{B}_2 - \mathbf{B}_1) \cdot \mathbf{n} = 0$, which is written as

$$[\mathbf{B}_n] = 0 \qquad (5.43)$$

The normal component of **B** is continuous across a boundary. The boundary condition for **H** is obtained by integrating $\nabla \times \mathbf{H} = \mathbf{J}$ along the path ABCD as was done for **E**. We then obtain

$$|\mathbf{H}_2 \cdot \Delta \mathbf{l} + \mathbf{H}_1 \cdot (-\Delta \mathbf{l})| = |\mathbf{K}_s \times \Delta \mathbf{l}| \qquad (5.44)$$

where $\mathbf{K}_s$ is the surface current density flowing at the interface. Equation (5.44) can be rewritten in terms of the unit normal **n** as

$$\mathbf{n} \times (\mathbf{H}_2 - \mathbf{H}_1) = \mathbf{K}_s \qquad (5.45)$$

$$[\mathbf{H}_t] = |\mathbf{K}_s| \qquad (5.46)$$

The discontinuity of the tangential component of **H** is related to the presence of a surface current $\mathbf{K}_s$. Surface currents can only exist in boundaries if the electrical conductivity is very large because in finite conductivity medium, currents will decay and so will the magnetic field.

We can use $\nabla \cdot \mathbf{D} = \rho$ and by a similar procedure derive the requirements for $\mathbf{D}$.

$$[\mathbf{D}_n] = \sum \qquad (5.47)$$

where $\sum$ is the surface charge density. The normal component of $\mathbf{D}$ is discontinuous at the surface density of free charges. However, we remind the readers that plasmas in steady state do not support free charges.

5.5.2 Moving Boundaries

If the boundary is moving, Eq. (5.45) must be modified (Noerdlinger 1971). Assume that in the laboratory frame S there is a stationary current sheet $\mathbf{K}_s$ in the Y-direction separating the two regions (Fig. 5.10). The magnetic field $|\mathbf{B}| = |\mathbf{B}_o|$ is in the Z-direction and $|\mathbf{E}| = 0$. The right side is vacuum and there is no magnetic or electric field there. Let S' be the moving frame (middle) and it is moving to the left with a velocity βc where $\beta = V/c$ so the surface current is moving to the right (middle figure). In the S' frame, the magnetic field $|\mathbf{B}'| = \gamma |\mathbf{B}_o|$ and motion across the magnetic field also induces an electric field $|\mathbf{E}'| = \gamma \beta |\mathbf{B}_o|$, where $\gamma = \left(1 - V^2/c^2\right)^{-1/2}$.

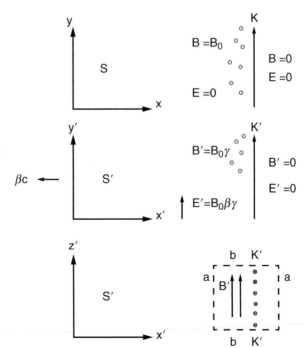

Fig. 5.10 A sketch showing moving boundary surfaces used to calculate the modification of the discontinuous relationship in Eq. (5.45) which is for a stationary boundary. The bottom plot is in S' frame with the Y-axis into the page and so are $\mathbf{E}'$ and $\mathbf{K}'$ (from Noerdlinger, *Am. J. Phys.*, **89**, 191, 1971)

To calculate the current sheet K_s in the S' frame, let us assume that the current sheet has finite thicknesses L and L' so $KL = J$ and $K' = J'L'$ where J and J' are current densities in the two frames. The velocity of the boundary is normal to J, hence Lorentz transformation shows that $J' = J$ (see Chap. 1). However, the thickness is Lorentz contracted so $L' = L/\gamma$. Consequently, the relationship of the current sheets in the two frames is

$$K' = K/\gamma \tag{5.48}$$

This result shows that the current sheet measured in the S' frame is smaller than K by a factor $1/\gamma$. The reason why the current densities are different is due to the induced electric field in the S' frame. The Maxwell equation to be considered across the boundary is now $\nabla \times H = J + \partial D/\partial t$. Integration of this equation over the rectangle (bottom, Fig. 5.10) yields

$$B'a = K'a + E'\beta a \tag{5.49}$$

where the last term comes from integrating $\int (\partial D/\partial t)dS$ around the rectangle abab in Fig. 5.10. Use now the relationship $|E'| = \gamma\beta|B_o|$ and rewriting the above equation yields

$$B_o = K' + \gamma\beta^2 B_o \tag{5.50}$$

The boundary condition in the frame of reference where the interface is moving with a velocity V written vectorially is

$$\hat{n} \times [B - V \times E] = K \tag{5.51}$$

This equation is much more difficult to handle than Eq. (5.45), suggesting current sheet K should be calculated in the frame moving with the boundary so the current sheet is stationary.

Note that Eq. (5.51) is different if the plasma moves with $E \times B/B^2$ as in fluids with infinite conductivity ($\sigma = \infty$). To see the difference, multiply both sides of the equation $E + V \times B$ with n and noting that $n \cdot [B] = 0$, one obtains

$$n \times [E] = n \cdot V[B] \tag{5.52}$$

The results are different because ideal plasmas do not induce electric field or current. Real plasmas include electric field $|E'|$ produced by the motion of the boundary in the S' frame as can be verified by integrating the equation $\nabla \times E = -\partial B/\partial t$ over the rectangle *abab* lying in the X'Y' plane (Fig. 5.10, bottom).

5.6 Kinetic Characterization of Boundaries

We view a boundary as a current layer that separates two different regions of plasmas. The dayside magnetopause separates magnetosheath and magnetospheric plasmas. Most models of dayside magnetopause assume that the magnetic fields on both sides are parallel and the solutions are obtained for stationary boundaries with straight magnetic field geometry.

Magnetic field on the two sides of a boundary can be in the same or opposite directions. We have already mentioned that when the B_{IMF} is southward, it is in the opposite direction of Earth's field. Another example of opposite fields is the geomagnetic tail, where magnetic field direction reverses across the plasma sheet. The currents produced by these magnetic field reversal geometries are more dynamic than those in parallel magnetic fields. Particle trajectories that include magnetic null regions are complicated and we discussed briefly some of the trajectories from the single particle point of view. The physics of reversed magnetic field geometry is more interesting as it can reveal important information on how particles are accelerated in the magnetic neutral region. Below, we will discuss the calculations for stationary boundaries with magnetic fields pointing in the same and opposite directions.

We begin with the Vlasov equation,

$$\frac{\partial f}{\partial t} + \mathbf{v} \cdot \frac{\partial f}{\partial \mathbf{r}} + \frac{q}{m}(\mathbf{E} + \mathbf{v} \times \mathbf{B}) \cdot \frac{\partial f}{\partial \mathbf{v}} = 0 \qquad (5.53)$$

This equation will be solved for a *time independent* boundary assuming collisions are absent. Hence we will ignore the first term on the left side. The charge and current densities are

$$\rho = q \int (f_+ - f_-)\, d^3 v$$

$$\mathbf{j} = q \int \mathbf{v}\,(f_+ - f_-)\, d^3 v \qquad (5.54)$$

where f_+ and f_- are distribution functions of ions and electrons. Maxwell equations can then be written as

$$\nabla \cdot \nabla \phi = -\rho/\epsilon_o$$

$$\nabla \times (\nabla \times \mathbf{A}) = \mu_o \mathbf{j} \qquad (5.55)$$

where ϕ and $\mathbf{A}$ are scalar and vector potentials. These partial differential equations are integrated and the solutions ϕ and $\mathbf{A}$ are used to determine the electric and magnetic fields

$$\mathbf{E} = -\nabla \phi$$

$$\mathbf{B} = \nabla \times \mathbf{A} \qquad (5.56)$$

We then insert the field equations back into the Vlasov equation to form closure and satisfy self-consistency.

5.6.1 Magnetic Fields Pointing in the Same Direction

Let us use a coordinate system with the magnetic field is in the z-direction and a straight plane boundary in the y-direction. The particles are approaching the boundary from the x-direction. For the steady state boundary under consideration, the distribution function is not only independent on time but also independent on y because of the symmetry of the problem. There are no forces along the magnetic field direction, hence, the distribution function in v_z is arbitrary and will not affect the physics. The distribution function can thus be represented by $f(x, v_x, v_y)$. The Vlasov equation for this problem is then reduced to

$$v_x \frac{\partial f}{\partial x} + \frac{qE}{m} \frac{\partial f}{\partial v_x} + \frac{qB}{m} \left(v_y \frac{\partial f}{\partial v_x} - v_x \frac{\partial f}{\partial v_y} \right) = 0 \qquad (5.57)$$

The vector and scalar potentials for this problem become

$$B = \frac{dA}{dx}$$

$$E = -\frac{\partial \phi}{\partial x} \qquad (5.58)$$

Note that we have assumed existence of a static potential difference across the boundary. Substitution of these into the above equation yields

$$v_x \frac{\partial f}{\partial x} - \frac{q}{m} \frac{\partial \phi}{\partial x} \frac{\partial f}{\partial v_x} + \frac{q}{m} \frac{\partial A}{\partial x} \left(v_y \frac{\partial f}{\partial v_x} - v_x \frac{\partial f}{\partial v_y} \right) = 0 \qquad (5.59)$$

The distribution function is constructed using constants of motion. That such a distribution function is a solution to the Vlasov equation can be easily illustrated. If $f(\mathbf{r}, \mathbf{v}, t) = f(c_1, c_2, c_3 \ldots)$ is a distribution function with the constants of motion c_i's as the variables, we see that

$$\frac{df}{dt} = \frac{\partial f}{\partial c_i} \frac{dc_i}{dt} = 0 \qquad (5.60)$$

since the constants of motion are independent of time. The most general solution is thus $f = f(c_1, \ldots c_6, t)$, and the possible set of constants can be the initial coordinates of the particles, for example. In this case, we have $f(\mathbf{r}, \mathbf{v}, 0) = f(\mathbf{r}_o, \mathbf{v}_o)$ which is a solution to any initial value problem. However, difficulty arises because this requires that we know $c_i = c_i(\mathbf{r}, \mathbf{v}, t)$ which is equivalent to solving the equations of motion for all the particles.

For the problem we are studying, we have two constants of motion, the total energy and the generalized canonical momentum,

$$w = \frac{1}{2}m(v_x^2 + v_y^2) + q\phi(x)$$

$$p_y = mv_y + qA(x) \tag{5.61}$$

The canonical momentum is a constant because the potentials are independent of y. The general solution of (5.59) can thus be written as

$$f = f(w, p_y) \tag{5.62}$$

As an example, consider

$$f = \delta(w - w_o)\delta(p_y - p_{yo}) \tag{5.63}$$

where w_o and p_{yo} are constants and the δ represents the usual delta function. The solution here is zero except for $w = w_o$ and $p_y = p_{yo}$. These two equations give the trajectories of the particle in the phase space. For a given x, we can determine v_y from the canonical momentum equation and v_x from the energy equation. Let us now examine the details of the momentum equations,

$$m_e v_y - eA(x) = m_e v_{yo}$$

$$m_i v_y + qA(x) = m_i v_{yo} \tag{5.64}$$

where m_e and m_i are the electron and ion masses, respectively, and v_{yo} is the velocity in the y-direction far in the plasma where the vector potential $A = 0$. If electrons and ions have the same initial velocity, the ion momentum is much larger than the electron momentum even if they have the same energy. It then turns out that the vector potential in the ion momentum equation can be ignored because $A(x)$ is large enough and comparable to $m_e v_y$ but small compared to $m_i v_y$. The approximation here is the same as the one made earlier when we neglected the magnetic force on the ions. With these assumptions, the general solution for the ions can be written as

$$f_+ = f\left(\frac{1}{2}m_i\left[v_x^2 + v_y^2\right] + q\phi, v_y\right) \tag{5.65}$$

For simplicity, consider only the x motion and set $v_y = 0$. Then

$$f_+ = f\left(\frac{1}{2}m_i v_x^2 + q\phi, v_y\right) \tag{5.66}$$

If now we assume that the ion and electron densities are equal, $n^+(x) = n^-(x)$ then this allows us to avoid solving the Poisson equation. The justification for this is simply that the difference of n^+ and n^- that will produce the electric field is much smaller than n^+ or n^- so that the dynamics will not be distorted much. The asymptotic solutions for the electron and ion distribution functions are

$$f_+ = n_o v_o \delta \left(v_x^2 + 2q\phi/m_i - v_o^2 \right) \delta(v_y)$$

$$f_- = n_o v_o \delta \left(v_x^2 + v_y^2 - 2e\phi/m_e - v_o^2 \right) \delta(v_y - eA(x)/m_e) \qquad (5.67)$$

The densities are obtained by integrating the equations in (5.67) over velocity space, $n_+ = \int f_+ dv_x dv_y$ and $n_- = \int f_- dv_x dv_y$. The result is

$$n_+ = n_- = \frac{n_o v_o}{\sqrt{v_o^2 - (m_e/m_i)(eA/m_e)^2}} \qquad (5.68)$$

Question 5.9 Derive the expression for the density shown in Eq. (5.68).

From the results of the densities, we can write the current density in the y-direction as

$$j_y = -n_- e v_y(x)$$

$$= -\frac{n_- e A(x)}{m_e}$$

$$= -\frac{n_o e^2}{m_e} \frac{A(x)}{\sqrt{1 - (m_e/m_i)(eA/m_e v_y)^2}} \qquad (5.69)$$

Substitution of this equation into one of the Maxwell equations will yield

$$\frac{d}{dA} \left(\frac{B^2}{2\mu_o} \right) = \frac{n_o e^2}{m_e} \frac{A}{\sqrt{1 - (m_e/m_i)(eA/m_e v_o)^2}} \qquad (5.70)$$

This problem can be made more general assuming the plasma distributions on the two side are Maxwell distributions. Use of such distributions will show that the presence of trapped particles becomes important. Interested readers are recommended to read Sect. 5.7 of Longmire (1963),

5.6.2 Magnetic Fields Pointing in Opposite Direction

We now proceed to the current sheath model of Harris (1962). The equations he solved are the same equations we presented earlier in Eqs. (5.53), (5.54), and (5.59). He assumed that **E**, **B**, and f depend on only x-coordinate and constructed the distribution function in terms of the constants of motion, the total energy, and the canonical momenta in the y and z directions. The equations that he obtained are

$$w = \frac{m}{2}\left(v_x^2 + v_y^2 + v_z^2\right) + q\phi(x)$$

$$p_y = mv_y + qA_y(x)$$

$$p_z = mv_z + qA_z(x) \tag{5.71}$$

Assume now that $\mathbf{E}$ has only an x-component, $\mathbf{B}$ has only a z-component, and $\mathbf{A}$ has a y-component. Then, the above equations can be rewritten as

$$\alpha_1^2 = v_x^2 - (2q/m)v_y A_y - \left(q^2/m^2\right)A_y^2 + (2q/m)\phi$$

$$\alpha_2 = v_y + (q/m)A_y$$

$$\alpha_3 = v_z \tag{5.72}$$

Note that there is a set of $(\alpha_1, \alpha_2, \alpha_3)$ for each particle set, electrons and ions. The solution to the Vlasov equation (5.53) is $f(\alpha_1, \alpha_2, \alpha_3)$. However, the nature of the solution depends on the choice of f and the boundary condition. Assume that at $x = 0$, the distribution function is a Maxwellian centered about a mean velocity drifting in the y-direction.

$$f_e = \left(\frac{m_e}{2\pi\kappa T}\right) n_0 \, \exp - \left[m_e\left(\alpha_1^2 + (\alpha_2 - V_e)^2 + \alpha_3^2\right)/2\kappa T\right]$$

$$f_i = \left(\frac{m_i}{2\pi\kappa T}\right) n_0 \, \exp - \left[m_i\left(\alpha_1^2 + (\alpha_2 - V_i)^2 + \alpha_3^2\right)/2\kappa T\right] \tag{5.73}$$

where the subscript e and i refer to electrons and ions and the mean velocities of electrons and ions are given by V_e and V_i. If these equations are substituted in the Maxwell equations (5.54), one obtains

$$\frac{d^2\phi}{dx^2} = -n_o q \left\{\exp\frac{q}{\kappa T}(V_i A_y - \phi) - \exp\frac{q}{\kappa T}(-V_e A_y + \phi)\right\}$$

$$\frac{d^2 A_y}{dx^2} = -n_o q \left\{V_i \exp\frac{q}{\kappa T}(V_i A_y - \phi) - V_e \exp\frac{q}{\kappa T}(-V_e A_y + \phi)\right\}$$

$$\tag{5.74}$$

Transform now to a moving coordinate system so that in this frame $V_e = -V_i = -V$. The fields in this moving frame are then obtained from

$$\frac{d^2\phi}{dx^2} = -n_o q \left\{\exp\left(\frac{qV}{\kappa T}A_y\right)\exp\left(-\frac{q\phi}{\kappa T}\right) - \exp\left(\frac{q\phi}{\kappa T}\right)\right\}$$

$$\frac{d^2 A_y}{dx^2} = -n_o q V \left\{\exp\left(\frac{qV}{\kappa T}A_y\right)\exp\left(-\frac{q\phi}{\kappa T}\right) + \exp\left(\frac{q\phi}{\kappa T}\right)\right\} \tag{5.75}$$

The first of these equations is satisfied by $\phi = 0$. Then the vector potential equation [second equation in Eq. (5.75)] becomes

$$\frac{d^2 A_y}{dx^2} = -2qn_o V \, \exp\left(\frac{qV}{\kappa T}\right) A_y \tag{5.76}$$

The solution of Eq. (5.76) is

$$A_y = -\frac{2\kappa T}{qV} \log \, \cosh\left(\frac{xV}{\lambda_D}\right) \tag{5.77}$$

where $\lambda_D = \sqrt{\kappa T / n_o q^2 \epsilon_o}$ is the deBye length (solution provided by Professor Michael McCarthy. The details of the calculations are shown in Solution 5.10).

Question 5.10 Solve the differential equation (5.76) with the boundary condition $A_y = 0$ and $B = 2A_y/2x = 0$ at $x = 0$ and show that the solution is given by Eq. (5.77).

The magnetic field is then

$$B = B_o \tanh\left(\frac{Vx}{\lambda_D}\right) \tag{5.78}$$

The density of electrons and ions is equal, $n_e = n_i = n$, and the expression of the density is

$$\begin{aligned} n &= n_o \exp\left(\frac{qV}{\kappa T} A_y\right) \\ &= \frac{n_o}{\cosh^2(Vx/\lambda_D)} \end{aligned} \tag{5.79}$$

where we have used (5.77). The thickness of the sheath is about λ_D/V. In this model, $n_e = n_i$ and since it has been assumed that electron and ion velocities are equal and opposite, there is also no electric field in this frame. However, the electric field in any frame moving relative to this frame can be obtained from the Lorentz transformation. If we now assume a coordinate frame in which the ions are at rest, the electric field in that frame is given by

$$\begin{aligned} E_x &= \frac{V/c}{\sqrt{1 - V^2/c^2}} B_z \\ &= \frac{\sqrt{2n_o\phi}}{\sqrt{1 - V^2/c^2}} \frac{V}{c} \tanh\frac{Vx}{\lambda_D} \end{aligned} \tag{5.80}$$

The charge density is obtained from $q(n_i - n_e) = \partial E_x/\partial x$. This yields

$$q(n_i - n_e) = -2n_o q \frac{V^2/c^2}{\sqrt{1 - V^2/c^2}} \tag{5.81}$$

The 1D Harris current sheet model is an equilibrium configuration. The model can be extended to include a small magnetic field in the z-direction. Analytical solution to a 2D current sheet with magnetic field reversal geometry has been derived in the direction perpendicular to the magnetic field. These self-consistent solutions are important as they represent the first step toward understanding the structure of boundaries in collisionless plasmas. Readers interested in this problem are encouraged to read the paper by Brittnacher and Whipple (2002).

The Harris current sheet is the starting point of many instability analyses for the geomagnetic tail. For example, soon after the geomagnetic tail and the neutral current sheet were discovered, Coppi et al. (1966) applied the Harris current sheet to the geomagnetic tail and showed that the geomagnetic tail is unstable to tearing mode instability. The question about the stability of the geomagnetic tail is fundamentally important for understanding the physics of how large-scale magnetic topologies change during auroral events.

5.7 Tearing Mode Instability

An important instability mechanism for the geomagnetic tail or the magnetopause is the tearing mode instability. This instability has interested space plasma physicists because it provides a mechanism by which magnetic energy is converted to particle energy. The tearing mode instability was first studied in 1966 (Coppi et al. 1966) shortly after the geomagnetic tail and the neutral sheet were discovered. Further interests in this instability followed focusing more on the dayside (Galeev et al. 1986). Laboratory experiments also showed that magnetic neutral sheet configurations are not stable, and the idea was to find if similar consequences would ensue in neutral sheets embedded in collisionless space plasmas. Because this instability is relevant for boundaries separated by magnetic fields of opposite polarity, we review below the material presented previously in Parks (2004).

Consider now the 1D Harris current sheet problem which starts with a Maxwellian distribution function with a constant drift in the current direction. The distribution function for either the electrons or ions takes the form

$$f_o = n_o \left(\frac{m_j}{2\pi\kappa T} \right) \exp\left(-\frac{mV}{2\kappa T^2} \right) \exp\left(-\frac{\epsilon}{\kappa T} + \frac{p_y V}{\kappa T} \right) \tag{5.82}$$

where V is the constant drift in the direction of the current, $\epsilon = mv^2/2$ is the energy, and $p_y = mv_y + qA_{oy}(z)$ is the generalized momentum. Here we assume quasineutrality, hence the electrostatic potential vanishes, $\phi = 0$.

The self-consistent equations of the magnetic field and the density variation for the Harris model are (see Eqs. (5.79) and (5.78)),

$$\mathbf{B} = B_o \tanh z/\Delta \tag{5.83}$$

and

$$n(z) = \frac{n_o}{\cosh^2(z/\Delta)} \tag{5.84}$$

Here Δ is a measure of the thickness of the current sheet and n_o is the density at the center of the neutral sheet.

Assume now that the current sheet can be represented by current filaments uniformly distributed along the sheet. Suppose a small perturbation alters the current distribution, bringing together adjacent pairs while separating the others. The strengthening and weakening of localized currents can produce magnetic nodes and null regions. This redistribution of currents will change the original magnetic field topology and the sheet can become unstable. The collisionless tearing mode instability is important for understanding how large-scale magnetic field reconfigurations occur in the geomagnetic tail during plasma disturbances such as the auroral events.

Consider a perturbation of the magnetic field represented by a vector potential of the form

$$\delta A = A_y(z) \exp i(kx - \omega t) \tag{5.85}$$

where δ represents perturbation. The perturbed vector potential satisfies

$$\nabla^2 \delta \mathbf{A} = -q \int \mathbf{v} \delta f d^3 \mathbf{v} \tag{5.86}$$

Both electrons and ions contribute to the perturbed vector potential. As mentioned earlier, we assume quasi-neutrality for this problem, hence the perturbed electro-static potential vanishes

$$\phi = q \int \left(\delta f^+ - \delta f^- \right) d^3 \mathbf{v} = 0 \tag{5.87}$$

Let us now consider a plasma that consists of one ion species and electrons. The linearized Boltzmann equation for the perturbed distribution function δf is

$$\frac{\partial \delta f}{\partial t} + (\mathbf{v} \cdot \nabla)\delta f + \frac{q}{m}(\mathbf{v} \times \mathbf{B_o}) \cdot \frac{\partial \delta f}{\partial \mathbf{v}} = -\frac{q}{m_j}(\delta \mathbf{E} + \mathbf{v} \times \delta \mathbf{B}) \cdot \frac{\partial f_o}{\partial \mathbf{v}} \tag{5.88}$$

where $\delta \mathbf{E} = -\partial \delta \mathbf{A}/\partial t$ and $\delta \mathbf{B} = \nabla \times \delta \mathbf{A}$.

The right side of (5.88) can be rewritten in terms of the equilibrium distribution function (5.82). This yields

$$\frac{\partial \delta f}{\partial t} + (\mathbf{v} \cdot \nabla)\delta f + \frac{q}{m}(\mathbf{v} \times \mathbf{B_0}) \cdot \frac{\partial \delta f}{\partial \mathbf{v}} = \frac{q f_o}{\kappa T}\left[\left(\frac{\partial}{\partial t} + \mathbf{v} \cdot \nabla\right) V \delta A_y - \frac{\partial}{\partial t}(v_y \delta A_y)\right]$$

(5.89)

Galeev and Zeleny (1976) have shown that the solution of this equation can be written in the form of the integral along the particle trajectories,

$$\delta f = \frac{q f_o}{\kappa T}\left(V \delta A_y + i\omega \int_{-\infty}^{t} v_y(t')\delta A_y[x(t'), t']dt'\right)$$

(5.90)

The first term is due to the adiabatic change of the distribution function and the second term is due to the unmagnetized particles moving in the neutral sheet region. Integration over t' is made along undisturbed particle orbits inside and outside the neutral sheet. These particles can be treated as free particles moving between two ideally reflecting walls standing at $z = \pm d$, where d is the half thickness of the layer where the magnetic field does not influence the particle motion. Then the time integration of the integral can be performed along the trajectory $x = v_x t$ and after lengthy and complicated algebra, one obtains

$$\delta f = \frac{q f_o}{\kappa T}\left(V \delta A_y - \frac{\omega v_y \delta A_y}{\omega - k v_x}\right)$$

$$= \delta f_{\text{adiabatic}} + \delta f_{\text{nonadiabatic}}$$

(5.91)

Consider now the energy balance between the magnetic field energy of fluctuation $|\delta \mathbf{B}|^2$ and the dissipated energy of the current $\delta \mathbf{j} \cdot \delta \mathbf{E}^*$ in the Harris sheet,

$$\frac{1}{\mu_o}\frac{\partial}{\partial t}\int |\delta \mathbf{B}|^2 dz = -\int \delta \mathbf{j} \cdot \delta \mathbf{E}^* dz$$

(5.92)

where the $*$ denotes complex conjugate. We now compute the right-hand term for the adiabatic contribution. Since the current $\delta \mathbf{j} = \int q \mathbf{v} \delta f d^3 \mathbf{v}$ and $\delta \mathbf{E} = -\partial \delta \mathbf{A}/\partial t$, we deduce that change of energy for the adiabatic contribution is

$$\int \delta j_y^{\text{adiab}} \delta E_y^* dz = -\frac{q^2 V^2}{2\kappa T}\frac{\partial}{\partial t}\int \frac{n(z)}{n(0)}|A(z)|^2 dz$$

(5.93)

To compute the contribution from the nonadiabatic part, note that the main interest for us is the low frequency mode. For phase velocities much less than the thermal velocity of ions, $\omega << k v_{\text{th}}$, one can approximate

$$\frac{1}{(\omega/k - v_y)} \approx P(1/v_x) + i\pi\delta(v_x)$$

(5.94)

where P is the principal value and the other term is the residue. The principal part does not contribute. Hence

$$\int \delta j_y^{\text{nonadiab}} \delta E_y^* dz d^3 \mathbf{v} = \frac{\pi q}{k\kappa T} \frac{\partial}{\partial t} \int f_o \delta(z) |\delta A_y| v_y|^2 dz d^3 \mathbf{v} \tag{5.95}$$

The right side of (5.92) then becomes

$$\int \delta \mathbf{j} \cdot \delta \mathbf{E}^* dz = \frac{\pi}{k} \frac{q}{\kappa T} \frac{\partial}{\partial t} \int dz \int f_o \delta(v_x)(\delta A_y v_y)^2 d^3 \mathbf{v} - \frac{q^2 V^2}{2\kappa T} \frac{\partial}{\partial t} \int \frac{n(z)}{n(0)} \left| \frac{\delta A_y}{\Delta} \right|^2 dz \tag{5.96}$$

The first term represents the change of the kinetic energy for the resonant particles contained in the region $|z| < \Delta$. The second term represents the change due to the loss of the macroscopic kinetic energy of the remaining particles. Coppi, Laval, and Pellat have shown the kinetic ion resonant term is smaller than the electron term by approximately the square root of the ratio of gyro-radii of electrons and ions. Hence, the final result only involves the electrons and the total energy of the resonant electrons is

$$\frac{1}{\tau} \int dz \int f_o \delta(v_x)(\delta A_y v_y)^2 d^3 \mathbf{v}$$

$$= -\frac{k\kappa T_e}{2e^2} \int dz \left[\left| \frac{\partial \delta A_y}{\partial z} \right|^2 + |\delta A_y|^2 \left(k^2 - \frac{2}{\Delta^2 \cosh^2(z/\Delta)} \right) \right] \tag{5.97}$$

We have used the density profile information to arrive at this equation. The wave number is represented by k and $1/\tau = -i\omega$ is the growth rate. The tearing instability is purely growing and it exists for long wavelengths, such that $k^2 \Delta^2 < 1$. The growth rate is estimated to be of the order

$$\frac{1}{\tau} = \frac{(\Delta/r_{\text{Le}})^2 d_e}{v_{\text{th}}} \tag{5.98}$$

where $\Delta = 2\kappa(T_i + T_e)/q B_o(V_i - V_e)$, $d_e = \sqrt{(r_L \Delta)}$, and r_L is the mean gyroradius of the particles. The tearing mode transfers the free macroscopic energy to resonant electrons within the sheet thickness $d_e = \sqrt{(r_{\text{Le}} \Delta)}$ centered around the neutral sheet. The tearing mode instability violates the frozen-in-field condition since it involves dissipative processes and generation of new currents. Using published data of the geomagnetic tail and plasma sheet, Coppi et al. (1966) have estimated the growth rate to be extremely fast, less than 60 s. If the tearing mode is the instability responsible for substorms, then the recovery time of the substorm can be interpreted as the time required for the solar wind-geomagnetic field interaction to reform the tail current sheet. Typical recovery time of the plasma sheet is observed to be $\sim$30–60 min.

While the Harris sheet is unstable, theory shows that the existence of a small magnetic field across the current sheet will stabilize the tearing mode. A two-dimensional geometry that included a small B_z-component shows that the presence of a small B_z quenches the wave-particle interaction that is needed to make the tearing grow.

For the geomagnetic tail instabilities the ion tearing mode has been considered. However, detailed theory and simulation results show that ion tearing is stabilized by electron dynamics which was ignored in the theory. Several papers attempted to revive the ion tearing in the geomagnetic tail but Pellat et al. (1991) have shown that ion tearing is not a viable candidate for the geomagnetic tail. Does the geomagnetic tail undergo tearing mode instability? How do we validate the theory with observations? This question is still not answered. Recent MMS results have shown that the magnetopause is quite turbulent and currents are small scaled micro structures. One could think about these micro-currents interacting and undergoing instability but this picture is qualitative and needs to be further developed.

Question 5.11 The readers are requested to summarize what s/he has learned about the use of magnetic field reconnection concept in space plasma physics.

5.8 Concluding Remarks

We have argued that the best way to look at magnetic reconnection is to realize that some process *as yet undefined* is inducing small- and large-scale currents whose magnetic field is strong enough to modify the ambient magnetic field. How Nature can induce such large currents remains a mystery. A way to study this problem is to ask what physics can self-consistently produce small- and large-scale electric fields and currents in space plasmas.

The environment of the plasma boundaries is dynamic and complex. For example, the magnetopause region includes boundary structures consisting of multilayer mixtures of magnetosheath, magnetospheric and ionospheric plasmas (Ogilvie et al. 1984; Song et al. 1993). The MMS results show that the boundary dynamics include formation of micro-scale current sheets, often turbulent. These boundary structures are formed by the collisionless solar wind plasma interacting with the geomagnetic field. However, to date, the physical picture of this interaction process is not very clear. Eriksson et al. (2016) show examples of electron phase space holes in the boundary region and interpreted the dynamics in terms of reconnection process. However, an alternative interpretation is that these electron phase space holes could have originated from the ionosphere and have nothing to do with the boundary dynamics. Nonlinear whistler mode waves have also been observed at the dayside magnetopause and they have also been interpreted to be associated with reconnection dynamics (Wilder et al. 2017). These whistler mode waves are produced by pitch-angle anisotropy and just how the electron dynamics are related to reconnection remains to be seen.

A case study Thomsen et al. (1987) studied distribution functions of particles in the flux transfer events (FTE) (Russell and Elphic 1979). They found that while the FTEs contained a mixture of magnetosheath and magnetospheric plasmas, the relation of the FTEs to low latitude boundary layer remains unsolved. Li et al. (2000) examined the distributions of particles in the magnetosheath, boundary layer and magnetosphere and used MHD simulation tool to study how the magnetopause boundary layer might be formed. They, as in Thomsen et al. (1987), did not find a definitive answer about the sources of particles in the boundary layer. However, their simulation results indicate that the newly identified cold dense plasma sheet plasmas (Fujimoto et al. 1998) could play an important role.

Magnetopause and magnetopause boundary layer have been studied for a long time, and while observations have improved considerably, many aspects of the physics remain not understood well. The motional aspects of magnetopause boundary can be determined by the four spacecraft Cluster mission (Oksavik et al. 2002), now also with the MMS mission. Interpretation of magnetopause observations must use kinetic theory as a foundation in order to identify the physical mechanisms. Using "cartoons" can help us visualize what may be going on but it is not physics and should be avoided in spite of the fact it gives a "nice" picture. The basic goal in boundary studies is to understand the physical mechanisms. Kinetic particle codes are important tools because they can provide important information about the different processes in boundary physics (Zhu and Winglee 1996). Recent PIC simulation results have shed further light on the processes showing that reconnection is reversible (Ishizawa et al. 2015)

5.9   Solutions to Questions in Chap. 5

Solution 5.1 In an arbitrarily shaped contour C that is bounded by a surface S with $\mathbf{n}$ as the unit normal vector, the magnetic flux Φ enclosed by the contour C is $\int \mathbf{B} \cdot \mathbf{n} dS$ and the $\mathcal{EMF}$ around the contour is $\mathcal{EMF} = \oint_C (\mathbf{E} + \mathbf{V} \times \mathbf{B}) \cdot d\mathbf{l}$ where the electric field is measured in the frame of the contour element $d\mathbf{l}$ moving with a velocity $\mathbf{V}$ (see Chap. 1). Now note that $\oint_C (\mathbf{E} + \mathbf{V} \times \mathbf{B}) \cdot d\mathbf{l} = \int \nabla \times \mathbf{E}' \cdot d\mathbf{S} = -\int (\partial \mathbf{B}'/\partial t) \cdot d\mathbf{S}$. Here $\mathbf{E}'$ and $\mathbf{B}'$ are measured in the moving frame S'. Using the Lorentz transformation equation, we see that $\mathbf{E}' \approx \mathbf{E} + \mathbf{V} \times \mathbf{B}$ and $\mathbf{B}' \approx \mathbf{B}$, where we have assumed $V \ll c$, hence $\gamma = (1 - V^2/c^2)^{-1/2} \approx 1$. The electric field is still dependent on the frame but the magnetic field is the same in both coordinate frames. Since $\int (\partial \mathbf{B}'/\partial t) \cdot d\mathbf{S} = (\partial/\partial t) \int \mathbf{B} \cdot d\mathbf{S}$, which is time variation of the total magnetic flux, Faraday showed

$$\mathcal{EMF} = -\frac{d\Phi}{dt}$$

Faraday's law shows $\mathcal{EMF}$ is induced by changing either the magnetic flux in time, motion or shape of the contour. Since $d/dt = \partial/\partial t + \mathbf{V} \cdot \nabla$, Faraday's law indicates that $\mathcal{EMF}$ is induced even in steady state if there is a gradient of the magnetic flux along $\mathbf{V}$.

Solution 5.2 We start with the equation $\nabla \times \mathbf{E} = -\partial \mathbf{B}/\partial t$. Eliminate the electric field using Ohms law, $\mathbf{J} = \sigma(\mathbf{E} + <\mathbf{V}> \times \mathbf{B})$ where $<\mathbf{V}>$ is the bulk speed of the fluid and then use $\nabla \times \mathbf{H} = \mathbf{J}$ and the constitutive relation $\mathbf{B} = \mu_o \mathbf{H}$.

Solution 5.3 The speed of 1 keV electron is $\sim 17{,}500\,\mathrm{km\,s^{-1}}$ and the electrons will zip out of the open flux tube. A flux tube of length $10\,R_E$ will be emptied in a few seconds. The known transport processes (gradient and curvature and $\mathbf{E} \times \mathbf{B}$ drifts) that can replenish the electron losses in the flux tube are a few $\mathrm{km\,s^{-1}}$, too slow to account for the presence of electrons in FTEs. These electrons are present for a long time in FTE events giving strong evidence that the open field line model of FTE is incorrect. Can the readers think of a way to replenish electrons of a flux tube, whose one end is connected to the interplanetary space and the other to the ionosphere? For example, if there were electric field parallel to open flux tube, how large should the electric field be?

Solution 5.4 The density time profile of the SW transient showed it has the same shape as at L1, (Fig. 5.4), (panel 2, black). However, the bulk ion parameters changed. The peak density increased to nearly $95\,\mathrm{cc^{-1}}$ and the transient temperature increased to $T_{\mathrm{msh}} \sim 200\,\mathrm{eV}$ (panel 2, red). The bulk speed was reduced to $V_x = -300\,\mathrm{km\,s^{-1}}$. The density $n_{\mathrm{msh}} \sim 95\,\mathrm{cc^{-1}}$ is about three times the value at L1. This increase is due to the slowing down of the solar wind and compression across the bow shock and is generally observed in all of the events we have studied. The increase of n_{msh} up to four times n_{sw} has been observed.

Solution 5.5 This plot shows that in the magnetosphere $V_{xz} \sim V_{\parallel} = 400\,\mathrm{km\,s^{-1}}$ and the flow lasted until 2116 UT when the density reached a maximum (panels 1 and 3). The measured bulk speed was $V_{\perp} \sim 100\,\mathrm{km\,s^{-1}}$. The ion temperature was $\sim 1\,\mathrm{keV}$ and the distributions were anisotropic with $T_{\parallel}/T_{\perp} \sim 3$ and β reached 1. These values are unusually high for the Cluster perigee altitudes. The heat flux $H_{\parallel} \sim 3\,\mathrm{erg\,cm^2\,s^{-1}}$ was extremely large and flowing into the ionosphere. The geomagnetic field underwent large deformation, reducing B intensity from 350 nT 200 nT. The $110\,\mathrm{cc^{-1}}$ peak density measured by SC1 is more than twice the $50\,\mathrm{cc^{-1}}$ at L1.

Solution 5.6 Shocked electron distributions of solar wind found in the magnetosheath have flattop shape. That such distributions can appear at mid-altitude cusp is consistent with Cluster traversing a region that was connected to the magnetosheath. The temperature of the cold electron component is $T_e \sim 10\text{--}20\,\mathrm{eV}$ and hot $T_e \sim 60\text{--}100\,\mathrm{eV}$. The current density $J_{\parallel}$ deduced from electrons along B shows a bipolar feature with $J_{\parallel} \sim 400\,\mathrm{nA\,m^{-2}}$. Heat flux $H_{\parallel}$ is $\sim 0.03\,\mathrm{erg\,cm^2\,s^{-1}}$ and was observed at the leading and trailing edges flowing into and out of the ionosphere.

Solution 5.7 The Bernstein mode is different from the extraordinary mode. Whereas the Bernstein mode has $\mathbf{k}$ parallel to $\mathbf{E}$ and $\mathbf{E}$ perpendicular to $\mathbf{B}_0$, the extraordinary mode has $\mathbf{k}$ perpendicular to $\mathbf{E}$ and $\mathbf{E}$ parallel to $\mathbf{B}_0$. The extraordinary wave is nearly a pure electromagnetic mode. Also, note in general that electrostatic waves in plasma are normally damped (Landau damping). However, the Bernstein waves propagating across $\mathbf{B}_0$ are not damped because the Landau damping mechanism is not effective for $\mathbf{k} = \mathbf{k}_\perp$.

Solution 5.8 In the example we showed, the nearly equal densities observed at mid-altitude cusps to those in the magnetosheath suggest that some of the particles may be *free streaming* along B. The estimated number of particles entering the cusp for a 1-h transient using a typical $R_E \times R_E$ cusp area yields 10^{30} particles. This large number is significant indicating the cusp entry is a major source of particles for the magnetosphere. However, our observations do not tell us what fraction remains in the magnetosphere or mirror back to the magnetosheath. Simulation modeling could help reveal what fraction of these particles become trapped and become part of the ring current.

Solution 5.9 Performing the integration of the above ion distributions over dv_x and dv_y yields

$$n_+ = n_o v_o \int \frac{\delta(v_x^2 + 2q\phi/m_i - v_o^2)}{v_x} v_x dv_x = \frac{n_o v_o}{v_x(\phi)} = \frac{n_o v_o}{\sqrt{v_o^2 - 2q\phi/m_i}}$$

The corresponding equation for the electrons is

$$n_- = \frac{n_o v_o}{\sqrt{v_o^2 + 2e\phi/m_e - (eA/m_e)^2}}$$

We can go one step further since these results were obtained assuming that $n_+ \approx n_-$. Equating the two densities assuming $m_e/m_i << 1$ yields

$$e\phi \approx \frac{m}{2} \left(\frac{eA}{m_e}\right)^2$$

Substituting this relation into the above equations will yield

$$n_+ = n_- = \frac{n_o v_o}{\sqrt{v_o^2 - (m_e/m_i)(eA/m_e)^2}}$$

Solution 5.10 This differential equation can be solved exactly. The procedure is to convert Harris' equation into a simpler form and then integrate it twice. First collect the constants together so that the equation becomes

$$\frac{d^2 A(x)}{dx^2} + k_1 e^{k_2 A(x)} = 0 \tag{5.99}$$

with initial conditions $A(0) = A'(0) = 0$ and where $k_1 = 8\pi NeV/c > 0$ and $k_2 = eV/\theta c > 0$. Next, define new variables $Y = k_2 A$ and $y = \sqrt{k_1 k_2}\, x$ which will change the above equation to

$$Y''(y) + e^{Y(y)} = 0 \tag{5.100}$$

with initial conditions $Y(0) = Y'(0) = 0$. Multiply this equation by $Y/(y)$ and integrate once to get the first order ordinary differential equation,

$$\frac{1}{2}\left(\frac{dY}{dy}\right)^2 + e^Y = c_1 \tag{5.101}$$

Apply the initial conditions to find that $c_1 = 1$, reorganize the equation and then integrate a second time to obtain

$$\int \frac{dY}{\sqrt{2(1 - e^Y)}} = c_2 \pm y \tag{5.102}$$

where c_2 is a second integration constant. A change of variable, using the substitution $y = \log(1 - \xi^2)$ where $|\xi| < 1$, simplifies the integral to

$$\int \frac{dY}{\sqrt{2(1 - e^Y)}} \to -\sqrt{2} \int \frac{d\xi}{1 - \xi^2}$$

$$= -\sqrt{2}\tanh^{-1}\xi$$

$$\to -\sqrt{2}\tanh^{-1}\sqrt{1 - e^Y} \tag{5.103}$$

Substitute this equation and rearrange to obtain an expression for $Y(y)$,

$$Y(y) = \log\operatorname{sech}^2\left(\frac{c_2 \pm y}{\sqrt{2}}\right)$$

$$= -\log\cosh\left(\frac{c_2 \pm y}{\sqrt{2}}\right) \tag{5.104}$$

to which the initial conditions can again be imposed so as to find that $c_2 = 0$. Finally, return to the original variables and write the solution of the last equation as

$$A(x) = \frac{2}{k_2}\log\cosh\sqrt{\frac{k_1 k_2}{2}}\, x$$

$$= -\frac{2\theta c}{eV}\log\cosh\frac{xV}{L_{DC}} \tag{5.105}$$

where $L_D = \sqrt{4\pi N e^2/\theta}$.

Solution 5.11 The shape of the planetary dipole in the outer regions is very much modified in the presence of dynamic IMF. Magnetic fields originate from currents which are charges in motion. For the purpose of visualizing magnetic fields, we have introduced the concept of magnetic field lines. However, while it works as a visualization tool, it is not possible to define a unique field line. Feynman et al. (1964) has cautioned us that field lines are not fundamental and one must exercise caution about their use. In plasmas, field lines are meaningful only in the context of ideal fluids with infinite conductivity but ideal plasmas do not exist in nature.

Merging of magnetic field lines is a picturesque way to describe how nature can induce large-scale electric fields to produce currents strong enough to modify the existing magnetic field topology. While this picture is useful, the reader is cautioned that Maxwell's equations of electrodynamics say nothing about merging of field lines. The main challenge is to understand how the induction process works to produce the $\mathcal{EMF}$ that drive large-scale currents.

The concept of magnetic field lines merging is widely used to describe how magnetic energy is converted to particle energy. Merging has been suggested as the origin of high-energy particles produced in solar flares and it has been suggested that merging in magnetic tails across the neutral sheet produces energetic auroral particles. Moreover, merging between an IMF and an oppositely directed planetary magnetic field on the sunward magnetopause boundary leads to an open magnetospheric configuration that permits the solar wind energy to be transported into the magnetosphere. These descriptions are not adequate and they do not give us the needed information about the physical mechanisms. A fundamental problem in space physics is to understand how collisionless plasmas produce and dissipate small- and large-scale currents, and how particles are accelerated by electric fields and time varying magnetic fields.

Sketches of any model are a visual aid. In the case of cutting and connecting different magnetic field lines the underlying physics involves the inductive processes which produce electric fields that drive currents. However, electric field and currents have not been stressed in reconnection models, probably because they are not as easily visualized as magnetic field lines. But as discussed in this chapter, interpretation of physics performed using magnetic field lines is limited and can give an incorrect picture.

Computer simulation tools have been used to help determine what processes are involved, but here again the emphasis has been to organize the results by magnetic field lines. One also needs to worry about convolving numerical noise to produce physics. For example, ideal MHD physics cannot produce reconnection. Yet there have been numerous ideal MHD simulation results that show reconnection. The reconnection here is produced by noise which mimics resistivity. Similarly, physics produced by particles in cell (PIC) simulation is also reversible. Dissipative processes such as reconnection and annihilation of magnetic fields are not incorporated into the code. Moreover, computer capabilities are still limited and 3D simulations with proper spatial and time scales have not been modeled. Although much effort has been expended to understand the reconnection process, the physics remains elusive to this day.

References

Ashour-Abdalla, M., Kennel, C.F.: Nonconvective and convective electron cyclotron harmonic instabilities. J. Geophys. Res. **83**, 1531 (1978)

Brinca, A., et al.: Stimulation of electron Bernstein modes by perpendicular ion beams. Geophys. Res. Lett. **30**, 2075 (2003)

Brittnacher, M., Whipple, E.: Extension of the Harris magnetic field model to obtain exact two dimensional, self-consistent x-point structures. J. Geophys. Res. **107**, A2 (2002). https://doi.org/10.1029/2001JA000216

Canu, P., et al.: Abstract. In: URSI XXVII, General Assembly. ESA SP-598. European Space Agency, Noordwijk (2002)

Canu, P., et al.: Observation of continuum radiation close to the plasmapause: evidence for small scale sources. In: Rucker, H., Kurth, W., Mann, G. (eds.) Planetary Radio Emissions VI, p. 209. Austrian Academy of Sciences Press, Vienna (2006a)

Canu, P., et al.: A search for electron scale structures close to the magnetopause. In: Fletcher, K. (ed.) Proceedings of Cluster and Double Star Symposium. ESA SP-598. European Space Agency, Noordwijk (2006b)

Cargill, P., et al.: Cluster at the magnetospheric cusps. Space Sci. Rev. **118**, 321 (2005)

Chen, J., Palmedaso, P.: Chaos and nonlinear dynamics of single particle orbits in a magnetotaillike magnetic field. J. Geophys. Res. **91**, 1499 (1986)

Chen, L.J., et al.: Electron energization and mixing observed by MMS in the vicinity of an electron diffusion region during magnetopause reconnection. Geophys. Res. Lett. **43**, 6036 (2016)

Chernikov, A., et al.: Chaos - How regular can it be? Phys. Today **41**, 27 (1988)

Coppi, B., Laval, G., Pellat, R.: Dynamics of the geomagnetic tail. Phys. Res. Lett. **16**, 1207 (1966)

Dungey, J.: Conditions for the occurrence of electrical discharges in astrophysical systems. Philos. Mag. **44**, 725 (1953)

Dungey, J.: Interplanetary magnetic fields and the auroral zones. Phys. Rev. Lett. **6**, 47 (1961)

Dungey, J.: Neutral sheets. Space Sci. Rev. **17**, 1173 (1975)

Dunlop, M., et al.: Coordinated cluster/double star observations of dayside reconnection signatures. Ann. Geophys. **23**, 2867 (2005)

Eastwood, J.: Consistency of fields and particle motion in the 'Speiser' model of the current sheet. Planet. Space Sci. **20**, 1555 (1972)

Eriksson, E., et al.: Strong current sheet at a magnetosheath jet: kinetic structure and electron acceleration. J. Geophys. Res. **121**, 9608 (2016)

Frank, L.: Plasma in the Earth's polar magnetosphere. J. Geophys. Res. **76**, 5202 (1971)

Fu, S.Y., et al.: Energetic particles observed in the cusp region during a storm recovery phase. Surv. Geophys. **26**, 71 (2005)

Fujimoto, M., et al.: Plasma entry from the flanks of the near-Earth magnetotail: geotail observation. J. Geophys. Res. **103**, 4391 (1998)

Galeev, A., Kuznetsova, M., Zeleny, L.: Magnetopause stability threshold for patchy reconnection. Space Sci. Rev. **44**, 1 (1986)

Giovanelli, R.: Magnetic and electric phenomena in the Sun's atmosphere associated with sunspots. Mon. Not. R. Astron. Soc. **108**, 163 (1949)

Hall, D., et al.: Electrons in the boundary layers near the dayside magnetopause. J. Geophys. Res. **96**, 7869 (1991)

Hapgood, M., Bryant, D.: Re-ordered electron data in the low-latitude boundary layer. Geophys. Res. Lett. **17**, 2043 (1990)

Harris, E.G.: On a plasma sheath separating regions of oppositely directed magnetic field. Nuevo Cimento **23**(1), 115–121 (1962)

Hasegawa, H., et al.: Reconstruction of the electron diffusion region observed by the magnetospheric multiscale spacecraft: first results. Geophys. Res. Lett. **44**, 4566 (2017)

Hekkila, W., Winningham, D.: Penetration of magnetosheath plasma to low altitudes through the dayside magnetospheric cusps. J. Geophys. Res. **76**, 883 (1971)

Ishizawa, et al.: Electromagnetic gyrokinetic simulation of turbulence in torus plasmas. J. Plasma Phys. **81**(2), 435810203 (2015)

Krall, N., Trivelpiece, A.: Principles of Plasma Physics. McGraw Hill, New York, NY (1973)

Lavraud, B., et al.: Cluster observes the high-latitude cusp region. Surv. Geophys. **26**, 135 (2016)

Lichtenberg, A., Lieberman, M.A.: Regular and Stochastic Motion. Applied Mathematical Sciences. Springer, New York (1983)

Longmire, C.: Elementary Plasma Physics. Interscience Publishers, New York, NY (1963)

Lui, A.T.Y., et al.: Search for the magnetic neutral line in the Near-Earth plasma sheet, 2. Systematic study of the IMP 6 magnetic field observations. J. Geophys. Res. **82**, 1547 (1997)

Mozer, F., et al.: Evidence of diffusion regions at a subsolar magnetopause crossing. Phys. Rev. Lett. **89**, 015002 (2002)

Noerdlinger, P.D.: Boundary conditions for moving magnetic fields and Lorentz transformation of surface currents. Am. J. Phys. **89**, 191 (1971)

Oksavik, K., et al.: Three-dimensional energetic ion sounding of the magnetopause using cluster/RAPID. Geophys. Res. Lett. **29**, 1347 (2002)

Parks, G.K.: Physics of Space Plasmas: An Introduction. Westview Press, A Member of Perseus Books, Boulder, CO (2004)

Parks, G.K., et al.: Magnetopause boundary structure deduced from the high-time resolution particle experiment on the Equator-S spacecraft. Ann. Geophys. **17**, 1574 (1999)

Parks, G.K., et al.: Transport of transient solar wind particles in Earth's cusps. Phys. Plasmas **15**(8), 080702 (2008)

Pellat, R., Coroniti, F.V., Pritchett, P.L.: Does ion mode tearing mode exist? Geophys. Res. Lett. **18**, 143 (1991)

Phan, T., et al.: MMS observations of electron-scale filamentary currents in the reconnection exhaust and near the x line. Geophys. Res. Lett. **43**, 6060 (2016)

Pickett, J., et al.: Isolated electrostatic structures observed throughout the cluster orbit: relationship to magnetic field strength. Ann. Geophys. **22**, 2515 (2004)

Rothwell, P.L., Yates, G.K.: Global single ion effects within the Earth's plasma sheet. In: Hones Jr., E.W. (ed.) Magnetic Reconnection in Space and Laboratory Plasmas. AGU Geophysical Monograph, vol. 30. American Geophysical Union, Washington, DC (1984)

Russell, C., Elphic, R.: ISEE observations of flux transfer events on the dayside magnetopause. Geophys. Res. Lett. **6**, 33 (1979)

Sauvaud, J.A., et al.: Intermittent thermal plasma acceleration linked to sporadic motions of the magnetopause, first cluster results. Anna. Geophys. **19**, 1523 (2001)

Sonnerup, B.U.Ö.: Adiabatic particle orbits in a magnetic null sheet. J. Geophys. Res. **76**, 8211 (1971)

Speiser, T.: Particle trajectories in model current sheets, 1. Based on the open model of the magnetosphere with application to auroral particles. J. Geophys. Res. **70**, 1717 (1965a)

Speiser, T.: Particle trajectories in model current sheets, 2. Analytical solutions. J. Geophys. Res. **70**, 4219 (1965b)

Thomsen, M., et al.: Ion and electron velocity distributions within flux transfer events. J. Geophys. Res. **92**, 12127 (1987)

Trines, R., et al.: Spontaneous generation of self-organized solitary wave structures at Earth's magnetopause. Phys. Rev. Lett. **99**, 205006 (2007)

Willes, A., Robinson, P.: Electron-cyclotron maser theory for extraordinary Bernstein waves. J. Plasma Phys. **58**, 1531 (1997)

Willis, D.M.: Structure of the magnetopause. Rev. Geophys. Space Phys. **9**, 953 (1971)

Zhu, Z., Parks, G.K.: Particle orbits in model current sheets with a nonzero B_y component. J. Geophys. Res. **98**, 7603 (1993)

Zhu, Z.W., Winglee, R.: Tearing instability, flux ropes, and the kinetic current sheet kink instability in the Earth's magnetotail: a three-dimensional perspective from particle simulations. J. Geophys. Res. **101**, 4885 (1996)

Additional Reading

Alfvén, H., Fälthammar, C.-G.: Cosmical Electrodynamics. Fundamental Principles, 2nd edn. Oxford University Press, Oxford (1963)

Bernstein, I., Green, J.M., Kruskal, M.D.: Exact nonlinear plasma oscillations. Phys. Rev. Lett. **108**, 546 (1957)

Büchner, J., Zeleny, L.: Regular and chaotic particle motion in magnetotail-like field reversals, 1. Basic theory of trapped motion. J. Geophys. Res. **94**, 11821 (1989)

Escoubet, P., et al.: Staircase ion signature in the polar cusp: a case study. Geophys. Res. Lett. **19**, 1735 (1992)

Feynman, R.P., Leighton, R., Sands, M.: The Feynman Lectures on Physics. Addison-Wesley, Reading, MA (1964)

Galeev, A., Zeleny, L.: Tearing instability in plasma configurations. Sov. Phys. JETP **43**, 1113 (1976)

Goldstein, M., et al.: Multipoint observations of plasma phenomena made in space by cluster. J. Plasma Phys. **81**, 325810301 (2015)

Jackson, J.D.: Classical Electrodynamics, 2nd edn. Wiley, New York, NY (1975)

Janhunen, P., et al.: The ion shell distribution in the nightside magnetosphere. Eur. Geophys. Soc. Abstract 8152 (2003)

Landau, D.D., Liftschitz, E.M.: The Classical Theory of Fields. Addison-Wesley, Reading, MA (1951)

Lee, L.C., Kan, J.R.: A unified kinetic model of the tangential magnetopause structure. J. Geophys. Res. **84**, 6417 (1979)

Lee, L.C., Kan, J.R.: Structure of the magnetopause rotational discontinuity. J. Geophys. Res. **87**, 139 (1982)

Lee, E.S., Wilber, M., Parks, G.K., et al.: Modeling of remote sensing of thin current sheet. Geophys. Res. Lett. **31**, L21806 (2004)

Lembège, B., Pellat, R.: Stability of a thick two dimensional quasineutral sheet. Phys. Fluids **25**, 1995 (1982)

Li, Q., et al.: The geopause in relation to the plasma sheet and the low-latitude boundary layer: comparison between wind observations and multi-fluid simulations. J. Geophys. Res. **105**, 2563 (2000)

Lui, A.T.Y., et al.: Critical issues on magnetic reconnection in space plasmas. Space Sci. Rev. **116**, 497 (2005)

Mjolsness, E.C., et al.: Self-consistent reversed field sheath. Phys. Fluids **4**, 730 (1961)

Morse, R.L.: Adiabatic time development of plasma sheaths. Phys. Fluids **3**, 308 (1965)

Nichollson, D.: Introduction to Plasma Theory. Wiley, New York, NY (1983)

Ohtani, S., et al.: Electron dynamics in the current disruption region. J. Geophys. Res. **107**, SMP22-1 (2002)

Ogilvie, K., et al.: Observations of electron beams in the low latitude boundary layer. J. Geophys. Res. **89**, 10723 (1984)

Pellat, R.: About reconnection in collisionless plasma. Space Sci. Rev. **23**, 359 (1979)

Pu, Z.Y., et al.: Multiple flux rope events at the high-latitude magnetopause: cluster/rapid observation on 26 January, 2001. Surv. Geophys. **26**, 193 (2005)

Shi, Q.Q., et al.: Simulation studies of high-latitude magnetospheric boundary dynamics. Surv. Geophys. **26**, 367 (2005)

Shi, Q.Q., et al.: Solar wind entry into the high-latitude terrestrial magnetosphere during geomagnetically quiet times. Nat. Commun. **4**, 1466 (2013)

Song, P., et al.: Structure and properties of the subsolar magnetopause for northward interplanetary magnetic field - multiple-instrument particle observations. J. Geophys. Res. **98**, 11319 (1993)

Trattner, K., et al.: Temporal vs spatial interpretation of cusp ion structures observed by two spacecraft. J. Geophys. Res. **107**, 1287 (2002)

Trattner, K., et al.: On the occurrence of magnetic reconnection equatorward of the cusps at the Earth's magnetopause during northward IMF conditions. J. Geophys. Res. **122**, 605 (2017)

Wilber, M., et al.: Cluster observations of velocity space-restricted ion distributions near the plasma sheet. Geophys. Res. Lett. **31**(24), Cite ID L24802 (2004)

Wilder, F., et al.: The nonlinear behavior of whistler waves at the reconnecting dayside magnetopause as observed by the magnetospheric multiscale mission: a case study. J. Geophys. Res. **122**, 5487 (2017)

Zenitani, S., et al.: Electron dynamics surrounding the x line in asymmetric magnetic reconnection. J. Geophys. Res. **122**, 7396 (2017)

Zong, Q.G., et al.: Energetic electrons as a field line topology tracer in the high latitude boundary/cusp region: cluster rapid observations. Surv. Geophys. **26**, 215 (2005)

Chapter 6
Electric Field and Current

6.1 Introduction

One of the most important discoveries made in space plasma physics is the existence of electric field parallel to the direction of the magnetic field, $\mathbf{E}_\parallel$. Hannes Alfvén developed an auroral model in 1939 predicting existence of electric fields directed along the magnetic field $\mathbf{B}$ driving a field-aligned current (FAC), $\mathbf{J}_\parallel$. But it was not until 1960 that the first evidence of $\mathbf{E}_\parallel$ was obtained. Electron beams were observed moving earthward along the geomagnetic field to produce the aurora while the ions moved outward in the opposite direction. These beams are produced when thermal particles are accelerated by $\mathbf{E}_\parallel$ along $\mathbf{B}$. The electron and ion beams are part of the global auroral current system induced by the solar wind interaction with the magnetosphere. We have since learned that all of the dynamics associated with the aurora can be organized within the framework of the global current systems.

The first observation that gave clues about the parallel electric field was from precipitated energetic electrons exciting auroral emissions by an experiment made on the Injun III satellite in 1964 (O'Brien and Taylor 1964). Figure 6.1 shows data that include trapped and precipitated energetic electrons with energies $\geq 40\,\mathrm{keV}$ at polar altitudes together with auroral N_2^+ emissions measured by a photometer looking downward.

Equally important is the observation that currents are flowing parallel to magnetic field $\mathbf{B}$. Kristian Birkland deduced in 1908 from ground based magnetic recordings that ionospheric currents may include a field-aligned component, $\mathbf{J}_\parallel$. However the ground observations were not validated by space measurements until 1966. Evidence of FAC came in 1966 when satellite measurements showed fluctuations of the transverse component of the geomagnetic field above the auroral ionosphere (Patel 1965; Zmuda et al. 1966). The data were initially interpreted as hydromagnetic waves. However, Cummings and Dessler (1967) reinterpreted the same data and argued that the transverse variations behaved more consistently with the presence of

© Springer Nature Switzerland AG 2018
G. K. Parks, *Characterizing Space Plasmas*, Astronomy and Astrophysics Library,
https://doi.org/10.1007/978-3-319-90041-4_6

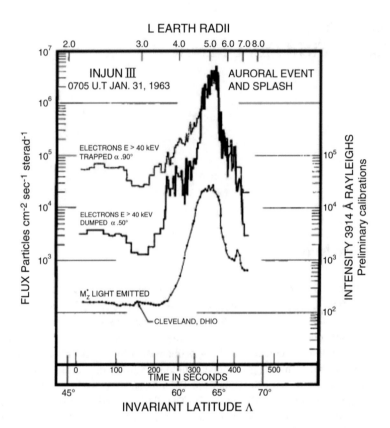

Fig. 6.1 Observations of an electron precipitation exciting atmospheric nitrogen producing auroral emissions. Both trapped and precipitated electron behavior are shown. From O'Brien and Taylor *J. Geophys. Res.*, **69**, 45 (1964)

FAC, which turned out to be correct. FACs have since been named Birkland currents. The history of FAC research is fascinating and good reading (Dessler 1983).

The data that measured magnetic disturbances transverse to the main geomagnetic field shown in Fig. 6.2 come from the Triad satellite as it crossed the polar regions in the different local time sectors. Using one of Maxwell's equations, $\nabla \times \mathbf{B} = \mu_o \mathbf{J}$ assuming steady state, the magnitude and directions of equivalent field-aligned currents are determined from the $\Delta \mathbf{B}$ variations using the approximation

$$J_\parallel = \frac{1}{\mu_o} \frac{\partial}{\partial x} \Delta B_y \tag{6.1}$$

where $J_\parallel$ is the field aligned current density, x is positive in the northward direction, and ΔB_y is the observed disturbance in the east-west direction. Equation (6.1) ignores the displacement current and further assumes that the observed disturbances in the geomagnetic north-south direction are always smaller than the east-west

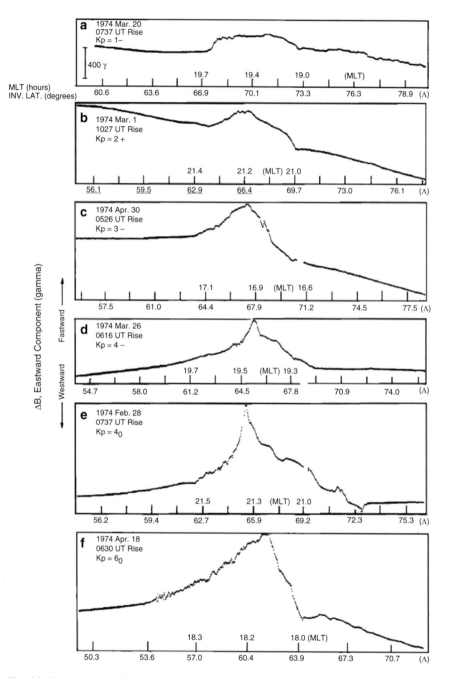

Fig. 6.2 The panels (**a**)–(**f**) show measurement examples of transverse deviations of the geomagnetic field at an altitude of 1200 km (from Iijima and Potemera, *J. Geophys. Res.*, **83**, 599, 1976)

disturbances and that the transverse magnetic disturbance is due to current sheets of infinite longitudinal extent.

This chapter will summarize the observations that established presence of $\mathbf{E}_\parallel$ and $\mathbf{J}_\parallel$ where the parallel currents are flowing in the same region where auroral beams and $\mathbf{E}_\parallel$ are observed. The existence of $\mathbf{J}_\parallel$ is important as it gives us insight into how the magnetosphere and ionosphere are connected. We examine how $\mathbf{E}_\parallel$ can be produced when the distribution functions of electrons and ions are different (Alfvén and Fälthammar 1963). We then discuss the behavior of the ring current and conclude with observations of electrostatic solitary waves and *double layers*, deBye scale structures, that may be the source of $\mathbf{E}_\parallel$. An important question still not answered is how the large-scale currents are maintained along the magnetic field direction. A question of fundamental importance is how the overall electric field and currents are tied to the global auroral current system of the magnetosphere.

6.2   Observations of Electron and Ion Beams

Parallel electric field $\mathbf{E}_\parallel$ is difficult to measure directly because any isolated charge that appears on magnetic field will be neutralized very quickly. The existence of $\mathbf{E}_\parallel$ was originally inferred from particle measurements. One of the first such observations came from a rocket experiment in 1960 (McIlwain 1960). Precipitating auroral electrons were detected along the magnetic field direction in a narrow energy range indicating they were *beams*. These beams were interpreted in terms of thermal electrons passing through a potential drop. Electric field experiments began reporting direct measurements of $\mathbf{E}_\parallel$ in 1976 corroborating the existence of the potential drops on geomagnetic fields in the auroral zone.

More evidence of $\mathbf{E}_\parallel$ came from *controlled experiments* (Wescott et al. 1976; Haerendel et al. 1976). In these experiments, shaped charged barium is released from a rocket at several hundred km altitudes. The released barium cloud, which is subsequently ionized by the solar UV radiation, could be observed and information on the local electric field is obtained from the structure and motion of the Ba^+ ions. On one occasion, Ba^+ ions were observed accelerated upward along $\mathbf{B}$ by $\mathbf{E}_\parallel$ and the estimated energy gained was $\sim 7\,\mathrm{keV}$.

A definitive observation of $\mathbf{E}_\parallel$ came from experiments on S3-3 satellite in 1976 when field-aligned ions were accelerated in the upward direction *opposite* to the direction of the electrons accelerated downward into the ionosphere to excite auroras. This observation clearly demonstrated that charged particles in auroras are accelerated by electric fields along the magnetic field (Shelley et al. 1976). These observations have since been verified by the more recent polar orbiting satellites, Frejer and FAST. Essentially all of the particle experiments have shown that $\mathbf{E}_\parallel$ is accelerating particles in the polar auroral ionosphere (Ergun et al. 2000).

The problem of how $\mathbf{E}_\parallel$ can be maintained by plasmas and what drives $\mathbf{J}_\parallel$ is still not completely understood. Space physicists have considered at least *six* different ways nature's plasma could support $\mathbf{E}_\parallel$. They include processes involving

Fig. 6.3 Observations of an electron beam in an auroral arc. The vertical scale multiplied by the geometrical factor yields the differential flux. The horizontal scale is energy in keV. This observation came from an electrostatic analyzer (ESA) above the aurora at 175 km altitude (from Albert, *Phys. Rev. Lett.*, **18**, 369, 1967)

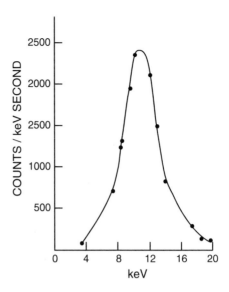

electrostatic shocks, double layers, pitch-angle anisotropy of ions and electrons, kinetic Alfvén waves, thermal potentials arising from contact of hot and cold plasmas, and anomalous resistivity. Some of these ideas will be discussed later in this chapter.

6.2.1 Electron Beams in Auroras

An example of electrons precipitating into the atmosphere over an aurora showing a beam structure is shown in Fig. 6.3. This observation comes from a rocket that flew over an aurora and included an ESA to detect electrons with an extremely good energy (E) and pitch-angle (α) resolutions: $\Delta E/E \sim 2\%$ and $\Delta\alpha/\alpha \sim \pm 3°$. The energy spectrum measured shows the electrons were nearly "*monoenergetic*." This beam is displaced from the origin (measurement platform) by ~ 10 keV, consistent with the interpretation that thermal electrons have gone through a potential drop of ~ 10 kV along the magnetic field direction.

6.2.2 Inverted-V Electrons

The existence of field-aligned electron beams was confirmed later by an experiment on the polar orbiting OGO-4 satellite above the aurora. The satellite experiment in addition showed a *new* feature: Precipitated electrons showed an *inverted* V signature as the satellite crossed the aurora (Frank and Ackerson 1971). An example

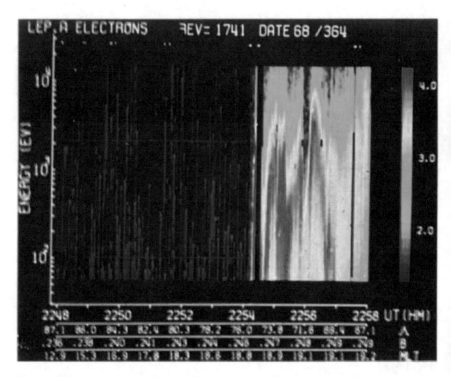

Fig. 6.4 The vertical axis is energy in eV and the horizontal axis Universal Time. The color bar on the right shows energy flux in units of (eV/cm²-s-ster-eV). This observation was made in the pre-midnight sector above quiet auroral arcs. The inverted-V electrons are directly connected to the auroral arcs (from Frank and Ackerson, *J. Geophys. Res.*, **76**, 3612, 1971)

of *inverted-V* feature in the energy flux spectrograms is shown in Fig. 6.4. The inverted-V feature refers to the shape of the electron energy flux vs time profile. The average energy of the inverted-V electrons starts out a few tens of eV, increases to a peak value of ∼1 keV, and then decreases back to a few tens of eV. Thus, a satellite passing through the auroral arc traces out the letter which resembles the Greek letter Λ, which is an upside down *V*, and has been called *Inverted-V*. The 10 min of electron data shown in Fig. 6.4 were obtained as the spacecraft crossed magnetic invariant latitudes ∼86° to 67°.

Figure 6.5 shows a 1D energy spectrum obtained by taking a cut through the energy spectrogram of Fig. 6.4. The measured data are shown as points and the solid line is a model fit of the data.

Question 6.1 Discuss the features shown in Fig. 6.5. What are the possible sources of the low and high energy components?

Fig. 6.5 Energy spectrum of inverted-V electrons. The measured data are dots and the modeling results are represented by the solid line. The Y-axis shows differential number flux and the X-axis energy in eV. The $\sim$1 keV peak is produced when the electrons pass through a potential drop $\Delta\phi$ of $\sim$1 kV (from Evans, *J. Geophys. Res.*, **79**, 2853, 1974)

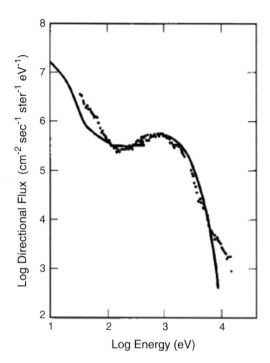

6.2.3   *Upward Moving Ions*

The most common ion species from Earth's ionosphere are H^+, O^+, N^+, and He^+. Note that He^+ from the ionosphere is singly charged whereas in the solar wind it is doubly charged, He^{++}. Also ionospheric oxygen is singly charged O^+ but the solar wind oxygen is O^{+6}, indicating the high coronal temperature exceeds the binding energies and has "stripped" all of the outer shell electrons. Ions with different charge states can be used to identify the source of the ions.

Figure 6.6 shows one of the first measurements of upward going ionospheric H^+ and O^+ ions. These measurements were made by a mass spectrometer on a polar orbiting satellite S3-3 in 1976. Data here come from the northern hemisphere on the dayside at a local time around 1430 h. The pitch-angle information determined from the onboard magnetometer is shown in the upper panel. The modulation represents the spin period of the rocket. The relative fluxes of the O^+ and H^+ ions are plotted versus time. The bottom panel shows the electron fluxes in the energy range $\sim$0.37 $\le E \le 1.28$ keV.

Question 6.2 Discuss what Fig. 6.6 is showing about the behavior of electrons and ions.

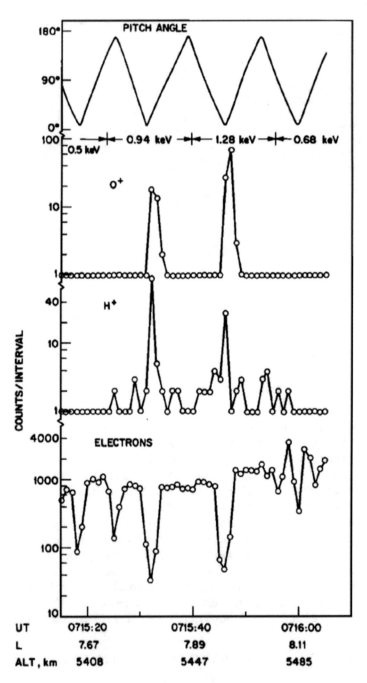

Fig. 6.6 Upward moving ions (H$^+$) and (O$^+$) in the polar region during a relatively quiet aurora. The energy per unit charge of the measured ions is indicated just below the pitch-angle information (top). Different pitch-angles are measured as the satellite rotates. Count-rate divided by the geometrical factor gives the differential flux (from Shelley et al., *Geophys. Res. Lett.*, **3**, 654, 1976)

6.2.4 Inverted-V Ions

Similar to electrons, accelerated ions also show inverted-V structure. An example from Cluster observations is shown in Fig. 6.7. The top three panels show H^+, He^+, and O^+ ion differential number flux spectrograms and the bottom three pitch-angle spectrograms of the same ions. These data were obtained when Cluster was crossing the southern auroral oval as indicated by the latitudes traversed by Cluster (bottom). The interval has been divided into four different sections according to the behavior of the data. Before 0042 UT, Cluster was in the polar cap. It then crossed into the outer boundary of the plasma sheet boundary layer. There it encountered a time dispersed injection event and also low energy (20–100 eV) field-aligned ions streaming outward from the ionosphere. Then Cluster encountered inverted-V structures in H^+, He^+, and O^+ ions. The peak energies of the inverted-V events are a few keV indicating the ions have gone through a several keV potential drop. The three ions species have been accelerated in a complex way indicating they may be accelerated from different heights. Readers interested in the details are encouraged to consult the original article by Cui et al. (2014). These ions are also field-aligned and streaming out of the ionosphere. After 0105 UT, the SC entered into the central plasma sheet where the ion fluxes were nearly isotropic and homogeneous. Observations of O^+ ion acceleration at 1700 km by Freja satellite have been discussed by André et al. (1998). For a statistical study showing ionospheric O^+ in the plasma sheet, see Kistler et al. (2006).

6.2.5 A Model of Auroral Potential Structure

A qualitative explanation of how the inverted-V feature arises assumes that there is a steady (or quasi-static) state U-shaped field-aligned potential structure originally proposed by Calqvist and Boström (1970) to accelerate auroral electrons. A sketch of such an electric potential connected to an auroral arc is shown in Fig. 6.8. The **B** field is in the vertical direction and since the electrical conductivity along **B** is very high, all equipotential surfaces are assumed parallel to **B**. The equipotentials are symmetric and at any altitude the potential drop $\Delta\phi$ is maximum at the center and decreases monotonically toward the sides. The potential ϕ decreases as a function of height and thus $\mathbf{E}_\parallel$ is directed upward. In this U-shaped potential, the electric field converges at the bottom and $\mathbf{E}_\parallel$ accelerates electrons downward and ions upward.

An electron detector moving across such a potential structure will detect electrons with increasing energy before the maximum and decreasing energy after the maximum. The bottom sketch shows a detector crossing the potential traces out an inverted-V feature in energy vs time. The downward accelerated inverted-V electrons by $\mathbf{E}_\parallel$ are presumably occurring within the upward flowing $\mathbf{J}_\parallel$ portion of the current system. A similar diagram can be constructed to explain the upgoing (downgoing) field-aligned electrons (protons).

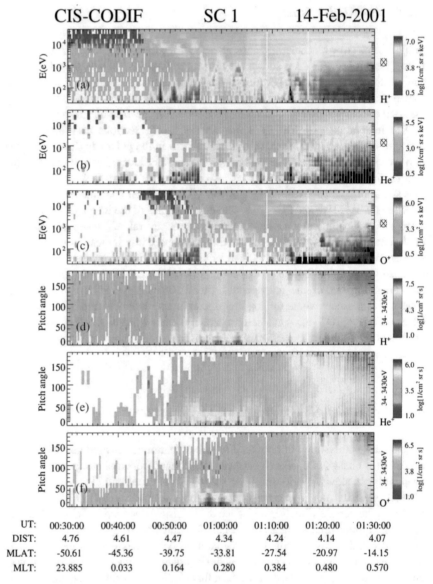

Fig. 6.7 Shown are H$^+$, He$^+$, O$^+$ ions obtained by the ion spectrometer on Cluster. The inverted-V feature arises as the spacecraft crosses a potential structure. Precipitating ions are accelerated upward going through the potential drop. These ions were detected at a height of about 21,000 km from the surface. The GSE position of the spacecraft location is shown on the bottom (from Cui et al., *Geophys. Res. Lett.*, **41**, 3752, 2014)

Fig. 6.8 One of the first schematic diagrams of a potential structure to explain the inverted-V feature based on suggestion by Gurnett (1972). The solid U-shaped lines represent the potential contours (**a**). The maximum electric field is at the center and decreases on both sides of the maximum (**b**). The inverted-V feature arises (**c**) as the spacecraft crosses the potential structure. Precipitating electrons are accelerated going through the potential drop (from Goertz, *Rev. Geophys. Space Phys.,* **17**, 418, 1979)

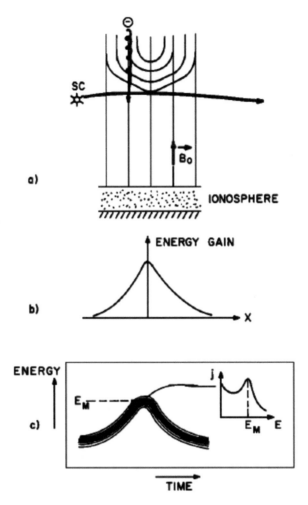

The model of Calqvist and Boström (1970) suggests that a double layer is located at the bottom of the U-shaped potential which is the source of $E_{\parallel}$. Above the double layer, plasma is convected perpendicular to **B** with $V_{\perp} = E \times B / B^2$. No information is available about the closure of the potential contours. One possibility is that the potential continues upward, parallel to **B** and closes in the southern hemisphere, maintaining symmetry about the geomagnetic equator.

The above model has assumed the potential is static or quasi-static. However, recent observations have shown that the potential can be time dependent (Marklund et al. 2011). The source of time variations is not known but suggestions have been made that the large-scale current system is possibly directly driven by the solar wind. However, at the present time, there are no models of time varying potential structures.

6.3   Acceleration Parallel to E and B Fields

For particles moving along the electric field **E** parallel to the magnetic field **B**, the scalar product **E** · **B** is a Lorentz invariant and electric and magnetic field will exist simultaneously in all Lorentz frames and the motion of particles must be considered in combined electromagnetic fields. We first look at the motion of a single particle in vacuum. Then the same problem is examined in plasma, originally discussed by Dreicer (1959), in Chap. 7.

6.3.1   Nonrelativistic Case

Choose $\mathbf{E} = (0, 0, E)$ and $\mathbf{B} = (0, 0, B)$, where the fields are homogeneous and uniform along the z-direction and parallel to each other. We begin with acceleration of nonrelativistic particles. The components of the Lorentz equation are

$$dv_x/dt = 0$$
$$dv_y/dt = 0$$
$$dv_z/dt = qE_z/m \qquad (6.2)$$

The third equation of Eq. (6.2) shows

$$v_z = (qE_z/m)t$$
$$z = (qE_z/2m)t^2 + v_{oz} \qquad (6.3)$$

We thus see the well-known result that the magnetic field has no effect on the motion along z and the velocity of particles can increase in time without bounds. This could lead to "run away" electrons first predicted by Giovanelli (1949) for acceleration of electrons on the Sun. The high energy X-rays observed in lightning discharges may be bremsstrahlung of runaway electrons accelerated by the electric field to relativistic energies (Parks et al. 1981; McCarthy and Parks 1985). These observations have been verified by balloon- and satellite-borne detectors (Eack et al. 1996; Smith et al. 2011) and recently by detectors flown on Airbus A340 from northern Australia (Kochkin et al. 2015). Runaway electron mechanisms in lightning discharges have been studied by Gurevich and Zybin (2005) and more recently by Oreshkin et al. (2017). Runaway electrons also occur in plasmas (Dreicer 1959) (Chap. 7).

The motion in the x and y directions are easily solved from Eq. (6.2) and we obtain

$$v_x = a \cos \omega t$$
$$v_y = -a \sin \omega t \qquad (6.4)$$

where a is generally a complex number that includes the phase of the particle. Since these functions are odd, the results indicate that the *average* velocity of the particles in the x and y direction vanishes. Integration of Eq. (6.4) yields

$$x = (a/\omega_c) \sin \omega_c t$$

$$y = (a/\omega_c) \cos \omega_c t \qquad (6.5)$$

where $\omega_c = qB/m$ is the cyclotron frequency (integration constants have been set to zero).

6.3.2 Relativistic Case

Relativistic electrons have been detected in the radiation belts during magnetic disturbed times (Baker et al. 2016). The presence of relativistic electrons is related to the solar Wind (Reeves et al. 2011). These electrons are likely accelerated by parallel **E** fields. Consider now relativistic particle acceleration discussed in Landau and Lifshitz (1962). As in the nonrelativistic case, the magnetic field has no effect on the motion along **B**, hence the governing equation along **B** is

$$\frac{dp_z}{dt} = eE_{\parallel} \qquad (6.6)$$

Note that here we let $E_z = E_{\parallel}$ denoting the electric field is parallel to the direction of **B**. From Eq. (6.6), we immediately obtain

$$p_z = qE_z t \qquad (6.7)$$

(we have set the integration constant $= 0$ assuming that at $t = 0$, $p_z = 0$). Similar to the nonrelativistic case, we see that the particle is accelerated by the electric field continuously in time without bounds.

Use now the relationship $\mathbf{v} = \mathbf{p}c^2/\mathcal{E}$ (from $\mathbf{p} = \gamma m_o \mathbf{v}$ and $\mathcal{E} = \gamma m_o c^2$), where $\mathcal{E}^2 = c^2 p^2 + m_o^2 c^4$, we can write

$$dz/dt = v_z$$

$$= c^2 p_z/\mathcal{E}_{kin}$$

$$= c^2 q E_z t/\mathcal{E}_{kin} \qquad (6.8)$$

where $\mathcal{E}_{kin} = [\mathcal{E}_o^2 + (q E_z t)^2]^{1/2}$ and $\mathcal{E}_o$ is the energy at $t = 0$. The integral is solved using the definition of the indefinite integral equation $\int x\,dx/(a + bx + cx^2)^{1/2} = \sqrt{x}/c - (b/2c) \int dx/(a + bx + cx^2)^{1/2}$. For our problem, $a = \mathcal{E}_o^2, b = 0, c = (cqE)^2$. Integration of Eq. (6.8) then yields

$$z = c^2 q E_z \int \frac{t\,dt}{\sqrt{\mathcal{E}_o^2 + (cq\,E_z t)^2}}$$

$$= \frac{\left[\mathcal{E}_o^2 + (cq\,E_z t)^2\right]^{1/2}}{q E_z}$$

$$= \frac{\mathcal{E}_{kin}}{q E_z} \tag{6.9}$$

The equations in the other two directions are

$$\frac{dp_x}{dt} = q B v_y$$

$$\frac{dp_y}{dt} = -q B v_x \tag{6.10}$$

which can be combined to give

$$\frac{d}{dt}(p_x + i p_y) = -q B (v_x + i v_y)$$

$$= \frac{i q B}{\mathcal{E}_{kin}}(p_x + i p_y) \tag{6.11}$$

where we have used the relation $\mathbf{v} = \mathbf{p} c^2 / \mathcal{E}$. This equation can be rewritten as

$$\int \frac{d(p_x + i p_y)}{p_x + i p_y} = -i e B c\, d\xi \tag{6.12}$$

whose solution can be written as

$$p_x + i p_y = p_o e^{-i\xi} \tag{6.13}$$

where p_o is the initial momentum in the XY plane and we have defined

$$d\xi = (e H c / \mathcal{E}_{kin}) dt \tag{6.14}$$

Integration of this equation yields

$$\xi = \int \frac{q B}{\mathcal{E}_{kin}} dt$$

$$= q B \int \frac{dt}{\sqrt{\mathcal{E}_o^2 + (q E_z t)^2}}$$

$$= \frac{B}{E} \sinh^{-1} \frac{(q E_z t)}{\mathcal{E}_o} \tag{6.15}$$

Rearranging Eq. (6.15) yields

$$ct = \frac{\mathcal{E}_o}{qE_z} \sinh \frac{E_z}{B}\xi \tag{6.16}$$

The equations for the trajectories are obtained noting that

$$p_x + ip_y = p_o e^{-i\xi}$$
$$= \frac{\mathcal{E}_{kin}}{c^2}\left(\frac{dx}{dt} + i\frac{dy}{dt}\right)$$
$$= qB\frac{d(x+iy)}{d\xi} \tag{6.17}$$

and integration of this equation yields

$$x = \frac{p_o}{qB} \sin \xi$$
$$y = \frac{p_o}{qB} \cos \xi \tag{6.18}$$

where we have used the relation $e^{-\xi} = \cos\xi + i\sin\xi$. Combining (6.16) and (6.18) yields a parametric equation

$$z = \frac{\mathcal{E}_o}{qB} \cosh \frac{E_z}{B}\xi \tag{6.19}$$

The trajectory is a helix with a radius p_o/qB and the particle moves with decreasing angular velocity $d\xi/dt = qBc/\mathcal{E}_{kin}$ along Z direction as the velocity approaches the speed of light c (since $\mathcal{E}_o$ is increasing). The motion of relativistic particles is much more complex than for nonrelativistic particles.

Question 6.3 Show that the relativistic expression (Eq. (6.19)) reduces to the nonrelativistic equation (Eq. (6.3)).

6.4   Electric Potential

The potential field which originates from charges is different from the induced electric field associated with the changing magnetic field. For electrostatic potential, $\nabla \times \mathbf{E} = 0$, hence $\mathbf{E} = -\nabla\phi$, where ϕ is the electrostatic potential. Although both kinds of electric fields exert forces on charges, there is a significant difference between them. Lines of $\mathbf{E}$ associated with charges always start from a positive charge and end on a negative charge whereas $\mathbf{E}$ associated with the changing magnetic flux is a force per unit charge on the particles moving along the periphery of an arbitrary closed contour.

Consider now the case of the electric field $\mathbf{E}$ associated with static charges. Since $\nabla \times \mathbf{E} = 0$, and $\mathbf{E}$ can be represented by the gradient of a scalar potential ϕ, $\mathbf{E} = -\nabla\phi$, we can define the potential difference between two points a and b as

$$\phi_b - \phi_a = -\int_a^b \mathbf{E} \cdot d\mathbf{l} \qquad (6.20)$$

However, if the path connecting a and b is a closed loop, then $\oint \mathbf{E} \cdot d\mathbf{l} = 0$. But note that the concept of potential difference could still be meaningful when inductive fields are involved. For example, if we take the potential difference of a limited path length on the periphery of a closed loop between points a and b, we will measure a finite value for the potential as in the static case. This example will become more clear when we discuss potential differences along the magnetic field direction above the aurora (also called field-aligned potential).

To clarify this point further, we mentioned earlier that the total force on any charge inside an equilibrium plasma must vanish, $\mathbf{F} = q(\mathbf{E} + \mathbf{V} \times \mathbf{B}) = 0$. If we integrate the force equation along a contour, it yields

$$\oint \mathbf{E} \cdot d\mathbf{l} = -\oint (\mathbf{V} \times \mathbf{B}) \cdot d\mathbf{l} \qquad (6.21)$$

This equation can be written as

$$\int \nabla \times (\mathbf{E} + \mathbf{V} \times \mathbf{B}) \cdot d\mathbf{A} = 0 \qquad (6.22)$$

using Stoke's law. The quantity $(\mathbf{E} + \mathbf{V} \times \mathbf{B})$ is just the electric field in the S' frame (nonrelativistic), and this equation is consistent with the fact that the force $q\mathbf{E}' = 0$. The particles experience no force. However, when $\nabla \times \mathbf{E} = -\partial \mathbf{B} \partial t$ and if the magnetic flux is changing, $\oint \mathbf{E} \cdot d\mathbf{l} \neq 0$ and Faraday's law shows $\oint \mathbf{E} \cdot d\mathbf{l} = -d\Phi/dt$ where $\Phi = \int \mathbf{B} \cdot \mathbf{n} dS$ is the magnetic flux enclosed inside the contour.

6.4.1   Auroral Potential Drop

In auroral observations, one often sees particles going through a potential drop along the magnetic field. Consider the case of the electric field $\mathbf{E}$ associated with static charges. As we have shown in Eq. (6.20), the potential difference refers to the potential between two points a and b along the magnetic field given by $\phi_b - \phi_a = -\int_a^b \mathbf{E} \cdot d\mathbf{l}$. If $\phi_a - \phi_b$ is positive, the point a is at a higher potential than the point b. If $\nabla \times \mathbf{E} \neq 0$ in a given region, $\phi_a - \phi_b$ at any two points is still given by Eq. (6.20) where $\mathbf{E}$ is now the electrostatic component of the total field (but not necessarily along $\mathbf{B}$). The induced electric field caused by the changing magnetic field *does not* contribute to the potential difference.

When there is a potential drop along a magnetic field, the particles going through this region will be accelerated or decelerated. If $\Delta\phi$ exists over a finite distance and is static (or quasi-static), we can let $E_\parallel = -d\phi/dz$ and the energy gained by the particle is then the total work done on the particle by the potential,

$$\Delta\mathcal{E} = \int \mathbf{F}_z \cdot \hat{\mathbf{z}}\,dz = \int_0^z q E_\parallel dz$$

$$= -q\int_{\phi=0}^{\phi=\phi} d\phi = -q\,\Delta\phi \qquad (6.23)$$

where we have assumed that the particle starts out at some reference point $z = 0$ where $\phi = 0$. Here $\hat{\mathbf{z}}$ is a unit vector along z and $\Delta\phi$ is the potential difference the particle has gone through from $z = 0$ to $z = z$. Depending on the sign of the charge, particles gain or lose energy by the amount equal to the potential drop. In the aurora, particles are accelerated typically to a few keV. However, in solar and astrophysical plasmas, particles along the magnetic field direction may be accelerated to relativistic energies.

6.4.2 Interpretation of Auroral Observations

We now interpret S3-3 observations in terms of a potential structure. The particles are traveling in the auroral potential structure as shown in Fig. 6.9. The left side shows a converging magnetic field toward Earth above an auroral arc. The right side shows the geometry used for calculating the effects of the electric fields.

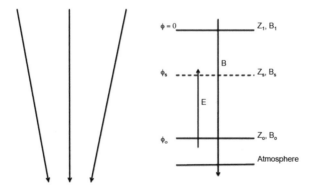

Fig. 6.9 A schematic diagram showing the potential structure in the auroral ionosphere to accelerate electrons downward and ions upward. The model shows B-field converging toward the atmosphere (left). The right side shows the potentials at various heights. The satellite is crossing the potential at height Z_s where the potential is ϕ_s and magnetic field B_s. Height Z_o is below the spacecraft where the accelerating potential is ϕ_o and the magnetic field B_o

Here $\mathbf{E}_{\parallel}$ points upward and accelerates electrons downward and ions upward. The equipotential contours are symmetric about $\mathbf{B}$ and the potential difference is maximum at the center and decreases toward the sides and the magnitude of the potential decreases toward higher altitudes. We let ϕ_o be at height Z_o where the atmosphere is dense so that collision becomes significant. The satellite is going across the height Z_s (dashed lines) where the potential is ϕ_s and $\phi = 0$ at height Z_1, which is far away in the distant region.

The presence of a potential drop will create many different types of particle distributions. For example, electrons moving upward from below the space will either pass through the potential or be reflected depending on the energy of the electrons. Similarly, ions moving downward from above the spacecraft will either pass through the potential or be reflected. The presence of the atmosphere will also alter the particle distributions. Electrons whose mirror points are below the atmosphere will interact with the atmosphere and some of the electrons will be backscattered and move upward. Interpreting the spacecraft data will thus require understanding the behavior of these particles.

Below, we begin first with a discussion of the quantities that a detector on a spacecraft will measure when there is no potential, that is $\mathbf{E}_{\parallel} = 0$. This will yield a simple picture of the physics of the particles in static inhomogeneous magnetic field. Then we extend the results including a potential drop.

Case $\mathbf{E}_{\parallel} = 0$ In this case, the differential flux measured by a particle instrument on the satellite is equal to the flux at height Z_o where we have assumed collisions become significant. Representing the differential flux as a function of the pitch-angle α, magnetic field B, and energy $\mathcal{E}$, we can write

$$j(\alpha_s, B_s, \mathcal{E}_s) = j(\alpha_o, B_o, \mathcal{E}_o) \tag{6.24}$$

With $\mathbf{E}_{\parallel} = 0$, the kinetic energy of the particles remains constant, and we have $\mathcal{E}_s = \mathcal{E}_o$ where

$$\mathcal{E}_o = \frac{m}{2}\left(v_{\parallel}^2 + v_{\perp}^2\right) \tag{6.25}$$

Here $v_{\parallel} = v_o \cos\alpha_o$ and $v_{\perp} = v_o \sin\alpha_o$. We also assume the magnetic moment of the particles is conserved, hence $\sin^2\alpha_s / B_s = \sin^2\alpha_o / B_o$. Here the subindex s represents an arbitrary point above Z_o.

All particles mirroring above the height Z_o are trapped. Those below height Z_o are precipitated. The pitch-angle of the loss cone at height Z_s can be defined in terms of mirroring particles $\alpha_o = \pi/2$ at height Z_o where the magnetic field B_o is (from first particle invariant equation)

$$\alpha_{s\,(\text{loss cone})} = \sin^{-1}\left(\frac{B_s}{B_o}\right)^{1/2} \tag{6.26}$$

Now, some of the precipitated flux will be scattered by the atmosphere at height Z_o. Let the backscattered flux going upward be denoted by $j_{bs}(\alpha_o, B_o, \mathcal{E}_{o(bs)})$ where the subscript "bs" denotes backscatter. Then the flux of backscattered particles measured by the satellite in the pitch-angle range $\pi/2 \leq \alpha_o \leq \pi$ produced at height Z_o is

$$j(\alpha_s, B_s, \mathcal{E}_s) = j_{bs}(\alpha_o, B_o, \mathcal{E}_{o(bs)}) \tag{6.27}$$

where α_s and α_o are related through the magnetic moment in the loss cone

$$\pi \geq \alpha_s \geq \pi - \alpha_s \text{ (loss cone)} \tag{6.28}$$

and we have assumed elastic scattering, hence $\mathcal{E}_s = \mathcal{E}_{o(bs)}$.

Case $\mathbf{E}_\parallel \neq 0$ When $\mathbf{E}_\parallel$ is present, the flux j is no longer invariant, but the phase space distribution is. The distribution function is related to the differential flux by $f = 2m\mathcal{E}j$ (see Chap. 1). In the presence of $\mathbf{E}_\parallel$, the equation of the flux $j^{E_\parallel}$ at height Z_1 above the satellite to the flux at Z_s which equals the undisturbed flux at Z_s for $\alpha_1 \leq \sin^{-1}(B_1/B_s)^{1/2}$ is transformed to

$$j^{E_\parallel}\left(\alpha_{s'}, B_s, \mathcal{E}'_s\right) = \frac{\mathcal{E}'_s}{\mathcal{E}_s} j(\alpha_s, B_s, \mathcal{E}_s) \tag{6.29}$$

where the energy $\mathcal{E}'_s$ is

$$\mathcal{E}'_s = \mathcal{E}_s + e\phi_1 \tag{6.30}$$

and the magnetic moment is

$$\sin^2 \alpha'_s = \frac{\mathcal{E}_s}{\mathcal{E}'_s} \sin^2 \alpha_s \tag{6.31}$$

for $0 \leq \alpha_s \leq \pi/2$. This equation relates the pitch-angle of the electron after it has gone through the potential drop in terms of the original pitch-angle before the potential drop. Note that if there were no potential drop, $\Delta\phi = 0$, we simply recover the equation of a particle in static magnetic field.

The effect of the potential drop is to lower the mirror point of the precipitating electrons. Suppose $\alpha_o = 90°$. Then $\sin^2 \alpha_o = 1$ and Eq. (6.31) shows α'_s will be less than $90°$. The *maximum* pitch-angle α'_s can have after going through the potential is

$$\sin^2 \alpha'_{max} = \frac{B'_s}{B_o}\left(1 - \frac{e\Delta\phi}{\mathcal{E}'_s}\right) \tag{6.32}$$

This equation shows no electrons will be found in the pitch-angle range

$$\alpha_{max} < \alpha'_s < 90° \tag{6.33}$$

In principle, one could determine the potential from measurements of the pitch-angle distributions of the particles. In practice, the problem is exacerbated by the fact that there are secondary particles produced by the primary accelerated particles.

6.4.3 Test of the Potential Model

Figure 6.10 shows 6 s of high resolution data obtained by the FAST spacecraft in a pre-midnight quiet auroral arc that included the inverted-V electrons. The top

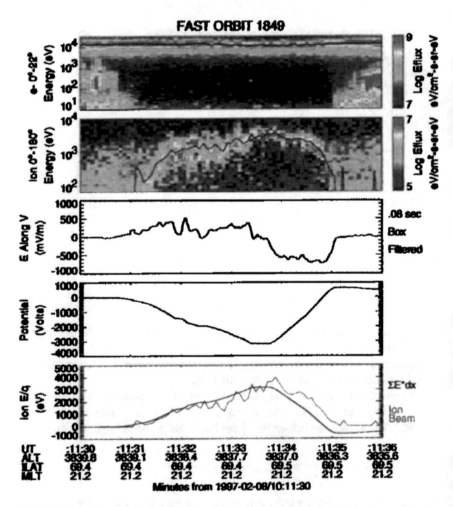

Fig. 6.10 The characteristic energy of the ion beam (black line, panel 2) is compared to the implied parallel potential below the spacecraft (black lines, panel 4 and 5) determined by integrating the measured electric field along the spacecraft trajectory (from McFadden et al., *Geophys. Res. Lett.*, **25**, 2021, 1998)

panel shows the energy spectrogram of inverted-V electrons precipitating into the atmosphere to produce the visible aurora. As can be seen, the electrons are nearly mono-energetic at ~ 1 keV. These electrons contribute to part of the upward going current of the auroral current system.

The second panel shows the spectrogram of upgoing ions. These upgoing ions are correlated with the measured electric field component perpendicular to **B** which at this time was nearly along the spacecraft velocity vector, V_{SC} (panel 3). The potential values (panel 4) are obtained by integrating the measured electric field along the spacecraft trajectory, $\Phi = -\int \mathbf{E}_\perp \cdot d\mathbf{S}$. The bottom panel shows the comparison of the characteristic energy of the ions, determined from the ratio of the energy flux to the particle flux of the beams (red), to the calculated potential (black). Note that the characteristic energy is also shown on the ion and electron spectrograms (black lines, panels 1 and 2). We see here that the integrated potential agrees remarkably well with the ion beam energy. The ions are accelerated upward by the ~ 800 mV/m electric field.

6.4.4 Theory of Potential Drop in Dipole Magnetic Field

The existence of a field-aligned electric field $\mathbf{E}_\parallel$ requires that space charges be maintained along **B**. However, space charges are difficult to maintain because the highly mobile electrons along the magnetic field will neutralize any excess charges. To show how this is achieved, consider a plasma immersed in a uniform magnetic field. If a potential is now applied along the magnetic field, electrons and ions will respond to this potential and they will move to cancel the $\mathbf{E}_\parallel$ (Fig. 6.11a).

Fig. 6.11 A schematic model showing how $E_\parallel$ could be produced (from Stern, *J. Geophys. Res.*, 5839, 1981)

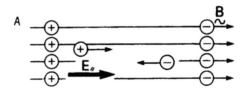

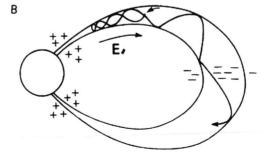

The physics is different if a plasma is in an inhomogeneous magnetic field, for example, a dipole field (Fig. 6.11b). If a potential is now applied between the equator and the ionosphere, the parallel motion of the particles will be affected by the combination of the electric $qE_\parallel$ and the magnetic mirror $-\mu \partial B/\partial s$ force. The mirror force will reinforce or counteract the electrical force and the dependence of the particle's speed on position will be different for positive ions and electrons. The net result is that the plasma charge distribution under the action of parallel electric fields in an inhomogeneous magnetic field is much more complicated than the case of a homogeneous magnetic field. However, a plasma in an inhomogeneous magnetic field supporting $\mathbf{E}_\parallel$ must still satisfy the plasma charge neutrality requirement. This problem has been investigated by Lennartson (1980). Here we show how $\mathbf{E}_\parallel$ arises in dipole magnetospheres using the single particle concept (Alfvén and Fälthammar 1963).

Consider the particles trapped in a planetary dipole magnetic field and assume that these particles are electrons and protons. The particles traveling along an inhomogeneous magnetic field experience a force,

$$F_\parallel = -\mu \frac{\partial B}{\partial s} + qE_\parallel \qquad (6.34)$$

where μ is the magnetic moment of the particle and the distance s is measured along the magnetic field. Now the force on a particle is $F_\parallel = m dv_\parallel/dt = m v_\parallel dv_\parallel/ds$. Hence, Eq. (6.34) for electrons and ions are

$$m_e \langle v \rangle_{\parallel e} \frac{d\langle v \rangle_{\parallel e}}{ds} = -\mu_e \frac{\partial B}{\partial s} - eE_\parallel$$

$$m_i \langle v \rangle_{\parallel i} \frac{d\langle v \rangle_{\parallel i}}{ds} = -\mu_i \frac{\partial B}{\partial s} + eE_\parallel \qquad (6.35)$$

where $|q| = e$ for protons and $-e$ for electrons ($e = 1.6 \times 10^{-19}$ C), the brackets $\langle \rangle$ mean the velocity has been averaged over the distribution function and the subscripts i and e denote ions and electrons, respectively.

For the particles that oscillate back and forth between the points s_1 and s_2 with a mean velocity $\langle v \rangle_\parallel$, note that $v_\parallel$ varies with position in the mirror field geometry. Since the amount of time particles spend on the line element ds is $dt = ds/\langle v \rangle_\parallel$, the average space charge in ds is $dQ = q ds/\tau \langle v \rangle_\parallel$, where τ is the half bounce-period of the particles travelling between s_1 and s_2. Let n_i and n_e be the number density of ions and electrons. The equation of the space charge density on the line element ds is then $n_i dQ_i + n_e dQ_e$. Since charge neutrality is required at every point in space we must set

$$n_i dQ_i + n_e dQ_e = 0 \qquad (6.36)$$

(Note that charge neutrality is also required for the entire plasma and thus the plasma must also satisfy $n_i q_i + n_e q_e = 0$.) This neutrality condition implies that $t_i \langle v \rangle_{\|i} = t_e \langle v \rangle_{\|e}$ and if we let $\alpha = t_i / t_e$, then

$$\langle v \rangle_{\|e} = \alpha \langle v \rangle_{\|i} \tag{6.37}$$

In general $\alpha \neq 1$ because electrons and ions in the dipole field do not necessarily have the same energy distributions or the same velocity. Using Eq. (6.37), we can rewrite Eq. (6.35) as

$$
\begin{aligned}
-\mu_e \frac{\partial B}{\partial s} - e E_{\|} &= m_e \langle v \rangle_{\|e} \frac{d \langle v \rangle_{\|e}}{ds} \\
&= \alpha^2 m_e \langle v \rangle_{\|i} \frac{d \langle v \rangle_{\|i}}{ds} \\
&= \alpha^2 \left(\frac{m_e}{m_i} \right) F_{\|i} \\
&= \alpha^2 \left(\frac{m_e}{m_i} \right) \left(-\mu_i \frac{\partial B}{\partial s} + e E_{\|} \right)
\end{aligned}
\tag{6.38}
$$

Note that in Eq. (6.38), the magnetic moment μ and the velocities $v_{\|}$ are quantities averaged over the distribution function. Solving for $E_{\|}$ yields

$$E_{\|} = -K \frac{\partial B}{\partial s} \tag{6.39}$$

Question 6.4   Show that Eq. (6.38) can be written as Eq. (6.39) and determine K.

   The results can be summarized as follows. Unless the distributions of electrons and ions are equal, plasmas in inhomogeneous magnetic fields will support parallel electric fields. For example, suppose the pitch-angle distributions of electrons and ions in the magnetosphere are different so that the electrons mirror closer to the planet than the ions. This will create a charge separation along **B** and the electric field will accelerate the ions downward and electrons upward (a downward current). If the source is static or quasi-static, the plasma will eventually neutralize itself. However, magnetospheric sources are dynamic and charge separation along **B** will be continuously created. This simple model qualitatively shows one way the parallel electric fields are produced in magnetospheres at auroral altitudes.

Question 6.5   Suppose the electrons have a Maxwellian distribution before entering a potential drop region. Derive the equation for the maximum pitch-angle of particles after the electrons have passed through the potential drop.

6.5   A Model of Double Layer

The parallel electric fields $\mathbf{E}_{\parallel}$ are the primary acceleration mechanism in the upward current region of the aurora. Although it has been established that a parallel electric field can drive a current against the magnetic mirror force, a theoretical understanding of how they are supported self-consistently in a collisionless plasma or how they are distributed along the magnetic field has not been fully established (Hudson and Potter 1981; Ergun et al. 2009; Lotko et al. 1998; Lotko 2004).

Double layers are ubiquitous in space plasmas. A DL consists of two ends with oppositely charged layers. Double layers are formed in regions that separate two plasmas of different potentials. In Earth's ionosphere, double layers are aligned along the magnetic field direction. Double layers might also exist in astrophysical plasmas (Borovosky 1986). The left side of Fig. 6.12 shows a schematic of a DL. The right side shows the potential, electric field, and charge distributions (Block 1978).

The formation of DLs requires (1) the potential drop $\Delta\phi$ through the layer be greater than the thermal energy/charge of the medium, $\geq kT_e/e$ where T_e is the temperature of the coldest plasma bordering the layer, (2) the electric field inside the DL is much stronger than outside, (3) quasi-neutrality is not maintained inside the DL, and (4) collision mean free path is much larger than the DL thickness. Experiments and theory show that if collisions are important, a DL will *not* be formed.

A typical auroral electron is accelerated to $\sim$1–10 keV. The potential structure is part of the auroral global current system and the diagram corresponds to the potential in the upward current region. For this potential structure, the high end of the potential is closer to Earth than the lower potential side. For the inverted-V potential, the plasma on the high potential side is the cold ionospheric thermal particles. The plasma on the low potential side includes the hotter plasma sheet particles.

The particles associated with DLs include four types, trapped ions and electrons and free or passing ions and electrons (see Fig. 6.13). The top panel shows the

Fig. 6.12 A schematic diagram of double layer (left). The right panel shows the potential, electric field, and charge density

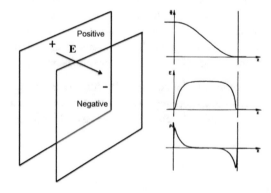

Fig. 6.13 A diagram of
double layer (**a**) potential, (**b**)
trapped and free ions, and (**c**)
trapped and free electrons
(from Block, *Astrophys.
Space Sci.*, **55**, 59, 1978)

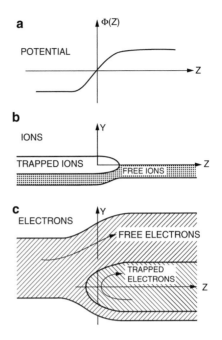

electrostatic potential along the magnetic field direction. The left side corresponds
to the high potential side and the right side shows the low potential side. Let the
potential difference be $\Delta\phi$. The ions starting on the left drift from the high potential
side initially and start with thermal energies but the particles are accelerated as they
pass through the potential forming a high energy beam as they emerge out of the low
potential side. Let the initial number flux be $n_o v_o$. This quantity is conserved going
through the potential. Hence the density will decrease as the velocity increases as the
particles are accelerated by the potential. Similarly for electrons starting on the low
potential side (right), their number density decreases as the particles are accelerated
as they drift toward the high potential side forming a beam.

If the energy of the ions on the high potential side do not have sufficient kinetic
energy to overcome the potential, these ions are reflected forming a positive charge
layer. Similarly, if the energy of the electrons on the low potential side do not
have sufficient kinetic energy to overcome the potential, they are also reflected
and form a negative charge layer. The charge density in the potential region is
$q\delta n_q = q(n_i - n_e)$.

6.5.1   Theory of Double Layers

The double layers can be formed by nonlinear waves (Bernstein et al. 1957;
Nicholson 1983; Turikov 1984; Tetreaualt 1988). There are many different kinds
of nonlinear waves. In this section, we discuss a class of nonlinear waves, the BGK

modes, named after Berstein, Green, and Kruskal. BGK waves involve spatially varying electrostatic potential and a distribution function in which a potential is produced self-consistently through the Poisson equation (Bernstein et al. 1957). For simplicity, consider the steady state case where the spatial variation is only in the x-direction. For each species, the one-dimensional Vlasov equation is

$$\left(v_x \frac{\partial}{\partial x} - \frac{q}{m} \frac{d\phi}{dx} \frac{\partial}{\partial v} \right) f(x, v) = 0 \tag{6.40}$$

The solutions of (6.40) are the equilibrium distribution functions represented in terms of the particle constants of motion. The two constants of motion in this case are the momenta $m v_y$ and $m v_z$ and the energy $m v_x^2/2 + q\phi(x)$ associated with the motion in the x-direction. The distribution functions of electrons and ions can be written as

$$f_e = f_e \left(v_x^2 - 2q\phi(x)/m_e, v_y, v_z \right)$$
$$f_i = f_i \left(v_x^2 - 2q\phi(x)/m_i, v_y, v_z \right) \tag{6.41}$$

where the subindices e and i denote electrons and ions. These distributions are used to solve for the potential ϕ which is determined self-consistently using the Poisson equation,

$$\frac{\partial^2 \phi}{\partial x^2} = q \left(\int_{-\infty}^{\infty} f_e(x, v) dv_x - \int_{-\infty}^{\infty} f_i(x, v) dv \right) \tag{6.42}$$

In this equation, it is understood that the dependences of $f(x, v)$ on the variables v_y and v_z have been integrated over. The Poisson equation is solved with appropriate boundary conditions. The solutions could involve, for example, periodic or localized solitary-like solutions. It turns out there are a large number of wave-like solutions that satisfy the above Eq. (6.42).

For simplicity, we study (6.42) assuming that the distribution function consists of cold beams. We let each species have the same speed at some given position. Then we can represent the distribution function for the electrons as

$$f_e(x, v) = 2n_o v_e \, \delta \left(v^2 - 2q\phi(x)/m_e - v_e^2 \right) \tag{6.43}$$

where we will choose only the positive roots inside the delta function. An important relation about the delta function we will need is

$$\delta f(y) = \frac{\delta(y - y_o)}{|df/dy|_{y=y_o}} \tag{6.44}$$

where y_o is the solution of $f(y_o) = 0$. Then (6.43) becomes

$$f_e(x, v) = n_o \frac{v_e}{v} \delta(v - \tilde{v}_e) \tag{6.45}$$

where $\tilde{v}_e = [v_e^2 + 2q\phi(x)/m_e]^{1/2}$. The same equation is obtained for ions. We take the distribution for ions as

$$f_i(x, v) = n_o \frac{v_i}{v} \delta(v - \tilde{v}_i) \tag{6.46}$$

where $\tilde{v}_i(x) = [v_i^2 - 2\phi(x)/m_i]^{1/2}$. The v_e and v_i are *arbitrary constants* chosen large enough so that (6.45) and (6.46) always yield real positive values for $\tilde{v}_e$ and $\tilde{v}_i$. The normalization constant n_o was chosen the same for electrons and ions so that the overall plasma remains neutral.

We now look for spatially periodic solutions to the Poisson equation (6.42) by integrating f_e and f_i over all velocity space. This yields

$$n_e(x) = n_o \frac{v_e}{\tilde{v}_e} \tag{6.47}$$

and

$$n_i(x) = n_o \frac{v_i}{\tilde{v}_i} \tag{6.48}$$

which reduces the Poisson equation (6.42) to

$$\frac{d^2\phi}{dx^2} = qn_o \left(\frac{v_e}{\tilde{v}_e} - \frac{v_i}{\tilde{v}_i} \right) \tag{6.49}$$

Substitution of the appropriate variables turns the Poisson equation to

$$\frac{d^2\phi}{dx^2} = \left[\left(1 + \frac{2q\phi(x)}{m_e v_e^2} \right)^{-1/2} - \left(1 - \frac{2q\phi(x)}{m_i v_i^2} \right)^{-1/2} \right] \tag{6.50}$$

This Eq. (6.50) is a potential equation often seen in classical mechanics whose general form is

$$\frac{d^2\phi}{dx^2} = -\frac{\partial V(\phi)}{\partial \phi} \tag{6.51}$$

where the potential is given by

$$V(\phi) = -n_o \left[m_e v_e^2 \left(1 + \frac{2q\phi}{m_e v_e^2} \right)^{1/2} + m_i v_i^2 \left(1 - \frac{2q\phi}{m_i v_i^2} \right)^{1/2} \right] \tag{6.52}$$

Equation (6.51) has the same form as Newton's equation of a particle in a potential, $md^2x/dt = -dV(x)/dx$. If the potential ϕ is small ($q\phi \ll mv_e^2/2$), we can expand (6.50) and obtain

$$\frac{d^2\phi}{dx^2} + \frac{8\pi n_o e^2}{T}\phi = 0 \tag{6.53}$$

which is a simple harmonic oscillator equation. The solution of (6.53) is

$$\phi(x) = \phi_o \sin \sqrt{2}x/\lambda \tag{6.54}$$

where $\lambda = v_e/\omega_e$. Here recall that v_e is a constant and not the thermal velocity of electrons. Readers are encouraged to think how the theoretical results could be applied to observations. A detailed discussion of BGK theory is found in Chapter 6 of Nicholson (1983).

6.6   Currents in the Magnetosphere and Ionosphere

Currents are flowing everywhere in space. The global Birkland current system circulates the polar ionospheres of Earth (Iijima and Potemera 1976) and they include field-aligned currents that connect the ionosphere to the distant magnetosphere. Both upward and downward currents are observed. Electrons (ions) are accelerated downward (upward) by $\mathbf{E}_{\parallel}$ in the upward flowing $\mathbf{J}_{\parallel}$ portion of the current system and electrons (ions) are accelerated upward (downward) in the downward portion of the $\mathbf{J}_{\parallel}$ current.

6.6.1   Observations of Field-Aligned Current

We now present observations of field-aligned currents and discuss their importance to coupling the magnetosphere and ionosphere (MI coupling). A fundamental question is how the large-scale currents are driven along the magnetic field direction. Although we know the auroral currents are intimately tied to the electron and ion beams and the parallel electric field, the physics of how the large scale electric field is sustained remains unknown.

Triad observations have established that currents run up and down the geomagnetic field. The sketch of the global auroral current system shown in Fig. 6.14 is an ionospheric "foot-print" of $\mathbf{J}_{\parallel}$ derived from magnetic field measurements.

The left plot shows the current pattern during quiet solar wind times and the right during moderately disturbed times. There are upward and downward current flowing regions. Note how the downward current region on the morning side spirals

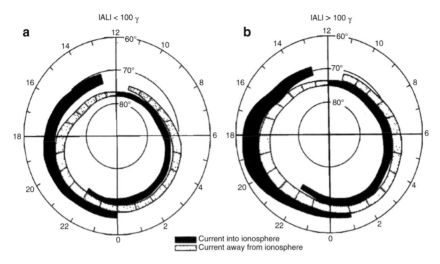

Fig. 6.14 A statistical global auroral field-aligned current system determined from magnetic measurements. The dark and light bands correspond to currents flowing downward and upward, respectively (from Iijima and Potemera, *J. Geophys. Res.*, **81**, 2165, 1976)

around and becomes an upward flowing current region on the evening side at lower latitudes. The current pattern is asymmetric: currents flowing on the noon side are located above 70° magnetic latitude and on the nightside, at 70° and slightly below (left). The term Region 1 current refers to currents flowing downward into the ionosphere and Region 2 currents flowing upward from the ionosphere. These currents are observed on the dawn side as well as the dusk side depending on the latitude.

The polar cap region becomes larger with increasing geomagnetic disturbances and the current regions move equatorward, which is more pronounced on the night side (right). These field-aligned currents $\mathbf{J}_\parallel$ are carried by downward moving electrons in the upward current region and upward moving electrons in the downward current region. The upward current region is associated with precipitating electrons and visible aurora maps to the auroral oval. An instantaneous snapshot of the auroral oval is shown in Fig. 6.15. This image was obtained by an ultraviolet imaging (UVI) camera on the Polar spacecraft from ~6 R_E above the surface of Earth.

The auroral emission here is produced by precipitating energetic electrons that excite the atmospheric constituents. The emissions come from the Lyman-Birge-Hopfield (LBH) bands of nitrogen in the ultraviolet wavelengths centered around 135.0 nm. This particular image shows the beginning of auroral brightening at midnight (substorm onset). Note the similarity of the auroral precipitation and the auroral current patterns. The electrons that excite the aurora are the same downward moving electrons in the upward current region accelerated by $\mathbf{E}_\parallel$ in Fig. 6.14. For a discussion of the dynamics of the aurora, see Lui (1991) and Liu et al. (2007). For inductive electric field along the magnetic field direction, see Murphy et al. (1974).

Fig. 6.15 An example of
global aurora taken by an
Ultra Violet Imager (UVI) on
Polar spacecraft. The
wavelength band of the filter
is centered around
~135.0 nm. Sunlight
contamination has been
removed. The truncation on
the left side is due to limited
field of view of the camera
(from Cardoso, et al. *J.
Geophys. Res.*, **122**, 6007,
2017)

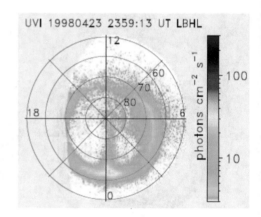

6.6.2   Relevant Equations for Currents

The formal definition for the current density at some point in space is $\mathbf{J}_{\parallel} = \int q\mathbf{v}_{\parallel} f(\mathbf{v})d^3v = qn\langle\mathbf{v}\rangle$. For a two-component plasma, the equation for the parallel current density along $\mathbf{B}$ reduces to

$$\mathbf{J}_{\parallel} = n^+ q^+ \langle\mathbf{v}\rangle_{\parallel}^+ + n^- q^- \langle\mathbf{v}\rangle_{\parallel}^- \qquad (6.55)$$

where $\langle\mathbf{v}\rangle_{\parallel}^{\pm}$ is the parallel component of the mean velocity of ions and electrons, $\langle\mathbf{v}\rangle^{\pm} = \int \mathbf{v}^{\pm} f(\mathbf{v}, \mathbf{r}, t)^{\pm} d^3 v$. From an experimental point of view, we can obtain $\mathbf{J}_{\parallel}$ by either computing the difference of the mean ion and electron velocities or applying the *curlometer* technique to the magnetic field measurements on the four Cluster spacecraft data which yields $\nabla \times \mathbf{B}$ where $\mathbf{B} = \mu_o \mathbf{H}$. In this case, we can use Maxwell's equations and write the parallel component of the current as (ignore time variations)

$$\mathbf{J}_{\parallel} = \mathbf{b} \cdot \nabla \times \mathbf{H} \qquad (6.56)$$

where $\mathbf{b}$ is a unit vector along the magnetic field direction and $\nabla \times \mathbf{H}$ is obtained from the curlometer results. The curlometer method measures the current at the center of a tetrahedron formed by the four spacecraft and represents the average current detected by the four spacecraft. Field-aligned currents are *force free*, $\mathbf{J}_{\parallel} \times \mathbf{B} = 0$.

6.6.3   Currents in Collisional Plasmas

When collisions are significant, the currents flowing in planetary ionospheres (and solar atmosphere) can be described using Ohm's law. For Earth's ionosphere, Ohm's law is given by

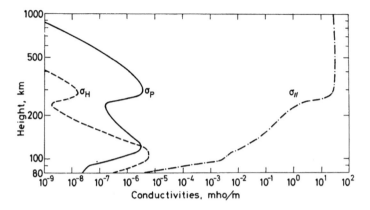

Fig. 6.16 Various conductivities as a function of height for Earth's ionosphere (from Boström, *Scandanavian University Books*, Oslo, Sweden, 1973)

$$\mathbf{J} = \sigma_p (\mathbf{E}_\perp + \mathbf{U} \times \mathbf{B}) + \sigma_H \frac{\mathbf{B}}{B} \times (\mathbf{E}_\perp + \mathbf{U} \times \mathbf{B}) + \sigma_\parallel \mathbf{E}_\parallel \qquad (6.57)$$

where σ_p, σ_H, and $\sigma_\parallel$ are the Pedersen, Hall, and parallel conductivities. Information on conductivity requires knowledge about the density and temperature of both neutral and ionized species. A model of these conductivities as a function of height has been obtained for Earth (Fig. 6.16). At heights where σ_H and σ_P are significant, Ohm's law adequately relates the current and the electric field. However, because the plasma is anisotropic, the simple Ohm's law becomes a tensor equation, $\mathbf{J} = \sigma \cdot \mathbf{E}$. At altitudes higher than ~ 100 km, $\sigma_\parallel$ dominates and one could calculate $\sigma_\parallel$ and use the simple form of Ohms law $\mathbf{J}_\parallel = \sigma_\parallel \mathbf{E}_\parallel$.

Question 6.6 At high altitudes where the plasma becomes *collisionless*, Ohm's law is not applicable. Derive the equation that describes the parallel current in collisionless plasmas.

6.7   Ring Current in Magnetospheres

The existence of ring current was predicted before the space age. However, the actual ring current in the Earth's magnetosphere was only discovered after a magnetometer was launched into space (Cahill and Amazeen 1963). Ions contributing to the ring current were subsequently detected by Frank (1967) and one of the first tests of the ring current theory was made by Berko et al. (1975). The ring current is intensified during magnetic storms and has been studied much recently by the RBSP (Radiation Belt Storm Probe) mission (subsequently renamed as Van Allen Probes) that were designed specifically to study the dynamics of the inner radiation belt particles.

The evolution of the ring current and the magnetic storm takes place over many hours to days mainly affected by ULF (ultra low frequency) waves interacting with the ring current particles. The sources of the ULF waves include the SW perturbing the magnetosphere or they are internally generated by instabilities (Zong et al. 2017). For example, these waves have periods close to the drift periods of the particles and they can grow when the drifting particle couples to the waves via the resonance mechanism. The ULF waves are important for the ring current as they can interact with the particles accelerating and diffusing the particle radially inward and outward. The magnetospheric medium in which the particles are modulated contains both the magnetic and pressure gradients. The particle modulation features are strongly dependent on the energy and pitch angle of the particles. Drifting particles can resonate with waves when phase velocities of the waves are close to the drift velocities allowing coupling to occur (Lin and Parks 1982)

Because there are several effects that can affect the ring current and the processes could be nonlinear, simulation models have been developed to help interpret the features associated with ring current particles (see, for example, Fok et al. 2014). Fok's simulation model solves the bounced-averaged Boltzmann equation for the distributions of ring current and radiation belt particles, and includes ionospheric current conservation equation for the ionospheric potential and the total plasmasphere ion content per unit magnetic flux tube. The model however does not incorporate the MI coupling physics. Interested readers are referred to the large number of papers published recently on this topic (see references at the end of the chapter).

6.7.1   Ring Current Ions

Until recently, it was thought that the ions in the ring current were all protons, H^+. It was thought that storms build up as a result of substorm injection of particles into the outer radiation belt. The plasma sheet (PS) and the plasma sheet boundary layer (PSBL) connected to the aurora and the PS could be the seat of instability responsible for substorms and storms. However, Kamide (1997) have argued that substorm injection occurs in the outer radiation zones while the ring current develops further inside.

The plasma sheet and plasma sheet boundary layer were once thought populated with only electrons and protons of the SW. However, we have since learned that a substantial fraction of the ring current ions are O^+ ions. For some magnetic storms, O^+ ions can even dominate the ring current population. Since the O^+ ions are mainly from Earth's ionosphere, it has become apparent that the ionosphere is playing an important role during magnetic storms.

Stevenson et al. (2001) have compared the field-aligned O^+ flows to auroral forms and suggest that a transition of downward and upflowing field-aligned O^+ ions occurs near boundaries of bright auroral arcs. O^+ ions have also been detected in the plasmasphere. Dandouras (2013) has shown that the plasmapause structure is complex and dynamic. For example, while the density profile of H^+ and He^+ is similar, O^+ is not observed as trapped population at the altitudes where measurements were made ($\sim 4\ R_E$). Low energy ($\sim 10\,eV$) O^+ ions are observed travelling only in one direction away from the ionosphere along the magnetic field (upwelling events). *Detached* plasmaspheric plasma events are observed within $0.5 R_E$ outside the plasmapause and these detached regions have bi-directional pitch-angle distributions, distinguishing them from the upwelling events. The dynamic nature of the plasmapause has been seen during moderate size storms in the extreme-ultraviolet images by Sandel et al. (2003) and Spasojevic et al. (2003). The plasmapause in the afternoon sector expands to the magnetopause and this could be the source of cold plasma seen by Cluster. We recommend the readers consult the review papers by Horwitz (1987) and Yau and André (1997) for further information on ionospheric ion sources and their implications about magnetospheric dynamics.

The high resolution instruments on the recent mission show the O^+ ions are abundant even with minor geomagnetic activity. Two populations of O^+ ions have been observed. One is the *freshly* accelerated O^+ ions of $\approx$ few tens of eV that are escaping the ionosphere along the direction of the magnetic field and the other is the energetic ions with a few tens of keV at larger pitch-angles. This latter component is thought to be *older* O^+ that have circulated the vast magnetosphere and subsequently accelerated. Cluster has also seen minor ions He^+ and O^{++} escaping the ionosphere along with O^+.

6.7.2   MHD Model of Ring Current

Ring currents are important for magnetospheric dynamics. We now summarize the equations on an MHD model to study the ring current. In MHD theory, the assumption that the fluid pressure is a scalar is only valid if there is equipartition of energy in all directions. We have to be cautious about this assumption for low density and collisionless plasmas because equipartition is not always achieved. Although in some situations equipartition of energy may be assumed for motions parallel to **B**, this cannot be assumed for particle motion in the perpendicular direction, since in a tenuous plasma the particles gyrate around a magnetic line of force many times between collisions. The perpendicular and parallel pressures are expected to be different.

In the ring currents problem, the pressure term must be treated as a tensor. For conducting plasmas, free charges vanish and the equation of motion for electron or ion fluids is given by

$$mn\frac{d\mathbf{U}}{dt} = -\nabla \cdot \mathbf{p} + \mathbf{J} \times \mathbf{B} \tag{6.58}$$

This equation looks identical to the one-fluid MHD equation, but the pressure is a tensor, instead of a scalar. The force involves the divergence of the pressure tensor, instead of the gradient of a scalar pressure. The equation for the pressure tensor is

$$p_{ik} = \pi_{ik} - U_i U_k \Sigma_\alpha m^\alpha n^\alpha, \tag{6.59}$$

where π_{ik} is the momentum transfer tensor $\pi_{ik} = \Sigma_\alpha m^\alpha \int v_i v_k f^\alpha(\mathbf{r}, \mathbf{v}, t) d^3 v$ (Chap. 1) and U_i and U_k are the ith and kth components of the center of mass velocity $\mathbf{U}$ given by $\mathbf{U} = \Sigma_\alpha m^\alpha \int \mathbf{v} f^\alpha(\mathbf{r}, \mathbf{v}, t) d^3 v / \Sigma_\alpha m^\alpha n^\alpha$. Here $\mathbf{p}$ in general has nine components in Cartesian coordinates. However, $\mathbf{p}$ is symmetric, $p_{ik} = p_{ki}$, and the number of variables is reduced to six. The computation of p_{ik} depends explicitly on the form of the distribution function.

In the coordinate system in which $\mathbf{B}$ is along the $\mathbf{z}$-axis, let $\mathbf{v}_\perp$ and $\mathbf{v}_\parallel$ represent the motions of the particles perpendicular and parallel to $\mathbf{B}$. In this frame of reference, we can write the pressure for a gyrotropic distribution as

$$\mathbf{p} = p_\perp \mathbf{I} + (p_\parallel - p_\perp) \mathbf{bb} \tag{6.60}$$

where $\mathbf{I}$ is the unit tensor (the Kronecker tensor) and $\mathbf{bb}$ is a second-order tensor (a dyadic) formed from the unit vectors $\mathbf{b}$ along the magnetic field direction. Here, $p_\perp$ and $p_\parallel$ are obtained from $p_\perp = m \int v_\perp^2 f(\mathbf{r}, v^2, t) d^3 v$ and $p_\parallel = m \int v_\parallel^2 f(\mathbf{r}, v^2, t) d^3 v$. If the magnetic field is curved, the orientation of $\mathbf{B}$ changes and this change must be taken into account.

We need to compute the divergence of $\mathbf{p}$ as required by the equation of motion. Consider $\mathbf{p}$ as given in the equation of motion. Define now $\nabla = \nabla_\parallel - \nabla_\perp$. It is important to note that while $\nabla \cdot \mathbf{B} = 0$, $\nabla \cdot \mathbf{b} \neq 0$. After a lot of algebra (Parks 2004), we find the momentum equation splits into two equations, one parallel and the other perpendicular to $\mathbf{B}$,

$$mn \frac{d\mathbf{U}_\parallel}{dt} = -\nabla_\parallel p_\parallel - (p_\parallel - p_\perp)(\nabla \cdot \mathbf{b})\mathbf{b}$$

$$mn \frac{d\mathbf{U}_\perp}{dt} = -\nabla_\perp p_\perp - (p_\parallel - p_\perp)(\mathbf{b} \cdot \nabla)\mathbf{b} + \mathbf{J} \times \mathbf{B} \tag{6.61}$$

An expression of the current density in the direction perpendicular to $\mathbf{B}$ is obtained by taking the cross-product of the second equation of (6.61) with $\mathbf{B}$. We find that

$$\mathbf{J}_\perp = \frac{\mathbf{B}}{B^2} \times \nabla_\perp p_\perp + \frac{(p_\parallel - p_\perp)}{B^2} [\mathbf{B} \times (\mathbf{b} \cdot \nabla)\mathbf{b}] + mn \frac{\mathbf{B}}{B^2} \times \frac{d\mathbf{U}_\perp}{dt} \tag{6.62}$$

where $\mathbf{B} = B\mathbf{b}$. This equation replaces the expression of the current density for scalar pressure when the pressure is anisotropic. Note that if $p_\parallel = p_\perp$, the scalar equation is recovered. The successive terms in (6.62) come from the gyromotion (first term), curvature and gradient drifts (second term), and inertial effects (last term). This expression for the current density is quite general and must satisfy the Maxwell equation $\nabla \times \mathbf{B} = \mu_0 \mathbf{J}$.

6.7.3 Magnetic Field of Ring Current

The above results have been applied to study the problem of ring current in the Earth's magnetosphere. A magnetic storm is accompanied by a worldwide decrease of the horizontal component of the geomagnetic field (not shown). A magnetic storm usually starts with a sudden commencement and includes the initial phase, the main phase, and the recovery phase. The sudden commencement is very rapid and the increase of the field intensity is accomplished in a few minutes. The field intensity remains increased during the initial phase and this phase lasts typically from tens of minutes to hours. The main phase is accompanied by a decrease in the geomagnetic field and this phase lasts typically 6–48 h. The decrease in the magnetic field is greatest at the geomagnetic equator and less at higher latitudes. The recovery of the field to the normal undisturbed value can take several days.

The sudden commencement and the initial phases can be explained in terms of compression of the planetary magnetic field by increased SW pressure, and the model of Chapman-Ferraro is applicable. However, the main phase requires a global ring current above the equatorial ionosphere flowing in the westward direction to account for the behavior of the main phase (a westward current decreases the magnetic field inside the ring). The recovery phase corresponds to the dissipation of this ring current.

The ring current is now known to be formed by a large number of trapped energetic particles drifting in the magnetosphere. The presence of ring current particles was identified by satellite-borne particle detectors in 1967 (Frank 1967). The behavior of the magnetic field in the ring current can be studied quantitatively by evaluating the Biot-Savart's equation, which relates the current and magnetic field. Ignoring time dependence,

$$\mathbf{B}(\mathbf{r}) = \frac{\mu_0}{4\pi} \int \frac{\mathbf{J}(\mathbf{r}') \times (\mathbf{r} - \mathbf{r}')}{|\mathbf{r} - \mathbf{r}'|^3} d^3 r' \tag{6.63}$$

where the magnetic field is evaluated at $\mathbf{r}$ due to the presence of a current at $\mathbf{r}'$ (for example, $\mathbf{r}$ could be the position of a magnetometer on the surface of a planet and $\mathbf{r}'$ is in the magnetosphere). To evaluate this integral equation requires knowledge about $\mathbf{J}$, which, for a plasma, comes from magnetization and drift currents.

We solve a simpler problem (Parks 2004) and consider the magnetic field at the center of a planet rather than at the surface. The solution to Biot-Savart's equation for a circular loop at the center is $\mathbf{B} = (\mu_0 I / 2r)\hat{\mathbf{z}}$. The current I is computed from the first-order particle orbit theory. The guiding center drift velocity equation of a charged particle trapped in a magnetic field is

$$\mathbf{W}_D = \frac{m v_\parallel^2}{q B^4}[\mathbf{B} \times (\mathbf{B} \cdot \nabla)\mathbf{B}] + \frac{m v_\perp^2}{2q B^3} \mathbf{B} \times \nabla B \tag{6.64}$$

Now substitute $v_\parallel = v \cos \alpha$, $v_\perp = v \sin \alpha$ where α is the pitch-angle of the particle, and $(\mathbf{B} \cdot \nabla)\mathbf{B} = \nabla B^2/2 = B\nabla B$ since $\nabla \times \mathbf{B} = 0$, and rewrite the above equation as

$$\mathbf{W}_D = \frac{mv^2}{2qB^3}(1 + \cos^2 \alpha)\mathbf{B} \times \nabla B \qquad (6.65)$$

To compute the $\mathbf{B} \times \nabla B$ term of a dipole field on the equator, first note that $B_r = -2\mu_0 M \cos\theta/4\pi r^3$ vanishes and the magnetic field has only a θ-component. $B_\theta = B_s(R_p/r)^3$ where $B_s = \mu_0 M/4\pi R_p^3$ is the field at the surface of a dipole planet of radius R_p and M is the magnetic moment. For a dipole on the equator, one finds $\mathbf{B} \times \nabla B = -(3B^2/r)\ \hat{\boldsymbol{\phi}}$. Use this result in the drift equation (6.65) and, for simplicity, consider only particles with $\alpha = 90°$ (this applies only to a ring current confined to the equator). This yields $\mathbf{W}_D = -3mv^2/2qBr\ \hat{\boldsymbol{\phi}}$ for the drift velocity on a dipole equator. Here $\hat{\boldsymbol{\phi}}$ is positive in the eastward direction, and positive and negative particles drift in the westward and eastward directions, creating a ring current flowing in the westward direction.

The current density is $\mathbf{J} = nq\mathbf{W}_D = -3nmv^2/2Br\ \hat{\boldsymbol{\phi}} = -3\mathcal{E}/Br\ \hat{\boldsymbol{\phi}}$ where n is the number density and $\mathcal{E} = nmv^2/2$ is the kinetic energy density of the drifting particles. The relationship between the current density $\mathbf{J}$ and the total current I is I $d\mathbf{l} = \mathbf{J}\ dV$. Define $\mathcal{E}_T = \int \mathcal{E} dV$ as the total energy of the particles and, noting that $\int d\mathbf{l} = 2\pi r\ \hat{\boldsymbol{\phi}}$, we can write the total current as $I = -3\mathcal{E}_T/2\pi r^2 B$. Use this result and obtain $\mathbf{B}_D = -3\mu_0 \mathcal{E}_T/4\pi r^3 B\hat{\mathbf{z}}$ as the magnetic field perturbation at the origin due to the drifting particles. The $(-)$ sign here denotes that the perturbation field is in the opposite direction of the planetary magnetic field.

The total perturbation of the planetary field by the ring current must also include the diamagnetic contribution (due to the cyclotron motion). Let $\mathbf{W}_p$ represent the diamagnetic drift term of the guiding center, given by $\mathbf{W}_p = -m/qB^2 d\mathbf{W}_E/dt \times$ $\mathbf{B} = m/qB^2 \partial\mathbf{E}_\perp/\partial t$, where $\mathbf{W}_E = \mathbf{E} \times \mathbf{B}/B^2$. Assume $\mathbf{E} \cdot \mathbf{B} = 0$ and let $\mathbf{E}_\perp$ be the electric field component perpendicular to the magnetic field direction. The equation for the polarization current is then reduced to $\mathbf{J}_p = \Sigma nmq\mathbf{W}_p = \rho_m/B^2 \partial\mathbf{E}_\perp/\partial t$. Here n is the number density, the summation Σ includes both ions and electrons, and ρ_m is the mass density, $n^+ m^+ + n^- m^-$. The electron mass density is much smaller than the ion contribution, hence the mass density is essentially due to the ions.

The diamagnetic current can be calculated noting that $\mathbf{J}_\mu = \nabla \times \boldsymbol{\mu}$ where $\boldsymbol{\mu}$ is the total magnetic moment of the particles ($\boldsymbol{\mu}$ is used instead of $\mathbf{M}$ so as not to confuse the magnetic moment of the particles with the magnetic moment of a planet, which is designated as $\mathbf{M}$ here). This problem can be solved by recognizing that the diamagnetic contribution can be physically modeled by a ring of dipoles of radius r and total magnetic moment $\boldsymbol{\mu}$. Since $r \gg r_c$, cyclotron radius, the estimate of the magnetic field at the center of a dipole ring is given by $\mathbf{B}_{\text{diamag}} = -\mu_0/4\pi r^3 \boldsymbol{\mu} = \mu_0 \mathcal{E}/4\pi r^3 B\hat{\mathbf{z}}$ where $\boldsymbol{\mu} = -mnv^2\mathbf{B}/2B^2 = -\mathcal{E}\mathbf{B}/B^2 = -(\mathcal{E}/B)\hat{\mathbf{z}}$

is the total magnetic moment of the ring current particles on the equator. Note that the individual dipoles are aligned along the planetary magnetic field direction. In contrast to the drift current, whose magnetic field is in the opposite direction of the planetary field the magnetic field of a ring of dipoles adds to the main planetary magnetic field.

The perturbation of the planetary magnetic field at the origin due to the ring current is obtained by adding the two contributions $\Delta \mathbf{B}_T = \mathbf{B}_D + \mathbf{B}_{\text{diamag}} = -2\mu_0 E/4\pi r^3 B\hat{\mathbf{z}} = -2EM/\hat{\mathbf{z}}$ where use was made of $B = \mu_0 M/4\pi r^3$. This simple result shows that the total magnetic perturbation at the center of a planet is related to the magnetic moment M of the planet and the total particle energy E of the ring current particles.

We now express the perturbation field in terms of the surface field B_s of a planet,

$$\frac{\Delta \mathbf{B}_T}{B_s} = -\frac{2\mathcal{E}_T/M}{\mu_0 M/4\pi R_p^3}\hat{\mathbf{z}} = \frac{-2\mathcal{E}_T \mu_0}{4\pi B_s^2 R_p^3}\hat{\mathbf{z}} \tag{6.66}$$

where $M = 4\pi B_s R_p^3/\mu_0$. This equation can be rewritten noting the total magnetic field energy of a dipole above the planetary surface is $E_M = \int B^2/2\mu_0 dV = [4\pi/\mu_0][B_s^2 R_p^3/3]$ where the integration limits are R_p to ∞. Use of this result in the above equation yields

$$\frac{\Delta \mathbf{B}_T}{B_s} = -\frac{2}{3}\frac{\mathcal{E}_T}{E_M}\hat{\mathbf{z}} \tag{6.67}$$

as the total decrease of the magnetic field at the center of a planet due to a ring current.

Figure 6.17 shows an example of the behavior of magnetic field during a magnetic storm that occurred on Earth. The magnetic field perturbation was compared to the particle energy of the ring current. The peak of the ring current was located between 3 and $4R_E$. Observation and theory were made to fit fairly well after three iterations.

Panaysuk et al. (2004) studied one of the largest magnetic storms using spacecraft and ground-based observations. It occurred in October 2003. The spacecraft data included electrons, protons, and ions over a wide energy range. The analysis included furthering understanding of the deformation of magnetosphere structure, boundaries of penetration of solar cosmic rays, boundaries of the auroral zone and polar cap, dynamics of the radiation belt, and the influence of substorm on the evolution of the current systems of a magnetic storm. Not all of the features are understood and we challenge interested researchers to model this event.

Question 6.7 Equation (6.67) was derived with many restrictions and assumptions. Discuss these restrictions and assumptions and how they affect the interpretation of the results.

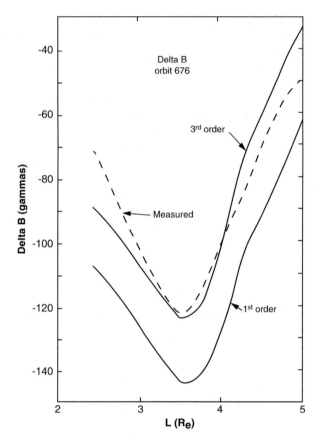

Fig. 6.17 Comparison of theory and observations for a ring current. The vertical axis δB and the horizontal axis L in R_E (from Berko et al., *J. Geophys. Res.*, **80**, 3549, 1975)

6.7.4   Ring Current and Field-Aligned Current

An equation of field-aligned current $\mathbf{J}_\parallel$ cannot be deduced from the macroscopic momentum equation as we have done for the perpendicular ring current because $\mathbf{J}_\parallel$ does not appear in the momentum equation, $d\langle\mathbf{v}\rangle/dt = \mathbf{J} \times \mathbf{B} - \nabla \cdot \pi$ (Chap. 1). Instead, we will examine what can be learned about $\mathbf{J}_\parallel$ from one of Maxwell's equations $\nabla \times \mathbf{H} = \mathbf{J} + \partial\mathbf{D}/\partial t$. If we take the divergence of this equation, we obtain the continuity equation, $\nabla \cdot \mathbf{J} + \partial\rho_c/\partial t = 0$ where we have used the relation $\nabla \cdot \mathbf{D} = \rho_c$. Since it is difficult to maintain free charges in plasmas, we let $\rho_c = 0$. Thus, the current density $\nabla \cdot \mathbf{J} = 0$ in steady state problems. Now let $\mathbf{J} = \mathbf{J}_\parallel + \mathbf{J}_\perp$. Then the continuity equation can be written as $\nabla \cdot (\mathbf{J}_\parallel + \mathbf{J}_\perp) = 0$ and the divergence of $\mathbf{J}_\parallel$ can be related to the divergence of $\mathbf{J}_\perp$,

$$- \nabla \cdot \mathbf{J}_\perp = \nabla \cdot \mathbf{J}_\parallel$$

$$= \nabla \cdot \mathbf{J}_\parallel \left(\frac{\mathbf{B}}{B} \right) = B \frac{\partial}{\partial s} \left(\frac{J_\parallel}{B} \right) \tag{6.68}$$

where $\partial/\partial s = (\mathbf{B}/B) \cdot \nabla$ is the gradient operator along the direction of $\mathbf{B}$. Given $\mathbf{J}_\perp$, $\mathbf{J}_\parallel$ can be obtained by integrating $\nabla \cdot \mathbf{J}_\perp$ along $\mathbf{B}$.

Here $\mathbf{J}_\perp$ is generated by the drift motion of trapped particles in the magnetosphere (ring current and tail current). Assume now a part of this current flows along $\mathbf{B}$ and closes in the ionosphere (how this current closes in the ionosphere will depend on the ionospheric conductivities, whose details need not concern us here). The perpendicular current $\mathbf{J}_\perp$ obtained from the macroscopic momentum equation is

$$\mathbf{J}_\perp = \left(\frac{\mathbf{B}}{B^2} \right) \times \left(\rho_m \frac{d\langle \mathbf{v} \rangle}{dt} + \nabla \cdot \boldsymbol{\pi} \right) \tag{6.69}$$

where $\langle \mathbf{v} \rangle$ is the macroscopic velocity averaged over the distribution function, $\langle \mathbf{v} \rangle = \int \mathbf{v} f(\mathbf{v}) d^3 v$, and $\boldsymbol{\pi}$ is the momentum transfer tensor. We can use the continuity equation for the current and obtain an equation for $J_\parallel$. The expression for the parallel current is

$$J_\parallel = \int 2\mathbf{J}_\perp \cdot \frac{\nabla B}{B^2} + \left(\frac{\mathbf{B}}{B^3} \right) \cdot \nabla \times \mathbf{F} \, ds \tag{6.70}$$

and we have asked the readers to derive this equation (Question 6.7).

Question 6.8 Take the divergence of Eq. (6.69) and use it with Eq. (6.68) to obtain the equation of the perpendicular current $\mathbf{J}_\perp$ in the guiding center approximation. Then show that the parallel current is given by Eq. (6.70).

6.7.5 Ring Current and Waves

Clues about how ionospheric ions are accelerated and transported out into the magnetosphere come from correlated observations of electric field fluctuations and upward flowing ions. Low frequency (frequency $\omega \leq \omega_{ci}$, ion cyclotron frequency) electric field fluctuations with amplitudes ≥ 0.1 V/m have been observed to accompany upward flowing ions. These waves include both ion cyclotron and Alfvén waves (Chaston et al. 2017). See also the statistical results of the low frequency waves by Nykyri and Dimmock (2015). Large-amplitude electromagnetic waves can give rise to significant ponderomotive force that can accelerate and transport ions out of the ionosphere (Allen 1992; Li and Temerin 1993; Guglielmi et al. 1993; Lee and Parks 1996; Lundin and Guglielmi 2006; Le et al. 2017; Nykyri and Dimmock 2015). These theories show that the ponderomotive force at low frequency may accelerate heavier ions more efficiently.

6.7.6 Ring Current Instability

The distribution function of the ring current particles is not isotropic in velocity space and moreover the ring current in real space can have sharp boundaries with density and temperature gradients. The ring current will dissipate giving rise to instabilities. An example is the stable auroral red (SAR) arcs that are diffuse and observed in the wavelength from $\sim$630.0 to 636.4 nm. This particular auroral form is observed in the mid-latitude regions of Earth following magnetic storms. Measurements of the frequency of occurrence lead to the conclusion that the phenomenon is due to the excitation of atomic oxygen by hot electrons in the neighborhood of the plasmapause region by precipitating electrons from the ring current. However, it has not been possible to determine whether it is the fresh precipitating or heated ambient local electrons that are responsible for the red emissions.

There are several suggestions about the mechanisms for producing the SAR arcs by the ring current. They include heat flow by transferring the kinetic energy of the ring current particles to the SAR arc region by Coulomb collisions, transfer of ring current proton energy to hydromagnetic waves, which are then damped by the electrons in the SAR arc region, and direct influx of energetic electrons into the SAR arc region. Which of these mechanisms is responsible is still not known at this time (Cornwall 1971; Rees and Roble 1975; Kozyra 1997).

The ring current is thought to form by a combination of the injected particles during substorms and subsequent radial diffusion (Schulz and Lanzerotti 1974). Some authors have suggested that ion acoustic waves generated by Kelvin-Helmhotz instability could be active in ring current dynamics (Moore et al. 2016).

6.8 Magnetosphere-Ionosphere Coupling

We showed earlier that electrons moving downward produce the aurora in the upward flowing current regions. There is also a current flowing downward where the electrons are moving upward. Although much is known about the currents in the ionosphere, not much is known about the behavior of currents in the magnetosphere. The four Cluster in tetrahedron configuration and Themis satellites (a five spacecraft mission with identical instrumentation spaced along a line in the X_{GSE}-direction) in the geomagnetic tail have now reported measurements of field-aligned electrons very likely to originate from the downward flowing current region (Zheng et al. 2012; Sun et al. 2013). These observations are *cutting edge* results because they are not only the first of such measurements in the distant plasma sheet but they can also shed light on the MI coupling of the magnetosphere and ionosphere. MI coupling will affect the dynamics of the ring current and thus can affect space weather. A 3D model for propagation of Alfvén waves through the ionosphere has been suggested by Lysak (1998). This model should probably be expanded to include the more recent results.

6.8.1 Field-Aligned Electrons in the Plasma Sheet

An example of the general behavior of magnetic field, ion, and electron fluxes from Earth's plasma sheet is given in Fig. 6.18. The data come from Cluster which was in the geomagnetic tail at a distance of $\sim 15\,R_E$.

Question 6.9 In Fig. 6.18, discuss the activities beginning around 1133 UT paying attention to the behavior of magnetic field, bulk velocities, and pitch-angle.

The differential energy spectra of electrons from $\sim 50\,eV$ to $26\,keV$ and pitch-angles ($0°$, $90°$, and $180°$) for a few selected energies are shown in Fig. 6.19. The energy spectra are for electrons with pitch-angles of $0°$, $90°$, and $180°$ (left panel). Except for the diminished intensity of the 1 keV peak at $0°$ as compared to $180°$, the form of these energy spectra is essentially the same. The spectrum of $90°$ pitch angle electrons (black crosses) remained essentially unchanged compared to the spectra at earlier times before the beam was detected (not shown). The differential flux of the beam at the peak is $\sim 8 \times 10^7\,cm^{-2}\,s^{-1}\,sr^{-1}\,keV^{-1}$ here and is comparable to the upward going fluxes observed in the auroral ionosphere (Carlson et al. 1998).

The behavior of the pitch-angle distributions for few selected energies (bottom right) shows "wings" at $0°$ and $180°$ for electrons up to $\sim 2\,keV$ indicating the presence of field-aligned electrons. To characterize the beam, the mean fluxes were subtracted and the beam was fit with a Maxwellian (top right). The estimated density of this beam is $n_b \sim 0.05 \pm 02\,cm^{-3}$ and using 0.6–$1\,cm^{-3}$ for the ambient density n_e (estimated from the plasma frequency, not shown), $n_b/n_e \sim 0.08 \pm 0.05$.

The temperature (kT) of the beam is $\sim 135 \pm 22\,eV$ and similar to temperatures of the beam observed in the auroral zone. The thermal velocity of the beam v_b is $\sim 6.89 \times 10^3\,km\,s^{-1}$. Since the drift speed v_D of 1 keV electrons is $\sim 1.89 \times 10^4\,km\,s^{-1}$, the ratio v_D/v_b is ~ 2.7. An estimate of field-aligned current carried by these electrons obtained by subtracting the electrons traveling along and against the direction of the magnetic field yielded $3 \times 10^{-8}\,A\,m^{-2}$. This current was most intense when the beam developed at $\sim 1134{:}47$ UT and it was measured on all 4 SC. The intensities were nearly the same on the four SC, but the features were different. For example, on Cluster 2, it was bipolar and on Cluster 4, there were two bursts (not shown). Note that the field-aligned current in the plasma sheet can be asymmetric (Shi et al. 2010).

6.8.2 Field-Aligned Currents in Unstable Current Sheet

We now give an example of FAC measured in the geomagnetic tail using the *curlometer* technique applied to the magnetic field measurements on the four Cluster spacecraft (Sun et al. 2013). The curlometer technique gives the total current $\mathbf{J} = \nabla \times \mathbf{H}$ and the relation $\mathbf{b} \cdot \nabla \times \mathbf{H}$ has been used to obtain the current along $\mathbf{B}$. Here $\mathbf{b}$ is the unit vector along the magnetic field direction.

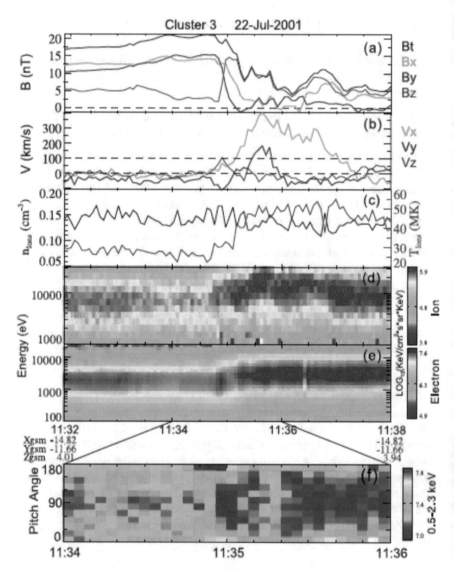

Fig. 6.18 Summary plot of 22 July 2001 from Cluster 3. Data are shown in the geocentric solar magnetospheric (GSM) coordinate system. Panel (**a**) shows components and the total intensity of the magnetic field, where XGSM is positive toward the Sun, ZGSM axis is the projection of the Earth's magnetic dipole axis on to the plane perpendicular to the X axis (positive North), and YGSM completes a right-handed coordinate system (YGSM is positive toward dusk). (**b**) components of the average plasma velocities from the velocity moments, (**c**) local density (black) and temperature (blue) deduced from the velocity moments, (**d**) and (**e**) show energy flux spectrograms of ions and electrons, and (**f**) pitch-angle spectrogram of electrons with energies 0.5–2.3 keV. The color bars on the right give the intensity (from Zheng et al., *Phys. Rev. Lett.*, **109**, 205001, 2012)

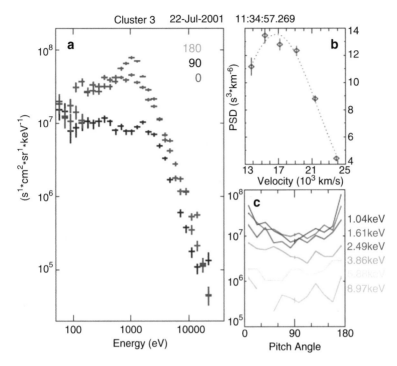

Fig. 6.19   (**a**) Differential energy spectra for electrons with pitch-angles of 0°, 90°, and 180° at 1134:57, 22 July 2001. (**b**) Phase space density for the "beam" with a Maxwellian fit. (**c**) Pitch-angle distribution for selected energy channels (from Zheng et al., *Phys. Rev. Lett.*, **109**, 205001, 2012)

The data in Fig. 6.20 come during a moderately disturbed time in the geomagnetic tail when the currents were enhanced. In particular, the time when the FACs were observed coincided with the onset of dipolarization. The term dipolarization front (DF) is used to describe the increase of the B_z component as a spatial structure. The mechanisms of dipolarization or how the DFs form are not known but they are tied to the dynamics of high-speed mean plasma flows and substorms. Some simulation models suggest DFs are spatial structures that propagate earthward with the plasma. During dipolarization disturbances, plasma current sheet can flap around (Sergeev et al. 2003).

Question 6.10   Study Fig. 6.20 and discuss the important features.

A summary of the field-aligned current system associated with the DFs in the geomagnetic tail is schematically shown in Fig. 6.21. The red arrows indicate FACs corresponding to region-1 sense and blue for region-2 sense. Black arrows show the magnetic field direction, and green arrows show directions of the DF normal. Multi-point Cluster observations of the DFs often observed together with earthward traveling high-speed flowing plasmas in the Earth's geomagnetic tail have shown

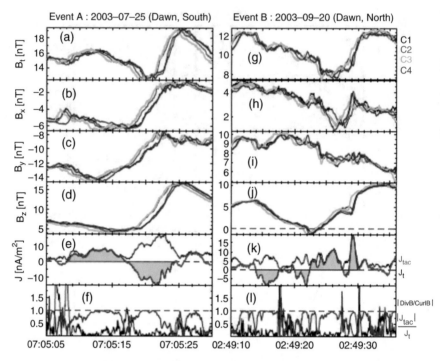

Fig. 6.20 Each side (left and right panels) shows 25s of data covering the time interval when the dipolarization fronts were encountered. Field-aligned current was measured in both cases using the curlometer technique. The quantity $|div B/curl B|$ in panels f and l (black) is small indicating the results are reliable (from Sun et al., *Geophys. Res. Lett.*, **40**, 4503, 2013)

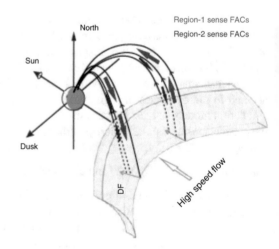

Fig. 6.21 A diagram illustrating the FAC system accompanying the DF structure (only Northern Hemisphere). Black arrows show the magnetic field direction, green arrows directions of the DF normal, red and blue arrows FACs correspond to region-1 and region-2 sense, respectively (from Sun et al., *Geophys. Res. Lett.*, **40**, 4503, 2013)

that FACs associated with DF are highly structured. Observations show the FAC in front of the DF are flowing in opposite directions than the FACs inside the DFs with FACs flowing downward into the ionosphere and then upward from the ionosphere. The DF events observed on the duskside and the dawnside show that FACs in the DF layer are downward on the dawnside and upward on the duskside, similar to the region-1-sense current and region-2-sense current of Iijima and Poterema system. Typical values of region-1-sense FACs flowing inside the DFs are $\sim$10–20 nA m^{-2} and region-2-sense FACs just in front of DF (in Bz dips) have values $\sim$5–10 nA m^{-2}.

6.8.3   A Simple Suggestion of MI Coupling

We now consider a simple example of plasma motion in the plasma sheet. A plasma blob is moving across a magnetic field with a velocity $\mathbf{V}$ (Fig. 6.22). There is no external applied electric field in laboratory S-frame but there is a steady uniform magnetic fled in the $+Z$-direction. The motion of the blob across $\mathbf{B}$ induces a motional $\mathcal{E}MF$. Electrons and ions in the S-frame are moving with a velocity $\mathbf{V}$ and the particles experience a $q(\mathbf{V} \times \mathbf{B})$ force. An observer in the S-frame thus sees positive and negative charges move to the opposite sides of the boundary of the blob.

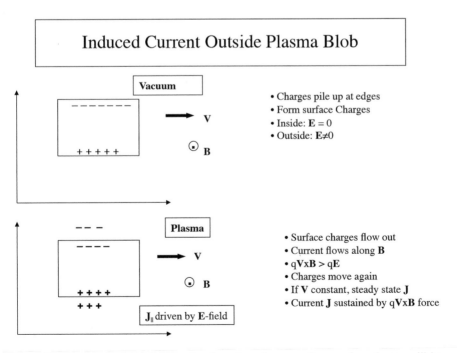

Fig. 6.22 When a plasma blob is moving through a plasma medium, surface charges will be conducted into the surrounding plasma and currents will flow. Here X-axis is horizontal, Y vertical, and Z comes out of the paper

These charges will move until the electric field due to the separation of charges is equal and opposite to the magnetic force $q(\mathbf{V} \times \mathbf{B})$. These surface charges will also produce electric fields *external* to the plasma blob.

Now we examine what will happen if the blob is surrounded by another plasma population (Fig. 6.22) in contact with the blob. Then surface charges from the blob will leak out and a *conduction current* will flow in the surrounding plasma. This induced current is driven by the electric field produced by the charges at the ends. The charge removal from the ends will *reduce* the number of the electric charges at the ends of the surface of the moving blob. Then the magnetic force inside the slab will exceed the electric force and the ions and electrons will start to move again to replenish the free charges that have been removed from the ends of the blob. The current flow is sustained by the magnetic force $q\mathbf{V} \times \mathbf{B}$ on the charges inside the blob. For a constant velocity blob, the current flow will become steady and a state of dynamic equilibrium will be reached. A fundamentally important point is that the presence of the moving plasma blob is essential for the current flow outside the blob.

This type of moving plasma blob is often seen in Earth's plasma sheet and very likely in the solar corona and elsewhere. The currents that flow outside the blob can subsequently flow along the magnetic field direction (because mobility is high along B) and produce field-aligned current. If the magnetic field is connected to another region, for Earth, this could be the ionosphere and for Sun, the chromosphere or lower, field-aligned currents will flow into these regions and vice versa. For Earth, this is how the magnetosphere and ionosphere are likely coupled. For the Sun, the concept of corona-chromosphere coupling has not been discussed but if currents are flowing in the solar regions, the topic of field-aligned currents should also be important on the Sun.

Question 6.11 A plasma blob is moving parallel to the X-axis observed in the S-frame and $\mathbf{B}$ is pointing in the positive Z-direction. Discuss what an observer sees in the S'-frame of reference.

6.9 Auroral Kilometric Radiation

A variety of electromagnetic waves are created by unstable plasma distributions. For example, the Earth produces very intense waves associated with the aurora on the nightside. These waves cover the frequency range about 100–600 kHz and correspond to wavelengths in the kilometric range, and thus have been called *auroral kilometric radiation* or AKR. For any event, the total radiated power is estimated to be about 10^7–10^9 W. This radiation represents the most intense radio power in this frequency range in our solar system. The AKR power is comparable to the power emitted by the decametric radiation from Jupiter. Satellite observations of AKR have established that the height of the AKR source lies between ≈ 1.3–$3.5 R_E$ and is located in geomagnetic latitudes of $\approx 70° \pm 3°$.

The AKR frequency is in the range of ordinary and extraordinary modes and the emission occurs close to the electron cyclotron frequency. Melrose (1976) originally proposed that the AKR radiation is generated by the Doppler-shifted cyclotron resonance mechanism. There have been other theories proposed for the AKR waves involving O-mode and X-mode waves. These same theories have been applied to waves observed in stellar and astrophysical objects. The details of these theories are different, but they all emphasize the induced radiation is intimately associated with precipitating electrons that produce the aurora. The theory that is most successful for the generation of AKR is the theory developed by Wu and Lee in 1965 which is now discussed. See also a theoretical review paper on emission mechanisms by Melrose (2017). Observations relating the behavior of auroras (Ieda et al. 2001) and AKR radiation have been studied by Morioka et al. (2008) and the behavior of waves associated with AKR emissions during auroras by Hanasaz (2001).

The theory is formulated using the relativistic Doppler shifted frequency resonance condition of O- and X-mode waves,

$$\omega - (s\,\omega_c^-)/\gamma - k_\parallel\,v_\parallel = 0 \tag{6.71}$$

where s is the harmonic number and is a positive integer. Positive s gives the normal Doppler shift and negative number gives the anomalous shift. The case $s = 0$ corresponds to the Landau or the Cerenkov resonance. What the relativistic factor does is to produce a resonance ellipse instead of a resonance line produced by the nonrelativistic cyclotron resonance theory. The growth-rate involves integration along the resonance ellipse in the region of the loss cone, and electrons in gyroresonance with the O- and X-mode waves can amplify these waves.

Observations have indicated that the AKR comes from a region depleted of the local ionospheric plasma density. A depleted region can have densities as small as $\approx 1\,\text{cm}^{-3}$ and thus the depletion region is called an auroral plasma cavity. Recent observations by the instrument on FAST show that the auroral cavity is observed on auroral lines of force where the electrons have a field aligned component precipitating into the atmosphere and upgoing collimated ion beams. The current interpretation is that the cavity is produced by the auroral field-aligned potential that sweeps out the low energy ionospheric electrons.

The excitation of AKR waves comes from the combination of precipitated electrons and the atmospheric loss cone which will be empty after a quarter bounce. That is, there is free energy in the velocity space, $\partial f/\partial v_\perp > 0$ after a quarter bounce assuming the electrons originate from the equator. The electrons outside the loss cone that are reflected by the more intense magnetic field (mirror effect) in the ionosphere now travel in the upward direction and will interact with the excited waves and amplify them. It is important in this induced *maser* action that the relativistic resonance condition be used, for without the relativistic correction, the instability is not excited.

Let f_- and f_+ be distribution functions of downgoing (precipitated) and upgoing electrons (not to be confused with electron and ion distributions that will be denoted as f_e and f_i). These distributions are truncated at low energies since the source region is depleted of low energy thermal electrons. Only high energy electrons are considered (≥ 1 keV). Note that f_+ includes an empty loss cone. The total electron distribution function is thus given by

$$f_e = n_+ f_+ + n_- f_- \qquad (6.72)$$

where $n_\pm$ are the densities and the distribution function $f_\pm$ here is normalized to unity. Note that in this model, we also require that $n_+ < n_-$ and that the fluxes

$$\left| n_+ \int f_+ v_\parallel d^3 v \right| < \left| n_- \int f_- v_\parallel d^3 v \right| \qquad (6.73)$$

indicating the net flux is downward. These fluxes can give rise to field-aligned currents but it will be assumed that they are extremely small and negligible. This allows the X-mode and O-mode to be decoupled with $k_\parallel^2 \ll k^2$. The theory thus assumes the wave propagation is in the perpendicular direction. Here $\omega^2 \approx k^2 c^2 \gg \omega_p^2$ where ω_p is the plasma frequency of electrons, $\omega_p^2 = n_e^2 / m_e \epsilon_o$ and $n_e = n_- + n_+$. With these assumptions, the dispersion relation of the ordinary mode derived from the full kinetic equations is

$$1 - \frac{c^2 k^2}{\omega^2} + \frac{\omega_p^2}{\omega^2 n_e^2} \int d^3 v \left(\omega_c^- \frac{\partial f_e}{\partial v_\perp} + k_\parallel v_\perp \frac{\partial f_e}{\partial v_\parallel} \right) \frac{v_\parallel^2 J_1^2(b)}{v_\perp (\omega - \omega_c^- / \gamma - k_\parallel v_\parallel)} = 0 \qquad (6.74)$$

and of the extraordinary mode is

$$1 - \frac{c^2 k^2}{\omega^2} + \frac{\omega_p^2}{\omega^2 n_e^2} \int d^3 v \left(\omega_c^- \frac{\partial f_e}{\partial v_\perp} + k_\parallel v_\perp \frac{\partial f_e}{\partial v_\parallel} \right) \frac{v_\parallel^2 J_1'^2(b)}{(\omega - \omega_c^- / \gamma - k_\parallel v_\parallel)} = 0 \qquad (6.75)$$

Here $J_1(b)$ is the Bessel function of order 1, $b = k_\perp v_\perp / \omega_c^-$ and $J_1'(b) = dJ_1/db$. It should be noted further that the relativistic effect is important only in the denominator, and elsewhere, $\gamma = 1$. The relativistic effect provides a mechanism which couples the perpendicular electron motion $v_\perp^2$ with the waves and gives the free energy available in the loss cone to the waves.

The growth rate is obtained in the usual way, letting $\omega = \omega_r + i \omega_i$ and solving for the imaginary part ω_i. If one assumes that $\omega_r \gg \omega_i$, the growth rate for the ordinary mode is

$$\omega_i = \frac{\omega_c^2 \pi^2 n_+}{4 \omega_r n_e} \int dv_\parallel \int dv_\perp \frac{v_\perp^2}{c^2} v_\parallel^2 \, \delta \left(\omega_r - \omega_c^- \left[1 - v^2 / 2c^2 \right] - k_\parallel v_\parallel \right)$$

$$\times \left(\omega_c^- \frac{\partial f_+}{\partial v_\perp} + k_\parallel v_\perp \frac{\partial f_+}{\partial v_\perp} \right) \qquad (6.76)$$

The growth rate for the extraordinary mode is

$$
\omega_i = \frac{\omega_c^2 \pi^2 n_+}{4\omega_r n_e} \int dv_\parallel \int dv_\perp v_\perp^2 \, \delta \left(\omega_r - \omega_c^- \left[1 - v^2/2c^2 \right] - k_\parallel v_\parallel \right)
$$

$$
\times \left(\omega_c^- \frac{\partial f_+}{\partial v_\perp} + k_\parallel v_\perp \frac{\partial f_+}{\partial v_\perp} \right) \tag{6.77}
$$

These equations have used the approximation $\sqrt{1 - v^2/c^2} \approx 1 + v^2/2c^2$ since the precipitated electrons have typical energies of only a few keV. Waves will be amplified if $\omega_i > 0$.

To solve these equations, several important features should be noted. First to note is that for the ordinary mode waves, $\omega_r \approx ck_\perp > \omega_c^-$. For the extraordinary mode waves, $\omega_r \approx kc > \omega_c^- (1 + \omega_p^2/\omega_c^2)$. Another point is that we are considering only waves with $\omega_r > \omega_c^-$. Hence, electrons in resonance, $\omega_r - \omega_c^- (1 - v^2/2c^2 - k_\parallel v_\parallel)$ can only occur for electrons with $k_\parallel v_\parallel > 0$. This means that for waves propagating upward ($k_\parallel > 0$), only the upgoing electrons ($v_\parallel > 0$) can resonate with the waves. This is the reason why only f_+ appears in the above equations. The waves that escape the source region without refraction have wave vectors pointing away from the ionosphere. The upgoing electrons play a decisive role in the amplification of these waves.

The γ for a few keV electrons is only about 1.01. It is quite surprising that this small correction can be so important. However, the relativistic effect provides a mechanism that couples the perpendicular motion $v_\perp$ with the X-mode waves that is the essence of the Wu-Lee theory. In this regard, the observations from VIKING and FAST are important. Louarn et al. (1990) and Delory et al. (1998) show that the electron distributions associated with the AKR radiation have positive gradients with respect to $v_\perp$ superimposed on a broad plateau covering a wide pitch-angle range. Thus, the source of free energy in the phase space for AKR generation is much larger than just the loss cone. This additional free energy can lead to much larger intensities of the electric field in the AKR radiation.

6.10 Concluding Remarks

Auroras have been seen by human beings since the dawn of civilization. However, the global images of the aurora were only imaged with modern technology. The imaging cameras used filters in the ultraviolet wavelengths and thus observed auroral emissions even on the dayside. An important aspect of global images is that they enable us to see what is going on in the more distant magnetospheric and plasma sheet regions because the auroral zone is connected to the distant regions by the geomagnetic field. The spatial and temporal variations in the aurora are reflecting the behavior of the dynamics that are active in these distant regions. The global images provide an instantaneous view of the auroral oval that is useful in dynamical studies of how the solar wind interacts with the magnetosphere.

Comparisons of features in auroral images with measurements made *in situ* by plasma detectors in the plasma sheet have enabled us to answer the long-standing question about the origin of aurora forms. Ground-based observations have raised the question whether the small-scale auroral brightening, called pseudo-breakup auroras, are different from the larger-scale *normal* auroral breakup events. This question has been studied by Fillingim et al. (2000) who showed that the electrodynamic activity associated with these two types of auroras is identical in the plasma sheet. This conclusion has since been verified by Lee et al. (2015).

The global images show complex spatial features including arcs across the polar cap. Sometimes there is a gap in the local midnight sector, which is thought to be Harrang discontinuity (Chua et al. 1998) The global images also indicate that the onset activity of a *normal* auroral breakup begins on the equatorward side of the auroral oval, confirming the ground-based observations made nearly 50 years ago (Akasofu 1968). This observation has important implications about the dynamics in the plasma sheet responsible for the substorm onset. Geomagnetic models map the auroral onset spot to the equatorial plane inside 6–10 earth radii (R_E).

New interesting physics has been revealed about the auroras by viewing the northern and southern hemisphere auroras at the same time. The simultaneous images show that in one case the onset of the breakup in the northern hemisphere occurred 1 min later than the onset in the southern hemisphere. The conjugate images can also be displaced from each other by as many as a few hours in longitude and the intensities are different. There are also seasonal effects that may arise because the electrical conductivities in the sunlit hemisphere are higher than in the dark hemisphere. Ionospheric conductivity affects current flow and consequently the MI-coupling.

The Viking and FAST observations at ionospheric altitudes have given us a better understanding of the electrodynamics of the aurora, although the measurements were localized. These observations have given us a clear picture about the role of the electric fields and currents associated with the aurora. New features included the discovery of electron solitary waves and that the electron distributions that drive the AKR radiation have a broad free energy source in the phase space.

Parallel electric fields are generally very small (a few tenths of milli-Volts m^{-1}), and their existence was deduced initially from observations of beams. This is unlike perpendicular electric fields that can reach 1 V/m in the neighborhood of the plasmapause (Kim et al. 2010). Most of our knowledge about parallel electric field has come from studying Earth's auroral beams. However, beams are ubiquitous in space plasmas. Auroral-like beams have been observed on the Moon, Venus, Mars, Jupiter and Saturn. Although the sources of these beams have not been identified, they are most likely produced by $\mathbf{E}_{\parallel}$ along the magnetic field direction. Whether these planets have large-scale current systems produced by the beams as on Earth is not known at the present time (See Sun et al. 2015).

Theories predict there are six different ways parallel electric field can be produced (Goertz 1979). The current view is that double layers are one of the most likely sources for the electric field. However, how the microscopic structures are related to the larger structure remains unknown. Other ways parallel electric fields are produced include sudden surges of ion convection in the plasma sheet (Mauk

1989). For general discussion of how $\mathbf{E}_\parallel$ affects particles, see Kaufmann (1976) and Whipple (1977).

The standard picture is that the ionospheric currents are part of the magneto-spheric current system. The tail current system is set up in the geomagnetic tail by the interaction of the solar wind with the magnetosphere (by mechanisms as yet not known) and these distant currents close along the magnetic field in the ionosphere. The upward ionospheric currents can be looked at as the return current. This picture is pieced together from different observations in the tail and ionosphere of Earth. Unfortunately, large-scale models are difficult to prove or verify by observations. However, global simulation models could help us see how the large-scale currents are set up.

Most of our knowledge of Earth's magnetic field has come from data obtained by magnetometers distributed over the surface of Earth. However, measurements from the surface alone are not adequate because external currents (external to the core) are important. Currents run laterally and vertically in the ionosphere, and this external contribution must be correctly accounted for. Researchers are now active in creating realistic models that can accurately predict the shape of the magnetosphere (including the tail) for various solar wind conditions. One of the main challenges is to construct a model that will correctly predict where the ionospheric field lines map to in the distant magnetosphere.

The final comment concerns the SW induced electric field in magnetospheres. Although it is assumed that this electric field originates in the solar wind, nothing was said about how this electric field appears inside magnetospheres. This is because we do not as yet fully understand the mechanisms by which the solar wind and the planetary boundaries interact. The role of electric fields induced by the rotation of the planet or star is important for this problem. In this regard, it is worthy of note that while only a portion of Earth's magnetosphere corotates, the entire magnetosphere of Jupiter may be corotating. On the other hand, none of Mercury's magnetosphere is believed to corotate because Mercury has no ionosphere. These problems are intimately related to how the solar wind mass, momentum, and energy are transported across the magnetopause boundary. Another way to study this problem is to ask how boundary currents are established by the solar wind interacting with planetary magnetic fields.

6.11 Solutions to Questions in Chap. 6

Solution 6.1 This plot shows differential number flux against energy in eV. This energy spectrum has a peak similar to the spectrum obtained by the rocket experiment shown in Fig. 6.3. However, unlike the rocket spectrum, the experiment also detected electrons below and above the peak energy. The peak energy can vary from a few hundred eV to $\sim$10 keV. The source of the lower energy electrons has been suggested to come from secondary electrons produced by the keV precipitating electrons interacting with the atmosphere. The higher energies are from the plasma sheet. The "mono-energetic" electrons at $\sim$1 keV are interpreted as electrons that have gained energy going through a field-aligned potential drop. When thermal electrons pass through the potential, they are accelerated and *beams* are formed. For the example shown, the peak corresponds to a streaming energy of the beam which is $E_o \sim 1000\,\text{eV}$ for this case. The thermal energy estimated by fitting the broad peak with a Maxwellian distribution function with a density $0.06\,\text{cm}^{-3}\text{-ster}^{-1}$ yields a typical energy of $\sim 100\,\text{eV}$. The low energy portion $E \leq 300\,\text{eV}$ estimated using a power law form, $dJ/dE \propto E^{-\gamma}$, shows γ for this example is ~ 2.2. Analysis of many examples shows the exponent ranges from 1.5 to 2.5. Simulation results (solid line, Fig. 6.5) show good agreement supporting the secondary electron interpretation. The high energy *nonthermal electrons* ($>8\,\text{keV}$) have been interpreted as plasma sheet electrons.

Solution 6.2 The measurements were made at a height of $\sim 5000\,\text{km}$. Comparison of the pitch-angles with both O^+ and H^+ shows the ions are sharply peaked along $\sim 0°$, indicating ions are streaming up along the geomagnetic field from below the spacecraft. The estimate of the absolute flux intensity corresponding to the peak H^+ is $\sim 10^8$ (cm²/s-ster-keV) at $1.28\,\text{keV}$. The deep minima in electrons correspond to the atmospheric loss cone at the satellite location. The ion peaks are clearly shown at the same locations as the loss cone for electrons. Note that when the ions are observed streaming outward, the loss cone is deeper. This is due to the presence of $E_\parallel$ as can be shown by a simple particle theory conserving the first particle invariant. During this particular pass, N^+ or He^+ was not detected but these ions have been observed streaming out of the ionosphere by other mass spectrometer experiments flown subsequently (not shown).

Solution 6.3 We begin with

$$z = \frac{\mathcal{E}_o}{qE_z} \cosh \frac{E_z}{B}\xi$$

and expand the cosh term, obtaining

$$z \approx \frac{\mathcal{E}_o}{qE_z}\left[1 + \frac{1}{2}\left(\frac{E_z}{B}\xi\right)^2\right]$$

Then, similarly, expand (6.16) and obtain

$$t \approx \frac{\mathcal{E}_o}{qE_z}\frac{E_z}{B}\xi$$

Inserting this equation into the above equation yields

$$z \approx \frac{\mathcal{E}_o}{qE_z}\left[1 + \frac{E_z^2}{2B^2}\frac{(qtB)^2}{\mathcal{E}_o^2}\right] = \frac{qE_z}{2m_o}t^2$$

where we have used the relation $\mathcal{E}_o = m_o c^2$. This equation is the same expression obtained for the nonrelativistic particles earlier (Eq. (6.3)).

Solution 6.4 The equation for K is given by

$$K = \frac{(\mu_e/m_e - \alpha^2 \mu_i/m_i)}{e(1/m_e + \alpha^2/m_i)} = \frac{W_{\parallel i} W_{\perp e} - W_{\parallel e} W_{\perp i}}{eB(W_{\parallel e} + W_{\parallel i})}$$

The above equation made use of the relations $\mu = W_\perp/B$ and $W_\parallel = mv_\parallel^2/2$ and $W_\perp = mv_\perp^2/2$. Equation (6.39) shows that $E_\parallel$ vanishes *if and only if* the magnetic field is homogeneous $\partial B/\partial s = 0$ or if

$$K = W_{\parallel i} W_{\perp e} - W_{\parallel e} W_{\perp i} = 0$$

Here, we remind the readers that

$$W_\parallel = \int \left(\frac{mv_\parallel^2}{2} \right) f(v_\parallel, v_\perp) d^3 v$$

$$W_\perp = \int \left(\frac{mv_\perp^2}{2} \right) f(v_\parallel, v_\perp) d^3 v \qquad (6.78)$$

are energies associated with particles moving in the parallel and perpendicular directions of the magnetic field averaged over the distribution function. Observations have shown that the distribution functions of electrons and protons in the magnetosphere are generally different. Hence $K \neq 0$.

Solution 6.5 A Maxwellian distribution is given by

$$f(v) = f_0 \exp \left(\frac{-mv^2}{2kT} \right)$$

where $mv^2/2$ is the kinetic energy of the electron and kT its characteristic thermal energy. Before entering the potential region, a Maxwellian distribution consists of isotropic circular iso-intensity contours in the velocity space centered about $v = 0$. If these particles pass through an accelerating potential ϕ drop along a constant magnetic field, while most of the phase space contours will remain circular (because of conservation of particle energy and phase space density; Liouville's theorem), the contours for $v_\parallel < (2e\phi/m)^{1/2}$ will vanish. If we further assume that the first invariant is conserved, this cutoff surface can be expressed as a maximum pitch-angle for any given contour given by,

$$\sin^2 \alpha_{max} = \frac{W}{W + e\phi}$$

where W is the energy of the contour corresponding to the original distribution. The pitch-angle distribution becomes narrower as $e\phi$ increases. However, since the acceleration takes place at high altitudes, the pitch-angles will increase as the electrons move into stronger magnetic fields at lower altitudes. If we let B_0 and B

be the magnetic field strength at the height where the potential is encountered and where the electrons are first detected, the above equation will be modified to

$$\sin^2 \alpha_{max} = \frac{B_0}{B} \frac{W}{W + e\phi}$$

This example demonstrates that the shape of the pitch-angle distribution function of the parallel electrons can be determined by the combination of electrostatic potential and the magnetic mirror forces. Detailed studies of these distributions can thus provide important information on the nature and structure of the potential and where the particle acceleration occurs.

Solution 6.6 When the collision mean free path is much larger than the dimension in which currents are flowing, one cannot use the standard calculation of the conductivity as is done when the plasma is collision dominated. When electrons are carrying the current the mean velocity of the nonrelativistic electrons along the magnetic field direction obeys the Lorentz equation,

$$m_e \frac{d}{dt} \langle v \rangle_{\|} = -e E_{\|}$$

The equation for the current density is obtained by multiplying both sides of this equation with $-en_e$, and defining $J_{\|} = -en_e \langle v_e \rangle$, the above equation can then be rewritten as

$$\frac{d}{dt} J_{\|} = \frac{ne^2}{m_e} E_{\|}$$

This equation is *not* Ohm's law and the classical definition of the conductivity has no meaning. If we integrate this equation, we obtain

$$J_{\|} = \frac{ne^2}{m_e} E_{\|} t + \text{constant}$$

In collisionless plasmas, one could still have $J_{\|}$ even when $E_{\|} = 0$. The relationship between $J_{\|}$ and $E_{\|}$ is obtained through the coupled Maxwell-Lorentz equations (Alfvén and Fälthammar 1963).

Solution 6.7 The pitch-angle $\alpha = 90°$, β of plasma is small and negligible. Additional assumptions are that the magnetic field is measured on the surface of a planet rather than at the center. This will result in an error. The above calculations also assume a fixed dipole geometry, which is not exactly correct. Intense currents in the magnetosphere will elongate the dipole geometry, making it more parabolic. This in turn will change the form of the current. In practice, the calculation must be iterated several times for the theory and observations to agree. Equation (6.67) can be derived more elegantly from the Virial theorem for steady-state plasmas (Carovillano and Maguire 1968).

Solution 6.8 The divergence of Eq. (6.69) is

$$\nabla \cdot \mathbf{J}_\perp = \nabla \cdot \left(\frac{\mathbf{B}}{B^2} \times \mathbf{F} \right)$$

$$= -2\mathbf{J}_\perp \cdot \frac{\nabla B}{B} - \left(\frac{\mathbf{B}}{B^2} \right) \cdot \nabla \times \mathbf{F}$$

where $\mathbf{F} = \rho_m d \langle \mathbf{v} \rangle / dt + \nabla \cdot \boldsymbol{\pi}$ and we set $(\nabla \times \mathbf{B}/B^2) \cdot \mathbf{F} = 0$. Inserting this equation into (6.68) yields

$$B \frac{\partial}{\partial s} \left(\frac{J_\|}{B} \right) = 2\mathbf{J}_\perp \cdot \frac{\nabla B}{B} + \left(\frac{\mathbf{B}}{B^2} \right) \cdot \nabla \times \mathbf{F}$$

and $J_\|$ is obtained by integrating along the length of the magnetic field.

$$J_\| = \int 2\mathbf{J}_\perp \cdot \frac{\nabla B}{B^2} + \left(\frac{\mathbf{B}}{B^3} \right) \cdot \nabla \times \mathbf{F} ds$$

Note that $J_\|$ depends on the combination of the magnetic topology and the properties of plasmas (through the pressure tensor if the distribution is assumed thermal). If it is assumed that $J_\perp$ is due to the ring current, $J_\perp$ is,

$$\mathbf{J}_\perp = \frac{\mathbf{B}}{B^2} \times \left[\nabla p_\perp + (p_\| - p_\perp) \frac{(\mathbf{B} \cdot \nabla)\mathbf{B}}{B^2} - \rho_m \frac{\partial}{\partial t} (\mathbf{W_D} - \mathbf{W_g}) \right]$$

where the pressure p is anisotropic, ρ_m is mass density, and $\mathbf{W}_D$ and $\mathbf{W}_g$ are guiding center drifts arising from electric field and gravity. The perpendicular current reduces to a much simpler form than the above equation if pressure is isotropic.

Solution 6.9 The geomagnetic activity on this day began around 1133 UT with Cluster (SC3) recording an increase in B_x and B_y while B_z decreased. The maximum values attained by B_x and B_y at ~1134:45 UT were 12 and 15 nT and the minimum by B_z ~3 nT. The total magnetic field intensity $|B|$ was 22 nT, which suggests Cluster was close to the high latitude plasma sheet boundary layer (PSBL). The energy flux spectrograms of ions (1–30 keV) and electrons (500 eV–6 keV) show typical behavior of plasma sheet particles (panels (d) and (e)). The slow build up of the magnetic field was observed on all four Cluster spacecraft indicating this disturbance covered a region larger than the SC separation of 1200 km in $X-Y$ and $X-Z$ directions (not shown).

An abrupt change of the magnetic field was observed around ~1134:47 UT with B_y decreasing to ~1 nT and B_x to 3 nT. The small value of B_x indicates Cluster was close to the current sheet (CS). The B_z component increased dramatically from 3 nT to 15 nT. This behavior is called *dipolarization* because the magnetic tail geometry becomes more dipolar. Dipolarizations occur when high beta plasma ($\beta \sim 0.4$) in the CS is abruptly "removed" or the tail current has been reduced, reconfiguring the geomagnetic tail. These current "disruptions" are accompanied by intense auroral activities and Earth becomes an intense radio emitter at kilometric wavelength (Gurnett 1972). The suggested mechanism for "substorm" onset involves plasma instabilities of the current sheet but the mechanism is not known.

Prior to 1134:47 UT, the plasma was relatively quiet with very small average mean velocity: V_x and $V_z \sim 0$ and V_y 25 km s^{-1}. Accompanying the onset of dipolarization, V_x increased and reached $\sim$400 km s^{-1} at 1135:35 UT, V_y changed from $\sim$25 to 125 km s^{-1} returning to 25 km s^{-1}, and V_z remained steady throughout the period except at the beginning when it increased to 100 km s^{-1}. Following the onset, Cluster encountered a hotter and more intense plasma as $|B|$ decreased to 5 nT (diamagnetic effect). The spacecraft was then located close to the central CS where the ion plasma temperature was $T \sim 5 \times 10^7$ °K.

The pitch-angle spectrogram (panel f) shows a field-aligned electron beam (0.5–2.3 keV) at 180° pitch angle. The maximum beam intensity peaked at 1134:47 UT just before the dipolarization and lasted for 12 s (three spins of the spacecraft). Cluster was in the northern hemisphere and 180° pitch-angle corresponds to electrons coming from the earthward direction. The beam was observed within one pixel, 15° of the detector field of view. The beam at 0° was observed 4–8 s later with diminished intensity. This time delay has been interpreted as either the time for the 180° beam to propagate to the opposite end of the field and mirror back or that the 0° electrons are coming from further down in the geomagnetic tail or the southern auroral ionosphere. The speed of 1 keV electrons is 18,700 km s^{-1}, and the time delay of 4–8 s corresponds to a distance of 11–22 R$_E$ (R$_E \approx$6400 km).

Solution 6.10 Observations in Fig. 6.20 were made on the duskside. The left plots show the data when Cluster was in the southern hemisphere (negative B_X) and the right in the northern hemisphere (positive B_x). The top panels of both sides (panels a and g) show the total magnetic field intensity with different Cluster spacecraft color coded. The subsequent panels show the components, B_x (panels b and h), B_y (panels c ansi), and B_z (panels d and i). The total current is shown in panels (e) and (k). Panels (f)and (l) show |div B/curl B| (black) and |J_{FAC}/J_t|(blue), where J_{FAC} is field-aligned current and J_t is the total current. The DFs are identified by a sharp increase of the Bz component ($>$10 nT). Plasma ion density and temperature were consistent with typical plasma sheet values ($n \sim$0.1 cm^3 and $T \sim$4.3 keV), and the mean earthward plasma speed was 50–100 km s^{-1} (not shown).

The total current and the FAC component show the average magnitudes are larger than 10 nA m^{-2}. The magnetic field direction used to obtain FAC is a spatial average of the four field orientations measured by the Cluster spacecraft. A FAC component is observed both before and inside the DF. For the southern hemisphere, the direction of FAC is positive (parallel to **B**) before the DF in the region where B_z is reduced (sometimes referred to as B_z dip) and negative (antiparallel) to **B** inside the DF. For the event in the northern hemisphere (panel k) the FAC first show negative (antiparallel) values and then turn positive (parallel), opposite to the behavior in the southern hemisphere. The ratios of |J_{FAC}|/J_t (blue lines in panels f and l) show that the FACs are the main component of the total current during these times. A similar analysis of observations made on the duskside shows identical behavior except the signs of the FACs are reversed from those of dawnside (Sun et al. 2013).

Solution 6.11 For the geometry indicated, $\mathbf{B}$ is in the $+Z$-direction and since $\mathbf{V}$ is in the X-direction, the $q\mathbf{V} \times \mathbf{B}$ is in the $+Y$ direction. So the positive charges move toward the positive Y-axis and electrons toward the $-Y$-axis. The same charge distribution is seen in the S'-frame since the physics must be the same as guaranteed by the principle of relativity. The blob is stationary in the S'-frame so the surrounding plasma will be moving toward the blob in the $-Y$-direction.

References

Akasofu, S.-I.: Polar and Magnetospheric Substorms. Springer, New York (1968)

Albert, R.: Nearly monoenergetic electron fluxes detected during a visible aurora. Phys. Rev. Lett. **18**, 369 (1967)

Alfvén, H., Fälthammar, C.-G.: Cosmical Electrodynamics: Fundamental Principles, 2nd edn. Oxford University Press, Oxford (1963)

Allen, W.: Ponderomotive mass transport in the magnetosphere. J. Geophys. Res. **97**, 8483 (1992)

Baker, D., et al.: Highly relativistic radiation belt electron acceleration, transport, and loss: large solar storm events of March and June 2015. J. Geophys. Res. **121**, 6647 (2016)

Bernstein, I., Green, J.M., Kruskal, M.D.: Exact nonlinear plasma oscillations. Phys. Rev. Lett. **108**, 546 (1957)

Berko, F., et al.: Protons as the prime contributors to storm time ring current. J. Geophys. Res. **80**, 3549 (1975)

Block, L.: A double layer review. Astrophys. Space Sci. **55**, 59 (1978)

Borovosky, J.: Parallel electric fields in extragalactic jets: double layers and anomalous resistivity in symbiotic relationships. Astrophys. J. **305**, 451 (1986)

Boström, R.: Electrodynamics of the ionosphere. In: Egeland, A., Holter, O., Omholt, A. (eds.) Cosmic Electrodynamics. Scandinavian University Books, Oslo (1973)

Cahill, L., Amazeen, P.: The boundary of the geomagnetic field. J. Geophys. Res. **68**, 1835 (1963)

Calqvist, P., Boström, R.: Space-charge region about the aurora. J. Geophys. Res. **75**, 7140 (1970)

Caravillano, R., Maguire, J.: Magnetic energy relationships in the magnetosphere. In: Carovillano, R., McClay, J., Radoski, H. (eds.) Physics of the Magnetosphere. D. Reidel Publishing/Co., Dordrecht (1968)

Cardoso, F., et al.: Auroral precipitating energy during long magnetic storms. J. Geophys. Res. **122**, 6007 (2017)

Carlson, C., Pfaff, R.F., Watzin, J.G.: The fast auroral snapshot (FAST) mission. Geophys. Res. Lett. **25**, 2013 (1998)

Chaston, C., et al.: Radial transport of radiation belt electrons in kinetic field-line resonances. Geophys. Res. Lett. **44**, 8140 (2017)

Chua, D., Brittnacher, M., Parks, G., et al.: A new auroral feature "The nightside gap". Geophys. Res. Lett. **25**, 3747 (1998)

Cornwall, J., et al.: Unified theory of SAR arc formation at the plasmapause. J. Geophys. Res. **76**, 4428 (1971)

Cui, Y.B., Fu, S.Y., Parks, G.K.: Heating of ionospheric ion beams in inverted-V structures. Geophys. Res. Lett. **41**, 3752 (2014)

Cummings, W.D., Dessler, A.J.: Field-aligned currents in the magnetosphere. J. Geophys. Res. **72**, 1007 (1967)

Dandouras, I.: Detection of a plasmaspheric wind in the Earth's magnetosphere by the Cluster spacecraft. Ann. Geophys. **31**, 1143 (2013)

Delory, G., et al.: FAST observations of electron distributions within AKR source regions. Geophys. Res. Lett. **25**, 2065 (1998)

Dreicer, H.: Electron and ion runaway in fully ionized gases, I. Phys. Rev. **115**, 238 (1959)

Eack, K., et al.: X-ray pulses observed above a mesoscale convective system. Geophys. Res. Lett. **23**, 2915 (1996)

Ergun, R., et al.: Parallel electric fields in discrete arcs. Geophys. Res. Lett. **27**, 4053 (2000)

Ergun, R., et al.: Observations of double layers in Earth's plasma sheet. Phys. Rev. Lett. **102**, 155002 (2009)

Evans, D.: Precipitating electron fluxes formed by a magnetic field aligned potential difference. J. Geophys. Res. **79**, 2853 (1974)

Fok, M.-C., et al.: The comprehensive inner magnetosphere-ionosphere model. J. Geophys. Res. **119**, 7522 (2014)

Frank, L.: On the extraterrestrial ring current during geomagnetic storms. J. Geophys. Res. **72**, 3753 (1967)

Frank, L., Ackerson, K.: Observations of charged particle precipitation into the auroral zone. J. Geophys. Res. **76**, 3612 (1971)

Giovanelli, R.: Electron energies resulting from an electric field in a highly ionized gas. J. Sci. **40**(301), 206 (1949)

Goertz, C.: Double layers and electrostatic shocks in space. Rev. Geophys. Space Phys. **17**, 418 (1979)

Guglielmi, A.V., et al.: Modifications of magnetospheric plasma due to ponderomotive force. Astrophys. Space Sci. **200**, 91 (1993)

Gurevich, A.V., Zybin, K.P.: Runaway breakdown and the mysteries of lightning. Phys. Today **58**, 37 (2005)

Haerendel, G., et al.: First observations of electrostatic acceleration of barium ions into the magnetosphere. In: European Programmes on Sounding Rockets and Balloon Research in the Auroral Zone, p. 203. ESA Special Publications, ESA SP-115, Paris (1976)

Hanasaz, J.: Wideband bursts of auroral kilometric radiation and their association with UV auroral bulges. J. Geophys. Res. **106**, 3859 (2001)

Horwitz, J.: Core plasma in the magnetosphere. Rev. Geophys. **25**, 579 (1987)

Hudson, M., Potter, D.: Electrostatic shocks in the auroral magnetosphere. In: Akasofu, A.I., Kan, J.R. (eds.) Physics of Auroral Arc Formation, p. 260. American Geophysical Union, Washington, DC (1981)

Ieda, A., et al.: Plasmoid ejection and auroral brightenings. J. Geophys. Res. **106**, 3845 (2001)

Iijima, T., Potemera, T.: The amplitude distribution of field-aligned currents at northern high latitudes observed by triad. J. Geophys. Res. **81**, 2165 (1976)

Iijima, T., Potemera, T.: Large-scale characteristics of field-aligned currents associated with substorms. J. Geophys. Res. **83**, 599 (1978)

Kamide, Y.: Magnetic storms: current understanding and outstanding questions. In: Magnetic Storms. Geophysical Monographs, vol. 98. American Geophysical Union, Washington, DC (1997)

Kaufmann, R.L.: Acceleration of auroral electrons in parallel electric fields. J. Geophys. Res. **81**, 1672 (1976)

Kim, K., et al.: Large electric field at the nightside plasmapause observed by the polar spacecraft. J. Geophys. Res. **115**, A07219 (2010)

Kistler, L., et al.: Ion composition and pressure changes in storm time and nonstorm substorms in the vicinity of the near-Earth neutral line. J. Geophys. Res. **111**, A11222 (2006)

Kochkin, P., et al.: Experimental study on hard X-rays emitted from metre-scale negative discharges in air. J. Phys. D: Appl. Phys. **48**, 025205 (2015)

Kozyra, J.: High-altitude energy sources for stable auroral red arcs. Rev. Geophys. **35**, 155 (1997)

Landau, L., Lifshitz, E.: The Classical Theory of Fields. Pergamon, Oxford (1962)

Le, G., et al.: Global observations of magnetospheric high-m poloidal waves during the 22 June 2015 magnetic storm. Geophys. Res. Lett. **44**, 3456 (2017)

Lee, N.C., Parks, G.K.: Ponderomotive acceleration of ions by circularly polarized electromagnetic waves. Geophys. Res. Lett. **23**, 327 (1996)

Lee, E., Parks, G.K., Fu, S.Y., et al.: Relating field-aligned beams to inverted-V structures and visible auroras. Ann. Geophys. **33**, 1 (2015)

Li, X., Temerin, M.: Ponderomotive effects on ion acceleration in the auroral zone. Geophys. Res. Lett. **20**, 13 (1993)

Lin, C., Parks, G.K.: Modulation of energetic particle fluxes by a mixed mode of transverse and compressional waves. J. Geophys. Res. **87**, 5102 (1982)

Liu, W.W., et al.: On the equatorward motion and fading of proton aurora during substorm growth phase. J. Geophys. Res. **112**, A10217 (2007)

Lotko, W.: Inductive magnetosphere-ionosphere coupling. J. Atmos. Sol. Terr. Phys. **66**, 1433 (2004)

Lotko, W., et al.: Electrostatic shocks and suprathermal electrons powered by dispersive anomalously resistive field line resonance. Geophys. Res. Lett. **25**, 4449 (1998)

Louarn, P., et al.: Trapped electrons as a free energy source for the auroral kilometric radiation. J. Geophys. Res. **95**, 5938 (1990)

Lundin, R., Guglielmi, A.: Ponderomotive forces in cosmos. Space Sci. Rev. **127**, 1 (2006)

Lui, A.T.Y.: A synthesis of magnetospheric substorm models. J. Geophys. Res. **96**, 1849 (1991)

Lysak, R.: The relationship between electrostatic shocks and kinetic Alfvén waves. Geophys. Res. Lett. **25**, 2089 (1998)

Marklund, G., et al.: Altitude distribution of the auroral acceleration potential determined from Cluster satellite data at different heights. Phys. Rev. Lett. **106**, 055002 (2011)

Mauk, B.: Generation of macroscopic magnetic-field-aligned electric fields by the convection surge ion acceleration mechanism. J. Geophys. Res. **94**, 8911 (1989)

McCarthy, M.C., Parks, G.K.: Further observations of X-rays inside thunderstorms. Geophys. Res. Lett. **12**, 393 (1985)

McFadden, J., et al.: Spatial structure and gradients of ion beams observed by FAST. Geophys. Res. Lett. **25**, 2021 (1998)

McIlwain, C.E.: Direct measurements of particles producing aurora. J. Geophys. Res. **65**, 2727 (1960)

Melrose, D.: Coherent emission mechanisms in astrophysical plasmas. Rev. Mod. Plasma Phys. **1**, 5 (2017)

Moore, T.W., et al.: Cross-scale energy transport in space plasmas. Nat. Phys. **12**, 1164–1169 (2016)

Morioka, A., et al.: AKR breakup and auroral particle acceleration at substorm onset. J. Geophys. Res. **113**, A09213 (2008)

Murphy, C.H., Wang, C.S., Kim, J.S.: Inductive electric field of a field-aligned current system. J. Geophys. Res. **79**, 2901 (1974)

Nicholson, D.: Introduction to Plasma Theory, p. 115. Wiley, New York (1983)

Nykyri, K., Dimmock, A.: Statistical study of the ULF Pc4-Pc5 range fluctuations in the vicinity of Earth's magnetopause and correlation with the low latitude boundary layer thickness. Adv. Space Res. **58**, 257 (2015)

O'Brien, B., Taylor, H.: High-latitude geophysical studies with satellite injun 3; 4. Auroras and their excitation. J. Geophys. Res. **69**, 45 (1964)

Oreshkin, E.V., et al.: Parameters of a runaway electron avalanche. Phys. Plasmas **24**, 103505 (2017)

Panaysuk, M., et al.: Magnetic storms in October 2003. Cosm. Res. **42**, 489 (2004)

Parks, G.K.: Physics of Space Plasmas, An Introduction, 2nd edn. Westview Press, A Member of Perseus Books Group, Boulder (2004)

Parks, G.K., et al.: X-ray enhancements detected during thunderstorm and lightning activities. Geophys. Res. Lett. **8**, 1176 (1981)

Patel, V.L.: Low frequency hydromagnetic waves in the magnetosphere: explorer 12. Planet. Space Sci. **13**, 485 (1965)

Rees, M., Roble, R.: Observations and theory of the formation of stable auroral red arcs. Rev. Geophys. Space Phys. **13**, 201 (1975)

Reeves, G., et al.: On the relationship between relativistic electron flux and solar wind velocity: Paulikas and Blake revisited. J. Geophys. Res. **116**, A02213 (2011)

Sandel, B.R., et al.: Extreme ultraviolet imager observations of the structure and dynamics of the plasmasphere. Space Sci. Rev. **109**, 25 (2003)

Schulz, M., Lanzerotti, L.: Particle Diffusion in the Radiation Belts. Physics and Chemistry in Space. Berlin, Springer (1974)

Sergeev, V., et al.: Current sheet flapping motion and structure observed by Cluster. J. Geophys. Res. **30**, 1327 (2003)

Shelley, E., et al.: Satellite observations of ionospheric acceleration mechanism. Geophys. Res. Lett. **3**, 654 (1976)

Shi, J.K., et al.: South-north asymmetry of field-aligned currents in the magnetotail observed by Cluster. J. Geophys. Res. **115**, A07228 (2010)

Smith, D., et al.: The rarity of terrestrial gamma-ray flashes. Geophys. Res. Lett. **38**, L08807 (2011)

Spasojevic, M., et al.: The global response of the plasmasphere to the geomagnetic disturbance. J. Geophys. Res. **70**, 1717 (2003)

Stevenson, B., et al.: Polar observations of topside field-aligned O+ flows and auroral forms. J. Geophys. Res. **106**, 18969 (2001)

Sun, W.J., Fu, S.Y., Parks, G.K., et al.: Field-aligned currents associated with dipolarization fronts. Geophys. Res. Lett. **40**, 4503 (2013)

Sun, W.-J., et al.: MESSENGER observations of magnetospheric substorm activity in Mercury's near magnetotail. Geophys. Res. Lett. **42**, 3692 (2015)

Stern, D.: One dimensional models of quasi-neutral parallel electric fields. J. Geophys. Res. **86**, 5839 (1981)

Tetreaualt, D.: Growing ion holes as the cause of auroral double layers. Geophys. Res. Lett. **15**, 164 (1988)

Turikov, V.A.: Electron phase-space holes as localized BGK solutions. Phys. Scr. **30**, 73 (1984)

Whipple, E.: The signature of parallel electric fields in a collisionless plasma. J. Geophys. Res. **82**, 1525 (1977)

Wu, C.S., Lee, L.: A theory of terrestrial kilometric radiation. Astrophys. J. **230**, 621 (1965)

Yau, A., André, M.: Sources of ion outflow in the high latitude ionosphere. Space Sci. Rev. **80**, 1 (1997)

Zheng, H., Fu, S.Y., et al.: Observations of ionospheric electron beams in the plasma sheet. Phys. Rev. Lett. **109**, 205001 (2012)

Zmuda, A., et al.: Transverse magnetic disturbances at 1100 kilometers in the auroral region. J. Geophys. Res. **71**, 5033 (1966)

Zong, Q., et al.: The interaction of ultra-low-frequency pc3-5 waves with charged particles in Earth's magnetosphere. Rev. Mod. Plasma Phys. **1**, 10 (2017)

Additional Reading

Alfvén, H.: Kgl. Svenska Vetenskapsakad. Handl. **18**(3) (1939)

Alfvén, H.: On the importance of electric fields in the magnetosphere and interplanetary space. Space Sci. Rev. **7**, 140 (1967)

Andersson, L., Ergun, R.: The search for double layers. In: Auroral Phenomenology and Magnetospheric Processes: Earth and Other Planets. Geophysical Monograph Series, vol. 197. American Geophysical Union, Washington, DC (2012)

André, M., et al.: Ion energization mechanisms at 1700 km in the auroral region. J. Geophys. Res. **103**, 4199 (1998)

Arnoldy, R., et al.: Field-aligned auroral electron fluxes. J. Geophys. Res. **79**, 4208 (1974)

Block, L., Falthämmar, C.-G.: The role of magnetic field aligned electric fields in auroral acceleration. J. Geophys. Res. **95**, 5877 (1990)

Boström, R., et al.: Characteristics of solitary waves and weak double layers in the magnetospheric plasma. Phys. Rev. Lett. **61**, 82 (1988)

Bryant, D., Courtier, G.M.: Electrostatic double layers as auroral particle accelerators - a problem. Ann. Geophys. **33**, 481 (2015)

Cardoso, F.: Auroral electron precipitating energy input estimate during magnetic storms with peculiar long recovery phase features. Doctorate Thesis submitted to INPE, Sān José (2010)

Cloutier, P.A., Anderson, H.R.: Observations of birkeland currents. Space Sci. Rev. **17**, 563 (1975)

Dessler, A., The evolution of arguments regarding the existence of field-aligned currents. In: Potemera, T.A. (ed.) Magnetospheric Currents. Geophysical Monograph, vol. 28. American Geophysical Union, Washington, DC (1983)

Fälthammer, C.-G.: Problems related to macroscopic electric fields in the magnetosphere. Astrophys. Space Sci. **55**, 179 (1978)

Fillingim, M., et al.: Coincident POLAR/UVI and WIND observations of pseudobreakups. Geophys. Res. Lett. **27**, 1379 (2000)

Gurnett, D.: Electric field and plasma observations in the magnetosphere. In: Dyer, E.R. (ed.) Critical Problems of Magnetospheric Physics, p. 123. National Academy of Sciences, Washington, DC (1972)

Hwang, K.-L., et al.: Test particle simulations of the effect of moving DLs on ion outflow in the auroral downward current region. J. Geophys. Res. **113**, A01308 (2008)

Lennartson, W.: On the consequences of the interaction between the auroral plasma and the geomagnetic field. Planet. Space Sci. **28**, 135 (1980)

Lin, Y., et al.: Investigation of storm time magnetotail and ion injection using three-dimensional global hybrid simulation. J. Geophys. Res. **119**, 7413 (2014)

Lysak, R.L.: Electrodynamic coupling of the magnetosphere and ionosphere. Space Sci. Rev. **52**, 33 (1990)

Melrose, D.: An interpretation of Jupiter's decametric and the terrestrial kilometric radiation as direct amplified gyroemission. Astrophys. J. **207**, 651 (1976)

Mizera, P.F., Fennell, J.F.: Signatures of electric fields from high and low altitude particle distributions. Geophys. Res. Lett. **4**, 311 (1977)

Mozer, F., Hull, A.: Origin and geometry of upward parallel electric fields in the auroral acceleration region. J. Geophys. Res. **106**, 5763 (2001)

Mozer, F., Kletzing, C.A.: Direct observations of large, quasi-static parallel electric fields in the auroral acceleration region. Geophys. Res. Lett. **25**, 1629 (1998)

Mozer, F., et al.: Observations of paired electrostatic shocks in the polar ionosphere. Phys. Rev. Lett. **38**, 297 (1977)

Østgaard, N, et al.: Observations and model predictions of substorm auroral asymmetries in the conjugate hemispheres. Geophys. Res. Lett. **32**, L05111 (2005)

Parks, G.K., Lee, E., Fu, S., et al.: Outflow of low-energy O^+ ion beams observed during periods without substorms. Ann. Geophys. **33**, 333 (2015)

Potemera, T.A.: Observation of birkeland currents with the TRIAD satellite. Astrophys. Space Sci. **58**, 1 (1978)

Potemra, T.A.: Birkeland currents in the Earth's magnetosphere. Astrophys. Space Sci. **144**, 155 (1988)

Reiff, P., et al.: Determination of auroral electrostatic potentials using high and low altitude particle distributions. J. Geophys. Res. **93**, 7441 (1988)

Shawhan, S., et al.: On the nature of large auroral zone electric fields at 1 R_E altitude. J. Geophys. Res. **83**, 1049 (1978)

Stevenson, B.A.: Relationship of O+ field-aligned flows and densities to convection speed in the polar cap at 5000 km altitude. J. Atmos. Sol. Terr. Phys. **62**, 495 (2000)

Temerin, M., et al.: Observations of double layers and solitary waves in the auroral plasma. Phys. Rev. Lett. **48**, 1175 (1982)

Tetreault, D.: Theory of electric fields in the auroral acceleration region. J. Geophys. Res. **96**, 3549 (1991)

Wescott., E.M., et al.: The Skylab barium plasma injection experiments, 2. Evidence for double layer. J. Geophys. Res **81**, 4495 (1976)

Wong, H.K., Wu, C.S., Gaffey, J.D., Jr.: Electron-cyclotron maser instability caused by hot electrons. Phys. Fluids **28**, 2751 (1985)

Yau, A., Andres, M.: Sources of ion outflow in the high latitude ionosphere. Space Sci. Rev. **80**, 1 (1997)

Chapter 7
Topics for Further Studies

7.1 Introduction

This book has presented a selected number of fundamental topics important for understanding the basic physics of space plasmas. We have come a long way since the early days of plasma physics (Grad 1969) when we began to quantify the physics. The purpose of this chapter is to discuss several topics that have not been adequately treated in space plasma physics textbooks. These topics include the concepts of collective interactions and synergistic effects responsible for the large-scale coherent structures, an example of instability that occurs within the large magnetospheric structure, heating of collisionless plasmas, and acceleration of electrons to runaway energies.

7.2 Collective Interactions and Synergistic Effects

A unique feature about space plasma systems is that the collision mean free path is very long thereby making them nearly *collisionless*. Collisionless plasmas behave differently from collision dominated plasmas found in laboratories and lower regions of atmospheres of planets and stars. Unlike collision-dominated plasmas whose physics is dictated by short-range forces, a particle in collisionless plasmas gyrates around a magnetic field "freely" without collisions. However, all of the gyrating particles are interacting *collectively* through the long-range electric and magnetic fields, giving rise to unanticipated effects.

Collective mechanisms are difficult to study because a complete description of a space plasma system requires as many equations of motion as there are particles that are coupled to each other by the electromagnetic fields. Even with low density collisionless space plasmas, the number of equations involved is so large and it is virtually impossible to solve the coupled equations analytically or with super

© Springer Nature Switzerland AG 2018   297
G. K. Parks, *Characterizing Space Plasmas*, Astronomy and Astrophysics Library,
https://doi.org/10.1007/978-3-319-90041-4_7

computers at this time. Analytical theory has thus far resorted to statistical methods using phase space densities and one-particle distribution functions. However, even then, the Boltzmann equation is nonlinear and only a few simple cases have been solved.

By processes as yet unidentified, the physics of collective interaction can become *synergistic* where the total effect produced by the interacting particles is greater than the effect produced by the sum of the individual particles. Many of the dynamics observed throughout the Universe likely come from such synergistic processes. Examples include the remnant of Crab Nebula, and coronal mass ejections (CMEs) which are huge explosions on the Sun that eject millions of tons of solar debris into space including electrons and protons of relativistic energies.

7.2.1 Large-Scale Current Structures

We have emphasized in this book that data are fundamentally important for understanding space plasmas and that the dynamics must be looked at from the concepts of particles. We have argued that electromagnetic induction drives large-scale dynamics and speculated that other planetary and stellar systems probably have similar large-scale electric field and current systems as on Earth. We know a lot about how the current systems work on Earth but unfortunately we have no details about the systems elsewhere. The challenging question is how we can transfer our earth-acquired knowledge to plasma systems elsewhere.

Particles and field structures cover a range of dimensions from galactic to planetary auroral scale lengths. In our solar system, particles contributing to large-scale structures have been routinely measured locally by spacecraft instruments. These observations have led to a picture of the heliospheric current sheet (Fig. 7.1). The solar wind (SW) beams are part of this current system as auroral beams are part of the auroral global current system. Auroral observations clearly show that the beams are accelerated by electric field parallel to the magnetic field direction and we have speculated that similar processes active on the Sun produce the SW beams. The spiraling heliospheric currents and magnetic fields in the ecliptic plane extend from the Sun to the edge of the heliosphere. This picture was originally proposed by Alfvén more than three decades ago and it needs to be improved and updated.

The ecliptic interplanetary current sheet can explain qualitatively many observations about the behavior of the IMF (interplanetary magnetic field). First note that whether the IMF is pointing toward or away from the Sun depends whether the observer is to the north or south of the ecliptic plane. The change of the IMF polarity can be explained by the up and down motion of the current sheet relative to the observer, so the observer initially located in the north will see the change of the polarity when the observer moves to the south. The up-down motion is produced by Alfvén waves propagating in the ecliptic plane. These waves could originate on the Sun or be locally excited. If solar rotation is included in the model, one would then see a wavy current sheet that resembles the skirt of a ballerina.

Fig. 7.1 A model of heliospheric current system in the ecliptic plane (top) (from Smith et al., *J. Geophys. Res.*, **83**, 717, 1978). The current sheet is associated with magnetic field lines (solid) in and out of the Sun. The Sun is at the center with a warped disk-like sheet in which electric currents flow azimuthally (bottom) around the Sun (from Alfvén, *D. Reidel Co.*, Dordrecht, Holland, 1981)

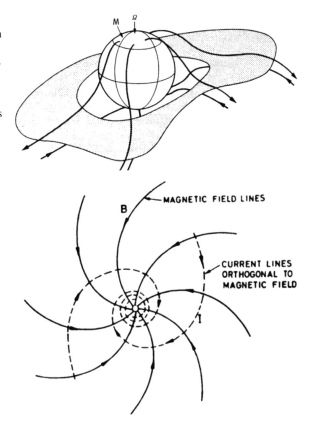

The heliospheric current sheet can be very dynamic as it is modulated by the solar activity. When there is activity on the Sun, the disturbance is directly transmitted to the interplanetary space in the current sheet. These disturbances can induce strong currents in the planets that are immersed in the current sheet. For example, ICMEs can induce currents in the magnetospheres, exciting magnetic storms. A sketch of what we know about the auroral and magnetospheric currents is shown in Fig. 7.2. The top sketch shows currents that flow in the magnetosphere and the bottom sketch shows currents in the ionosphere. The details of how the ionospheric currents are distributed in the more distant regions of the magnetosphere and how they connect to magnetospheric currents or where they close are not understood at this time. Some models show a portion of the auroral currents closes in the inner boundary of the flank magnetopauses or in the SW, but there is no unanimous agreement at the present time.

However, a coherent picture that has emerged from piecing together various observations is that the solar wind interaction with the geomagnetic field induces electric fields that drive currents. The auroral particle and field dynamics occur within the global Birkland current system. What we see in the aurora and the auroral oval are "foot prints" of the dynamics initiated in the distant magnetospheric regions

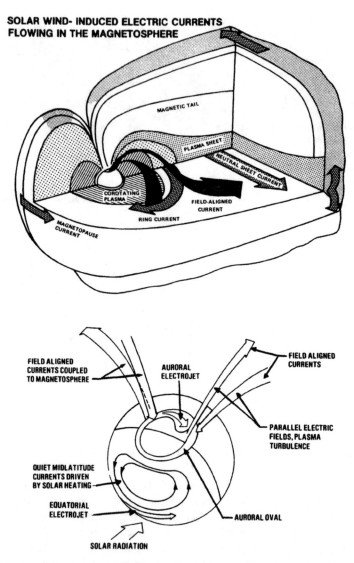

SOLAR WIND- INDUCED ELECTRIC CURRENTS
FLOWING IN THE MAGNETOSPHERE

MAGNETIC TAIL

PLASMA SHEET

NEUTRAL SHEET CURRENT

COROTATING
PLASMA

FIELD-ALIGNED
CURRENT

RING CURRENT

MAGNETOPAUSE
CURRENT

FIELD ALIGNED
CURRENTS COUPLED
TO MAGNETOSPHERE

AURORAL
ELECTROJET

FIELD ALIGNED
CURRENTS

PARALLEL ELECTRIC
FIELDS, PLASMA
TURBULENCE

QUIET MIDLATITUDE
CURRENTS DRIVEN
BY SOLAR HEATING

EQUATORIAL
ELECTROJET

AURORAL OVAL

SOLAR RADIATION

Fig. 7.2 A schematic diagram shows where different currents flow in the magnetosphere and ionosphere. The two regions are connected by field-aligned currents (courtesy of NASA)

by the solar wind. There are many details about the induced currents and boundaries inside the magnetosphere produced by the different types of plasmas existing there. New knowledge needs to be unveiled about how large-scale currents interact in nature producing parallel electric field and currents in large-scale plasma structures.

7.2.2 Dynamics of the Large-Scale Structures

The dynamics of the heliosphere and magnetosphere have been studied since the beginning of space age and while much is known about them the physics of what is going on is only understood qualitatively. Consider for example the problem of how the magnetospheres are populated by particles. While we have learned there are two possible sources of particles for the magnetosphere, the SW and ionosphere, we still do not know what processes can allow the SW particle to enter the magnetosphere and how the particles are accelerated to produce the radiation belts with energetic particles. Moreover, while we know that the parallel electric field can accelerate ionospheric particles outward to populate the magnetosphere, we do not have any information about the details of the SW interaction with the magnetosphere that induce parallel electric field. Although recently we have started to ask questions about how the various regions might be coupled, we have treated the problem piecewise and not in a global sense that includes the SW, planetary magnetic field, and the ionosphere.

7.3 Magnetospheric Instability

Solar flares, magnetic storms, and auroras are instabilities produced by a complex system of interacting plasmas, electric and magnetic fields, and electrical currents. On the Sun, the instabilities during solar flares accelerate particles to relativistic energies at the same time exciting complex electromagnetic waves, for example, type 2 and type 3 emissions. On Earth, we have a magnetospheric substorm instability that affects the entire magnetosphere. Auroras produced by precipitation of particles accompanied by the auroral kilometric radiation are major products of substorms. Substorms involve both macroscopic and microscopic instabilities and they are examples of macroscopic phenomena "controlled" by microscopic processes. We discuss below first macroscopic features associated with substorms and then show an example of microscopic instability associated with substorms.

7.3.1 Magnetospheric Substorm

Observations often show that a substorm starts with a chain of interacting events beginning in the solar wind with the turning of the IMF from a northward to a southward direction (Sauvaud 1987; Le et al. 2002). In response to this IMF turning, the magnetospheric topology changes and the solar wind is efficiently coupled across the magnetopause. Magnetospheric convection then enhances and plasma from the geomagnetic tail is transported closer toward the planet. This phase has been called *growth phase* (McPherron 1972). Some of the particles are accelerated in this process, and a small amount of particles near the loss cone will precipitate into the ionosphere because these particles mirror below the atmosphere. The phase space is no longer uniform and as the plasma is transported closer to Earth the nonuniformity increases further because the transport is pitch-angle dependent. There have been many studies about the onset mechanism. For example, Ohtani et al. (2002) studied possible causes of substorm onset processes using electron data but no conclusion was reached about the mechanisms.

Another possible mechanism is that a substorm is initiated by enhanced particle precipitation (Parks et al. 1972). Enhanced precipitation will increase ionospheric conductivity, permitting more currents to flow between the ionosphere and the magnetosphere. What follows subsequently leading to substorm trigger depends on the SW coupling of the magnetosphere and ionosphere and how the local phase space distributions evolve throughout the magnetosphere and ionosphere. At some point in time, one could anticipate plasma distributions to reach a state suitable for triggering many types of instabilities. This point could correspond to when the plasma distributions become so inhomogeneous that fluctuations will spread across them producing complex filamentary structures. A substorm can thus be looked at as a runaway situation with different plasma instabilities whose footprints are observed as different types of auroral structures in the ionosphere.

The geomagnetic tail reconfigures and the tail current sheet changes the geo-magnetic field to a more dipole-like configuration injecting and accelerating and precipitating the outer radiation belt particles including the particles in the loss cone. Subsequently, the magnetosphere recovers and the whole process repeats itself. An average-size substorm on Earth dissipates about 10^{21} ergs of energy in ≈ 10–30 min and it occurs once every 1–3 h (Parks and Winckler 1968), depending on the level of solar wind disturbance. The meaning of 1–3 h periodicity is not fully understood but it could represent the amount of time it takes for the SW-magnetosphere-ionosphere coupling to establish the conditions required to trigger a substorm. Note that the statistical results by Lin et al. (2009) are interesting and important because they show the location where a substorm is triggered is spread over a large distance in the geomagnetic tail.

The actual boundary interaction that initiates the efficient entry and coupling of the solar wind and magnetosphere and the mechanisms that "trigger" a substorm instability are also a mystery. However, readers interested in the general features of a substorm development in space and time that occurred on March 4, 1979 are encouraged to examine the observations made by Sauvaud (1987) reproduced in Chapter 11 of Parks (2004). This event was observed by five spacecraft, fortuitously situated in the "right regions" of space at the "right time." Additional data were also available from a high-altitude balloon-borne experiment instrumented to study the aurora, from an all-sky camera that provided pictures of the aurora from stations located in northern Scandinavia, and from auroral zone magnetometers that provided information on auroral ionospheric currents. See Gjerloev and Hoffman (2000) for the behavior of the current system during substorms.

Several other mechanisms have been suggested for substorms. They include the tearing mode instability (Coppi et al. 1966), drift ballooning mode instability (Roux et al. 1993), cross tail disruption instability of the geomagnetic tail (Yoon and Lui 1991), current driven Alfvén wave instability (Le Contel 2000; Perrault 2000), and gyroviscous magnetotail current layer model (Stasiewicz 1987). A model by Horton et al. (1999) gives the conditions required for fast onsets of substorms. The importance of $\mathcal{EMF}$ in substorm dynamics has been suggested by Lundin et al. (1991). These models can be tested in principle using Wind, Polar, Cluster, Themis, and MMS missions, but progress has been slow. Chen et al. (2003) have indicated that the behavior of particles in the tail plasma sheet is consistent with the ballooning mode. Much more work needs to be done validating the models using data from the various spacecraft. For more recent results on substorms using data obtained on multiple spacecraft, see Nakamura et al. (2000) and Lyons et al. (2011). For simulation results, see Sitnov et al. (2017) and Raeder et al. (2010).

7.3.2 Microscopic Instabilities

Global substorm instability is intimately tied to microscopic instabilities that precipitates particles and enhances magnetosphere-ionosphere coupling. This section will review briefly microburst precipitation phenomena, a dominant form of precipitation observed during substorms (Fig. 7.3). "Microburst" is a term used by Anderson and Milton (1964) to describe the short, impulsive auroral zone electron precipitation of durations $\sim$0.25 s. Microbursts tend to occur in groups of two or more spaced about 1 s apart. A systematic study of auroral X-rays from precipitated energetic Van Allen electrons shows that during daylight hours electron precipitation occurs primarily in the form of microbursts (Fig. 7.3). Microburst precipitation represents a major perturbation in the magnetospheric electron population. Precipitation bursts with less than 1 s duration of electrons with $\sim$40 keV energies have also been seen on Injun 3 experiment (O'Brien 1964). These bursts have subsequently been identified as microburst electrons (Milton and Oliven 1967; Oliven and Gurnett 1968).

Microburst electron precipitation on a rocket experiment was first detected from Fort Churchill ($L = 8$) by Lampton (1968). Later rocket measurements (Datta et al. 1957) provided more detailed information about the energy spectra and pitch-angle distributions. The rapid precipitation bursts of microbursts could be indicating that the first and second adiabatic invariants are violated. The study of microbursts could lead to important clues about the fundamental wave-particle and plasma interaction processes of relevance to understanding the particle loss mechanism in the magnetosphere.

Balloon-borne observations have shown that relativistic electron precipitation occurs primarily in the evening local hours (Parks et al. 1979; Millan and Thorne 2007). On the other hand, SAMPEX (Solar Anomalous and Magnetospheric Particle Explorer) has shown that relativistic precipitation occurs primarily on the morning side (Lorentzen et al. 2000; Nakamura et al. 2000). The electron precipitation in the morning hours is impulsive and bursty, and it is called *relativistic microburst*, named after the similarly structured microburst precipitation first observed by balloon-

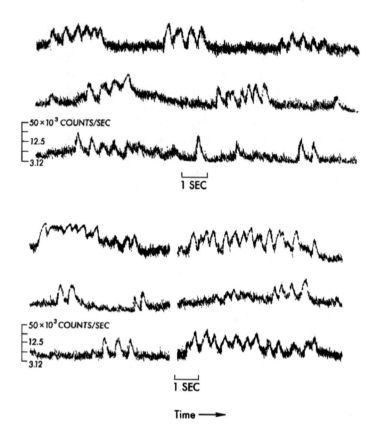

Fig. 7.3 Examples of microburst precipitation detected by a balloon-borne experiment. From Anderson and Milton, *J. Geophys. Res.*, **71**, 4617 (1964)

borne X-ray detectors in 1964 at lower energies. We do not know much about these relativistic electrons. What mechanisms can accelerate electrons to relativistic energies? What mechanisms precipitate relativistic electrons? These questions are of interest to space weather and remain unanswered.

The first measurements of energy spectra of microbursts with energies up to ~350 keV were made during the recovery phase of a magnetic storm by a Korean ScienTific SATellite (STSAT-1) (Lee et al. 2005). Figure 7.4 shows about 30 s of data obtained by the solid-state detectors (SST) (top two panels) and ESA (bottom two panels) during the recovery phase of a magnetic storm. STSAT-1 was at an altitude of ~680 km where the size of the loss cone is ~60°. One detector was aligned along the magnetic field direction looking upward and measured precipitated electrons (second panel). The other detector perpendicular to **B** measured trapped mirroring particles (top panel). The total integrated electron fluxes measured by the SSTs are shown in panel 3. Microburst precipitation is clearly evident in the SST data. Note that the loss cone is completely empty of

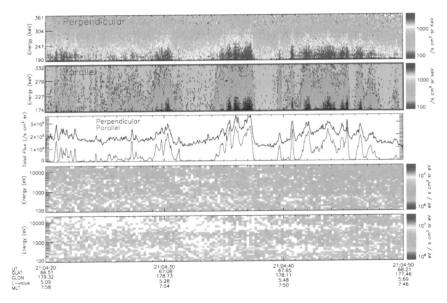

Fig. 7.4 Shown are 10 ms averages of electron fluxes in the energy range ∼170–360 keV measured by STSAT-1. Top panel, electron energy flux spectrogram perpendicular to the magnetic field. Second panel, energy flux spectrogram of a detector looking upward measuring precipitated microbursts. Third panel, total fluxes of the two SSDs integrated over the energy range. Bottom two panels show ESA data of electrons ∼100 eV and 20 keV/charge in the perpendicular and parallel directions (from Lee et al., *Geophys. Res. Lett.*, **32**, 13106, 2005)

electron fluxes except when the short duration impulsive microbursts are observed. The fluxes in the trapped particles exceeded the parallel fluxes by about a factor of two or slightly less. The lower energy electrons measured by ESA (∼100 eV–20 keV) do not show microbursts structure (bottom two panels) confirming previous observations (Lampton 1968).

Enhancements of relativistic electron fluxes in the outer radiation belt observed by Lee et al. (2005) and Hwang et al. (2004) may be the seed electrons of relativistic microbursts (Lee et al. 2005). Microbursts appear promptly over a large pitch-angle range with no measurable time delay of onsets in the perpendicular and parallel directions (∼50 ms). The relativistic microbursts have similar local time distributions as the chorus emissions (Lorentzen et al. 2000), suggesting that microburst precipitation is due to cyclotron resonance scattering of electrons by VLF waves (West and Parks 1984; Skoug et al. 1996). To explain the coincidence measurements of parallel and perpendicular microbursts observed by STSAT-1, microburst electrons would have to diffuse about 2.4° in 50 ms at the equatorial region. Estimating the pitch-angle diffusion coefficient (D_α) from $\Delta \alpha = (D_\alpha \Delta t)^{1/2}$ shows $D_\alpha \geq 3.5 \times 10^{-2}$ rad^2/s (Lee et al. 2005). This observation is a strong constraint for any theoretical models of microbursts.

The mechanism of relativistic electron precipitation in the evening hours is also due to wave-particle interaction. In this case, the responsible waves are the electromagnetic ion cyclotron (EMIC) waves (Lorentzen et al. 2000). According to cyclotron resonance theory, relativistic precipitation observed in the evening hours is due to EMIC waves interacting with electrons with energies ≥ 1 MeV. Relativistic electron microbursts have been reproduced in simulation results (Omura and Zhao 2013). See also the theory of wave-particle interaction to produce microbursts by Summers et al. (2007).

Outstanding questions about microbursts are: What mechanisms can cause such impulsive microburst precipitation to relativistic energies? How are the low energy and relativistic microbursts related? Balloons have not yet detected relativistic microbursts. Can the relativistic microbursts SAMPEX detected simply be an extension of the low energy microbursts with fluxes of relativistic microburst fluxes too low to be detected by balloon-borne X-ray detectors? Simulation model suggests this is a possibility (Lee et al. 2015). Why are microbursts observed only at energies ≥ 20 keV? Balloon-borne detectors show low energy microbursts can be as small as ~ 15–20 km (Parks 1967). What is the scale size of relativistic microbursts? How can one measure the scale size of relativistic microbursts? We hope future experiments will be able to answer some of these questions.

7.4 Heating Space Plasmas

Space plasmas have been treated as collisionless and the Vlasov equation has been used to study the dynamics. While for many cases the collisionless assumption is acceptable, data show space plasmas are often heated from the foreshock region in the SW, across the bow shock, in the magnetosheath and the magnetosphere and far down the plasma sheet in the geomagnetic tail.

Heating is due to particles scattered from one region of velocity space into another region, increasing the volume in velocity space. Two ways that have been considered for moving particles in velocity space are scattering by wave-particle interaction and collisions among the particles. We have discussed wave-particle interaction briefly in Chap. 2 but not wave-particle interaction leading to heating. Collisions among the particles under Coulomb force can also scatter particles in velocity space and this topic will be introduced below.

The dynamics of heating are irreversible and therefore the Vlasov description is not adequate to explain the physics. However, some papers have claimed that Landau damping can heat particles. Concepts such as *collisionless damping* and *phase mixing* have been used by some authors to explain heating collisionless plasmas.

7.4.1 *Landau Damping*

Landau damping is produced by the interaction of waves and particles derived from the Vlasov equation (Landau 1946). Landau damping is discussed in many textbooks (Krall and Trivelpiece 1973; Ichimaru 1973), hence we will only discuss the main conclusions. According to Landau, macroscopic quantities such as electric field and charge densities are damped exponentially but the perturbations in the electron phase space distributions $f(\mathbf{r}, \mathbf{v}, t)$ oscillate indefinitely. We can think of damping arising from phase mixing of various parts of the distribution function. The reversibility of Landau damping has been shown by the echo experiment (Malmberg et al. 1968). This experiment showed that macroscopic physical quantities could reappear in the plasma if we reverse the direction of phase space evolution of the microscopic elements.

Landau damping has been studied mostly at the linearized level and the behavior is valid for $t \rightarrow \pm\infty$. Many papers have devoted to understanding the Landau damping mechanism (Twiss 1952; van Kampen 1955; Case 1959). However, a complete solution of Landau damping requires solving the nonlinear equations. A comprehensive description of energy exchanges between particles and fields requires solving the complete nonlinear equation which is beyond the scope of the present book. The dynamics of nonlinear Landau damping has been a topic of intense mathematical research (Mouhot and Villani 2011).

7.4.2 *Phase Mixing*

To understand collisionless damping, the concept of phase mixing is important. This concept was introduced by Hasegawa and Chen (1974) for heating fusion plasmas and has also been applied to possible heating of solar coronal plasmas (Heyvaerts and Priest 1983; Tsiklauri and Haruki 2008; Sarkar 2003). Phase mixing arises when waves are in inhomogeneous medium. Consider an Alfvén wave whose local wave velocity is a function of position in the direction of the nonuniformity. Thus, Alfvén waves next to each other can have different velocities and as different phases are developed in time, the neighboring waves can couple by mixing the phases. Phase mixing can produce smaller scale structures that spread over a larger velocity space. The second velocity moment thus yields a higher temperature, and we conclude plasma has been heated.

This concept has not been applied as a possible heating mechanism in space plasmas. However, given that space plasmas depart from uniformity very often, our opinion is that it needs to be studied seriously. Moreover, Alfvén waves permeate the vast regions of space and phase mixing could be an important process. Phase mixing has also been suggested as a possible mechanism for mode coupling and amplification mechanism of electric field parallel to the magnetic field (Bian and Kontar 2011). In this regard we can ask the question, how is the mode coupling achieved between Alfvén and compressional waves (Lin and Parks 1978)? Is phase mixing involved? Note that phase mixing may also give rise to artificial heating of plasmas in simulation codes which is discussed further below.

7.4.3 Issues with Heating Collisionless Plasmas

Theorists have studied mechanisms that generate waves and heat plasmas since the early sixties. A usual approach used in these studies is to perturb the equilibrium distribution and examine the behavior of the plasmas and waves. If the perturbation is small, one finds the plasma oscillates around the equilibrium point and eventually damps out. If the perturbation is large, waves can grow and one needs to look at the nonlinear behavior of the plasma.

The usual way to study nonlinear equations is to expand the dynamic quantities in terms of perturbed quantities,

$$f = f_o + f_1 + f_2 \ldots$$
$$\mathbf{E} = \mathbf{E}_1 + \mathbf{E}_2 .. \tag{7.1}$$

where the subindices higher than 0 denote perturbed quantities. One finds the perturbed Vlasov equation to first order becomes

$$\partial f_1 / \partial t + (q/m)\mathbf{E}_1 \cdot \partial f_o / \partial \mathbf{v} = 0 \tag{7.2}$$

where the second term $(q/m)\mathbf{E}_1 \cdot \partial f_o / \partial \mathbf{v}$ is nonlinear. Bernstein and Tehran (1960) have studied the nonlinear term and showed that in addition to damping solution (Landau damping), there are also growing solutions depending on the initial conditions. For the growing solutions, the amplitude can become large (Fig. 7.5), suggesting that the nonlinear waves can break as water waves do and the energy of ordered wave motion can go into heating of particles.

However, space plasmas are collisionless and the particles must conserve entropy (the phase space volume must remain constant, $(df/dt) = 0$). Schmidt (1966) studied this problem and shows an example in which a cold plasma beam distribution *somehow* gets scattered into a spiral shape whose particles occupy the same volume as the initial cold beam (Fig. 7.6) since conservation of entropy requires the shaded areas in the two distributions be equal. But the particles in the spiral are spread out into a larger region of the velocity space and while the entropy remains unchanged the second velocity moment will yield a higher temperature than the original cold beam. This is because the second moment integrates over all velocity space smoothing over the empty space. The spiral feature has been observed in hybrid simulation of dispersive Alfvén waves and filamentation instability (Laveder et al. 2002; Borgogno et al. 2009).

Fig. 7.5 Three solutions are shown for the nonlinear waves depending on the initial conditions. The case of the solution $A > 1/k$ resembles a breaking wave solution (from Bernstein and Tehran, *Nucl. Fusion*, **1**, 3, 1960)

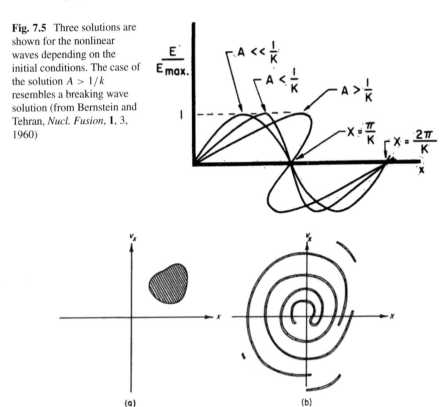

Fig. 7.6 The cold plasma beam (left) is somehow transformed to a spiral (right) maintaining the same volume (shaded region) (from Schmidt, *Academic Press*, New York, NY, 1966)

Schmidt (1966) summarizes that nonlinear interactions can destroy ordered motion by generating large amplitude waves and the initial distribution can end up a thermal distribution even in the absence of collisions. In Bernstein and Schmidt studies, they looked at only the nonlinear term arising from the electric field term and thus Schmidt states that "*in collisionless plasmas collective electrostatic interactions in a sense produce the same effect as collisions.*" However, this conclusion is paradoxical and we need to ask the question, is the temperature increase as described physically real?

7.4.4 Simulating Collisionless Shocks

One of the first simulations of Earth's bow shock was done in 1981 using a hybrid code (Leroy et al. 1982) and they showed SW plasma was heated going across the shock. Recently, Yang et al. (2014) applied a full 1D Particle in Cell (PIC) simulation study of supercritical and subcritical bow shocks.

Figure 7.7 shows the simulation results of a subcritical shock. The Alfvén Mach number for this shock was $M_A = 1.65$, and the setup of plasma parameters was kept unchanged from the supercritical shocks. For other examples of bow shock simulation, see Scholer and Matsukiyo (2004), Lembège (2004) and others. For an example of simulation of shock reformation observed by Cluster, see Mazelle et al. (2010).

This simulation shows that at the shock, the magnetic field jump was roughly in the form of a tangent function due to the limited number of reflected ions (Fig. 7.7 panel a). The main point of this figure is to indicate that entropy change at the shock front was small for both ions and electrons compared to supercritical shocks (not shown). For electrons, the velocity distribution (not shown) was still nearly Maxwellian and the temperature increase was much smaller in this subcritical case. The ions at the shock transition have a Maxwellian-like velocity distribution but with a weak high-energy tail (panel e). This observation leads to some entropy generation at the shock but not very much. This comparative study shows that entropy generation rate decreases as the shock M_A is reduced across the bow shock. Readers interested in the details of the two types of shocks are encouraged to read the original article by Yang et al. (2014).

PIC simulation results showing plasma can be heated presents a dilemma because the particles in the code move under the Lorentz force. One idea to possibly resolve this dilemma is that the particles and waves in PIC codes are interacting with inhomogeneous plasmas arising from fluctuations existing in the velocity space and can *phase mix* producing small-scale structures. These small-scale structures are unstable and can dissipate and produce even smaller structures spreading over a large velocity space. Hence the suggestion here is that phase mixing can mimic diffusion in velocity space resulting in higher temperatures. But note that this heating mechanism is not physical but an artifact of the code. How the PIC code produces heating is still a mystery. Whether heating in PIC codes is due to phase mixing needs to be investigated.

7.5 Boltzmann Collision Term $(\partial f / \partial t)_c$

Collisions among particles will change the velocities of the particles and diffuse particles in velocity space. Although collisions are rare in space plasmas, they can be important in some cases. We now discuss briefly how collisional effects are computed. The Boltzmann equation for a system of n particles is given by the distribution function $f_n(\mathbf{v}_1, \mathbf{r}_1, \mathbf{v}_2, \mathbf{r}_2, \ldots \ldots \mathbf{v}_n, \mathbf{r}_n, t)$ where n is the total number of particles. The one-particle distribution function $f(\mathbf{v}_1, \mathbf{r}_1)$ which we normally use in the Vlasov equation is obtained by integrating $f_n(\mathbf{v}_1, \mathbf{r}_1, \mathbf{v}_2, \mathbf{r}_2, \ldots \mathbf{v}_n, \mathbf{r}_n, t)$ over velocities of all of the particles except one, particle 1. To consider collisions, at least two particles are involved and the two-particle distribution function $f(\mathbf{v}_1, \mathbf{r}_1, \mathbf{v}_2, \mathbf{r}_2)$ is obtained by integrating over all particles except two. If collisions occur among three populations, we need a three-particle distribution function which is obtained by integrating over all velocities except for three particles.

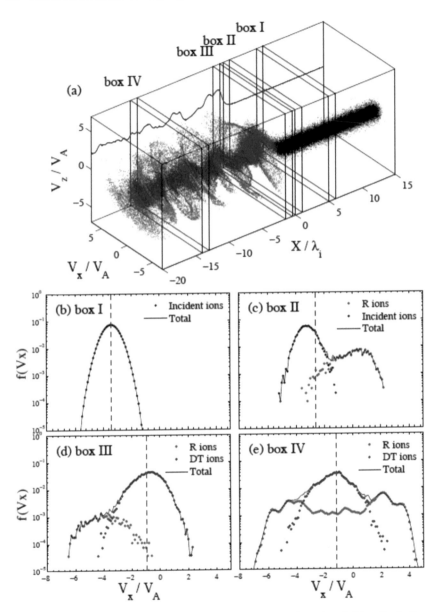

Fig. 7.7 PIC simulation of a subcritical bow shock crossing showing solar wind ion distributions. (**a**) Space velocity distribution (X-vx - vz) of incoming solar wind ions from right. The ions are separated into three groups: incident ions (black dots), reflected ions (R, red dots), and directly transmitted ions (DT, blue dots). The data in panels (**b**), (**c**), (**d**), and (**e**) are ion velocity distributions in the upstream region [box I as marked in panel (**a**)]. Ion velocity distributions right before and after the shock ramp in the shock transition are shown in boxes II and III, marked in panels (**a**) and (**e**), and ion velocity distributions in the far downstream in box IV, as marked in panel (**a**). The vertical blue dashed line shows the local bulk speed of the ions. The distribution function of incident, R, and DT ions are black, red, and blue dots, respectively. The total distribution (black solid curve) in each box is also shown for reference (from Yang et al., *Astrophys. J. Lett.*, **793**, L11, 2014)

Theoretical treatment of collisions has taken two routes, one looking at the problem as *time independent* (Spitzer and Härm 1953) and the other *time dependent* (Dreicer 1957). For the time independent case, the collisions are considered either strong or weak depending on whether the interaction involves large or small angle scattering. The Rutherford scattering model is used and for the small scattering angles, one can look at the scattering as analogous to the Brownian motion. For recent work on the Fokker-Planck equation, see Shizgal (2007 and references therein).

7.5.1   Time Independent Scattering

We follow the approach used by Longmire (1963). Consider two types of particles, and let us refer to them with subscripts 1 and 2. These particles can be, for example, electrons and ions or protons and helium. We let the distribution function of the two types be $f_1(\mathbf{v}, \mathbf{r})$ and $f_2(\mathbf{v}, \mathbf{r})$. The collision rate of particle 1 with particle 2 at the coordinate $\mathbf{r}$ will be proportional to $f_1(\mathbf{v}_1, \mathbf{r}) f_2(\mathbf{v}_2, \mathbf{r})$. Let the collisions change the initial velocities of particles 1 and 2 from $\mathbf{v}_1$ and $\mathbf{v}_2$ to $\mathbf{v}_1'$ and $\mathbf{v}_2'$. The differential scattering cross section which is a measure of the collision probability then depends only on the magnitude v and the angle of scattering given by $\mathbf{v} \cdot \mathbf{v}'$ and we can write the scattering cross section as $\sigma = \sigma(v, \mathbf{v} \cdot \mathbf{v}')$.

The collision term in Boltzmann's equation can then be represented by

$$\left(\frac{\partial f_r}{\partial t}\right)_s = -J(f_r f_s) - K(f_r f_s) \tag{7.3}$$

where the effects produced by $J(f_r f_s)$ are

$$J(f_r f_s) = \int_{v_s=0}^{\infty} \int_{\epsilon=0}^{2\pi} \int_{b=0}^{bc} \int (f_r' f_s' - f_r f_s) g b\, db\, d\epsilon\, d\mathbf{v}_s \tag{7.4}$$

Here $g = |\mathbf{v_r} - \mathbf{v_s}|$, b is the distance of closest approach and ϵ is the angle between the orbital plane and the plane containing the velocities of the two particles before encounter. The equation describing $K(f_r f_s)$ (weak scattering) is

$$K(f_r f_s) = \Sigma_i (f_r \langle \Delta v_{i,s} \rangle) - \frac{1}{2}\Sigma_{i,j} \frac{\partial^2}{\partial v_i \partial v_j}(f_r \langle \Delta v_i \Delta v_{j,s} \rangle) \tag{7.5}$$

where the two terms on the right are the diffusion coefficients. Here the quantity in the brackets $\langle\ \rangle$ is defined by

$$\langle x_s \rangle = \int_0^{\infty} g f_s dv_s \int_0^{2\pi} d\epsilon \int_{bc}^{bm} x b\, db \tag{7.6}$$

Note that $\langle x_s \rangle$ represents the mean value of x resulting from all encounters of particles of type s during the time interval dt.

7.5.2 Scattering Term

The equation for the rate at which particle 1 is scattered out of the volume d^3v_1 due to collisions with particle 2 in the volume d^3v_2 is

$$f_1(\mathbf{v}_1, \mathbf{r}) f_2(\mathbf{v}_2\mathbf{r}) d^3 v_1 d^3 v_2 v\sigma(v, \mathbf{v}\cdot\mathbf{v}')\delta(v-v')(d^3v'/v^2)\delta(\mathbf{V}-\mathbf{V}')d^3v' \quad (7.7)$$

where the $\mathbf{V}$'s are velocity of the center of mass frame, $\mathbf{V} = (m_1v_1 + m_2v_2)/(m_1 + m_2)$. The collisional term in the Boltzmann equation for $f_1(\mathbf{v}_1, \mathbf{r})$ then becomes

$$\left(\frac{\partial f_1(\mathbf{v}_1, \mathbf{r})}{\partial t}\right)_{1,2} = \int \int \int [f_1(\mathbf{v}'_1, \mathbf{r}) f_2(\mathbf{v}'_2, \mathbf{r}) - f_1(\mathbf{v}_1, \mathbf{r}) f_2 f(\mathbf{v}_2, \mathbf{r})]$$

$$\times (1/v)\sigma(v, \mathbf{v}\cdot\mathbf{v}')\delta(v-v')d^3v'\frac{1}{v^2}\delta(\mathbf{V}-\mathbf{V}')d^3v_2 d^3v'_1 d^3v'_2$$

$$(7.8)$$

The scattering terms are those in the square bracket. To examine the scattering terms in the square bracket, let the particle 1 be initially a collimated mono-energetic beam, defined by

$$f_1(\mathbf{v}_1) = \delta(\mathbf{v}_1 - \mathbf{v}_o) \quad (7.9)$$

This distribution describes a cold beam and all particles are traveling with the same velocity $\mathbf{v}_o$. To learn about the collisional effects on particle 1 by particle 2, and how particle spreads in velocity space, we use the Rutherford model. We remind the readers that the cross section for Rutherford scattering is

$$\sigma = Z_1^2 Z_2^2 e^4 / 4m^2 v^4 \sin^4(\theta/2)$$
$$= 4Z_1^2 Z_2^2 e^4 / m^2 |\mathbf{v}' - \mathbf{v}|^4$$
$$= (m_2/M)^4 (4Z_1^2 Z_2^2 e^4 / m^2 |\Delta\mathbf{v}_1|^4) \quad (7.10)$$

where $\Delta\mathbf{v}_1$ is the change of the velocity of particle 1 due to scattering by particle 2. To calculate the rate at which particle 1 is scattered by particle 2, we examine the scattering terms represented by

$$I(\mathbf{v}_o, \Delta\mathbf{v}_1) = \frac{4m_2 Z_1^2 Z_2^2 e^4}{Mm^2 |\Delta\mathbf{v}_1|^4} \int f_2(\mathbf{v}_o - \mathbf{v}')\frac{\delta(v-v')}{(v')}d^3v' \quad (7.11)$$

where $\Delta\mathbf{v}_1 = (\mathbf{v}_1 - \mathbf{v}')$. Assume now that f_2 is a Maxwellian given by

$$f_2(\mathbf{v}'_2) = N_2 \left(\frac{m_2}{2\pi kT}\right)^{3/2} \exp -[(v_2'^2)/(2kT)] \quad (7.12)$$

where N_2 is the density of particle 2 and kT is the temperature in energy units. Inserting this equation into Eq. (7.11) and performing the integration, we arrive at the final expression,

$$I(\mathbf{v_o}, \Delta\mathbf{v_1}) = \frac{4Z_1^2 Z_2^2 e^4}{m_1^2 |\Delta\mathbf{v_1}|^5} N_2 \left(\frac{m_2}{2\pi kT}\right)^{1/2}$$

$$\times \exp -(m_2/2kT)[v_{o\parallel} + (1/2)(M/m_2)\Delta v_1]^2 \qquad (7.13)$$

where $M = m_1 + m_2$, and $v_{o\parallel}$ is the component of $\mathbf{v_o}$ parallel to $\Delta\mathbf{v_1}$. Longmire (1963) has shown that in collisions of two particles via the Coulomb potential, the beam spreads in velocity space and the velocity distribution approaches equilibrium with a collision rate proportional to Δv^{-5}, where Δv is the change of velocity of particle 1 as a result of scattering by particle 2.

Equation (7.13) gives the probability of various velocity changes $\Delta\mathbf{v_1}$ per unit time. We now examine Eq. (7.13) more closely by noting that the exponential terms can be viewed as three separate factors

$$\exp[-(M^2 \Delta v_1^2 / 8m_2 kT)]$$

$$\exp[-(M \Delta v_1 v_o \cos \xi / 2kT)]$$

$$\exp[-(m_2 v_o^2 \cos^2 \xi / 2kT)] \qquad (7.14)$$

where ξ is the angle between $\mathbf{v_o}$ and $\Delta\mathbf{v_1}$. The second factor shows that the velocity changes in the backward direction ($\cos \xi$ is negative) are more probable than those in the forward direction. The third factor shows that if $\mathbf{v_o}$ is much larger than the thermal velocity of particle 2 ($m_2 v_o^2 / 2kT \ll 1$), velocity changes to the side ($\cos \xi \sim 0$) are much more probable than those in the forward or backward directions. This fact is responsible for observing meandering of fast electrons passing through matter. Note finally that in Eq. (7.13) we see the term $1/|\Delta v_1|^5$ shows that small changes of the velocity are most probable. We leave to the readers to estimate how much velocity is exchanged by collisions between H^+ and He^{++}, for example, in the solar wind.

In deriving Eq. (7.13), we assumed the f_2 was a Maxwellian. If we assume the distribution is a delta function, the exponential term on the right reduces essentially to 1 and Eq. (7.13) becomes

$$I(\mathbf{v_o}, \Delta\mathbf{v_1}) = \frac{4Z_1^2 Z_2^2 e^4}{m_1^2 |\Delta\mathbf{v_1}|^5} N_2 \left(\frac{m_2}{2\pi kT}\right) \qquad (7.15)$$

Note that -5 index of Δv is also a feature of the suprathermal SW (Gloeckler et al. 2000). The -5 spectral index is a fundamentally important observational constraint for understanding the dynamics of the SW and the origin of the suprathermal component. Fisk and Gloeckler (2008) have proposed that this index is a consequence of stochastic acceleration due to compressional turbulence in the SW. However, what is indicated here is the possibility that this unique index may come from collisions of SW beam with the ambient plasma.

7.5.3 Diffusion Coefficients

We now determine the mean rates of the velocity and energy of the initial mono-energetic beam. For convenience drop the subscript 1 on Δv_1 and remember that Δv is the change of velocity of a particle with initial velocity v_o. We need to calculate

$$\langle \Delta v_i \rangle = \int \Delta v_i I(v_o, \Delta v) d^3 \Delta v \qquad (7.16)$$

and

$$\langle \Delta v_i \Delta v_j \rangle = \int \Delta v_i \Delta v_j I(v_o, \Delta v) d^3 \Delta v \qquad (7.17)$$

After a lot of algebra, we obtain

$$\langle \Delta \mathbf{v} \rangle = \frac{4\pi Z_1^2 Z_2^2 e^4}{(mm_1)} \ln \left(\frac{2}{\theta_m} \right) \nabla_{vo} \int \frac{f_2(\mathbf{v_o} - \mathbf{v'})}{v'} d^3 v' \qquad (7.18)$$

where $m = m_1 m_2/M$ and θ_m is the minimum angle of deflection. The equation for the minimum deflection is estimated from

$$\theta_m = 2|Z_1 Z_2| e^2 / \lambda_D m v^2 \qquad (7.19)$$

which is based on classical physics. More sophisticated quantum estimation can be made but this is not needed for our problem. Note that θ_m enters the equation logarithmically and is roughly equal to ~ 10.

The expression for $\langle \Delta v_i \Delta v_j \rangle$ is

$$\langle \Delta v_i \Delta v_j \rangle = \frac{4\pi Z_1^2 Z_2^2 e^4}{(mm_1)} \ln \left(\frac{2}{\theta_m} \right) \frac{\partial}{\partial v_{oi}} \frac{\partial}{\partial v_{oj}} \int v' f_2(\mathbf{v_o} - \mathbf{v'}) d^3 v' \qquad (7.20)$$

This equation can be evaluated when f_2 is given. It is now possible to calculate the mean rate of energy change. After a lot of algebra, which is omitted here, the final result obtained is

$$\left\langle \frac{dW}{dt} \right\rangle = -N_2 \frac{4\sqrt{\pi} Z_1^2 Z_2^2 e^4}{\sqrt{m_1} v_o} \ln \left(\frac{2}{\theta_m} \right) \frac{m_1}{m_2}$$
$$\times \left[\int \exp(-y^2) dy - \left(1 + \frac{m_2}{m_1} \right) \alpha \exp(-\alpha^2) \right] \qquad (7.21)$$

where W is the energy, $\alpha^2 = m_2 v_o^2/2kT = (m_2/m_1)(W/kT)$, and $y = u - \alpha$ and $u^2 = (m_e/2kT')v'^2$. This equation gives the mean rate of change of energy of particle 1 with velocity v_o colliding with a set of particle 2 having a Maxwell distribution.

We now show the mean rate of change of energy transfer between two Maxwell distributions. To do this, we integrate the above equation over the distribution function

$$\left(\frac{m_1}{2\pi k T_1}\right)^{3/2} \exp -[(m_1 v_o^2 / 2k T_1)]4\pi v_o^2 dv_o \tag{7.22}$$

and obtain

$$\left\langle \frac{dW_1}{dt} \right\rangle = -N_2 4\sqrt{2\pi} Z_1^2 Z_2^2 e^4 \ln\left(\frac{2}{\theta_m}\right) \frac{\sqrt{m_1 m_2}(kT_1 - kT_2)}{(m_1 kT_2 + m_2 kT_1)^{3/2}} \tag{7.23}$$

This equation gives the average energy transfer rate from particle 1 to particle 2. Note that the rate vanishes when $T_1 = T_2$. We can compare the transfer rates when the two species are electron–electron, electron–ion, and ion–ion. The rates are determined by the mass factors and we obtain

electron–electron $\rightarrow 1/\sqrt{m_e}$
electron–ion $\rightarrow (1/\sqrt{m_i})\sqrt{m_e/m_i}$
ion–ion $\rightarrow 1/\sqrt{m_i}$

One sees here that the electron–ion exchange is the slowest. The reason is that the transfer of energy in collisions between particles with widely different masses is difficult. This also means that electrons and ions in contact with each other can have different temperatures. Electrons and ions will each come into equilibrium with themselves, much faster than the energy transferred from electrons (ions) to ions (electrons). If ion temperature is much less than m_i/m_e times the electron temperature, the above equation simplifies to

$$\left\langle \frac{dW_1}{dt} \right\rangle = -N_2 4\sqrt{2\pi} Z_i^2 e^4 \ln\left(\frac{2}{\theta_m}\right) \frac{\sqrt{m_e}(kT_i - kT_e)}{m_i (kT_e)^{3/2}} \tag{7.24}$$

Let us now apply this result to the solar wind. We let m_1 be H^+ and m_2 be He^{++}. The density N_2 of He^{++} at the coronal heights is $\sim 2\times 10^{20}$ cc^{-3}. $Z_1 = 1$ and $Z_2 = 2$. We let $kT_1 = 1.66 \times 10^{-10}$ ergs for a temperature of 1.2×10^6 K. We let kT_2 be four times the value of H^+. Substituting these values in the equation yields a value

$$\left\langle \frac{dW_1}{dt} \right\rangle \simeq 1.39 \times 10^{-20} \text{ergs/s/particle}$$

$$\simeq 1.7 \times 10^{-8} \text{eV/s/particle}. \tag{7.25}$$

This number is very small as expected. We can also calculate the mean rate of energy change per unit length of path. Dividing the Eq. (2.37) by $1/v_o$, we obtain

$$\left\langle \frac{dW}{dx} \right\rangle = -N_2 \frac{4\sqrt{\pi} Z_1^2 Z_2^2 e^4}{\sqrt{WkT}} \ln\left(\frac{2}{\theta_m}\right) \sqrt{\frac{m_2}{m_1}} \left(\frac{W}{3kT/2} - 1\right) \tag{7.26}$$

This formula gives the amount of energy loss per unit path and is useful in some applications, like when a charged particle penetrates an absorber.

7.5.4 Spitzer Conductivity of Plasmas

The electrons are interacting with ions at rest. If the interaction of the electrons with the ions occurs close to the Debye sphere, the Rutherford model is applicable (strong electric field case). For interactions occurring far away from the Debye sphere, the electrons are not scattered very much and as mentioned earlier the small angle scattered motions can be looked at as similar to Brownian motion (weak electric field case). Spitzer and Härm (1953) worked with the above equations and the equation of conductivity of plasmas they derived is

$$\sigma = 1.97 \left(\frac{Ne^2}{m_e c v_c} \right) \tag{7.27}$$

where the collision frequency v_c

$$v_c = \frac{4\sqrt{2\pi}}{3} \frac{Ne^4}{(kT_e)^2} \left(\frac{kT_e}{m_e} \right) \ln \left(\frac{2}{\theta_m} \right) \tag{7.28}$$

This is the famous equation that shows the conductivity σ has $T_e^{3/2}$ dependence. In deriving this equation, two assumptions have been made: (1) a steady state electron velocity distribution is obtained several mean free collision times after the electric field is applied and (2) the terminal electron drift velocity is small compared to the average random electron speed.

7.6 Runaway Electrons

In Chap. 6, we discussed particles' motion in the presence of an electric field parallel to the magnetic field and showed that *runaway* electrons can be produced. Here we will discuss the same problem for electrons in plasmas. The problem assumes electrons are moving through an infinite plasma of positive ions under the influence of static uniform electric field of arbitrary strength Dreicer (1957, 1959, 1960). The formulation begins with the Boltzmann transport equation,

$$\frac{\partial f}{\partial t} + \frac{eE}{m} \frac{\partial f}{\partial v_z} = \frac{\partial f}{\partial t}_c \tag{7.29}$$

in homogeneous plasmas and without the magnetic field term. The electric field is applied in the negative z-direction. The initial condition at $t = 0$ is

$$f(v, v_z, 0) = (m/2\pi kT_o) \exp -[(m/2\pi kT_o)(v^2 + v_z^2)] \tag{7.30}$$

The solution at $t = t$ is the displaced Maxwellian distribution

$$f(v, v_z, t) = (m/2\pi kT_o)^{3/2} \exp -(m/2\pi kT_o)(v^2 + [v_z - v(t)]^2) \tag{7.31}$$

where the electron drift velocity $v(t)$ is determined from the Lorentz equation of motion

$$dv/dt + eE_c\Psi(Z)/m = eE/m \tag{7.32}$$

The additional term arises from the friction exerted on the electrons by the ions. In this equation the various terms have the following meaning:

$$eE_c/m = 4\pi n(Z_i^2 e^4/m^2)(m/2kT_e) \ln \Lambda$$

$$n = \text{electron particle density}$$

$$Z_i = \text{ion charge}$$

$$Z = (2/kT_e)^{3/2}v$$

$$\Psi(Z) = E_2(Z) - Z(dE_2/dZ)/Z^2$$

$$E_2(z) = (2/\sqrt{\pi}) \int_0^Z e^{-t^2} dt \tag{7.33}$$

Here Z is a dimensionless measure of the electron velocity v and E_c is the critical electric field whose meaning will become clearer below. This formulation indicates that the problem of electrons passing through a plasma has a time-independent solution. The reason is that in Rutherford scattering, the electron drift is not bounded by a terminal value, and the velocity can grow monotonically with time. As illustrated in Fig. 7.8, the frictional force follows Stokes law for small Z but for $Z \gg 1$, the Rutherford scattering law shows a Z^{-2} dependence for the

Fig. 7.8 The velocity dependence $\Psi(Z)$ of the frictional force as a function of Z (from Dreicer, *Phys. Rev.*, **117**, 329, 1957)

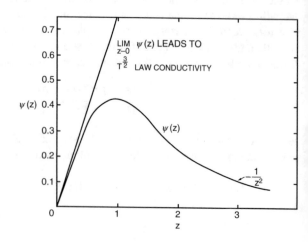

decay. The friction force is a maximum at $Z = 1$. There is no static solution when E exceeds $E_c \Psi(1)$. The value E_c thus plays the role of a critical field in the theory and sets the criterion for separating the weak and strong field cases.

The rate of Joule heating deduced from Eq. (7.30) is given by

$$\frac{d}{dt}\left(\frac{3}{2}kT\right) = eE_c(2kT_e/m)^{1/2}Z\Psi(Z) \tag{7.34}$$

The dimensionless variables used for this problem are E/E_c, T/T_o, Z, and $\tau = (eE_o/m)(m/2kT_o)^{1/2}t$, where E_c and Z are defined in terms of the initial electron temperature T_o and τ is the mean free time for collisions between electrons moving with the thermal speed $(2kT_e/m)^{1/2}$. Dreicer (1957) has computed E_c for various plasma densities and thermal energies and Z. He has solved the above equations numerically for a variety of applied fields (see Fig. 7.9).

An important point is that electrons in the high energy tail of the distribution make very few collisions and for these electrons almost any applied field may be considered strong. This suggests that the velocity space can be roughly divided into a runaway region where the applied field plays the role of a strong field and into a non-runaway region where the same field is weak. There, the time scale of appreciable depletion of the original distribution is determined by the diffusion of electrons into the runaway region. The time-dependent Boltzmann equation supplemented by Fokker-Planck collision terms shows the probability that all electrons have crossed into the runaway region in the time τ. This probability is given by

$$Q(\tau) = 1 - \exp^{-\lambda r} \tag{7.35}$$

where τ and E_c are defined in terms of the initial temperature of the electrons. Dreicer's work indicates that runaway in the weak field limit proceeds under the combined action of Joule heating and diffusion into high energy tail of the distribution.

7.7 Concluding Remarks

Space plasma physics will continue to receive attention as new missions obtain new data with instruments more capable than those on Cluster, Wind, Polar, FAST, Geotail, MMS, and other missions. To connect data to the real world, we have characterized space plasmas using a *data driven approach*. This approach has taken full advantage of the instrument's capability. Knowing what instruments can and cannot measure is critical to developing models consistent with the real behavior of space plasmas. We have argued that since plasma instruments measure particles, not fluids, observations must be interpreted with particle theories and concepts.

Even though space plasmas have been observed for more than 50 years, most of the fundamental space plasma phenomena are still understood only partially. This

Fig. 7.9 A plot showing λ as a function of E/E_C (from Dreicer, *Phys. Rev.*, **117**, 329, 1957)

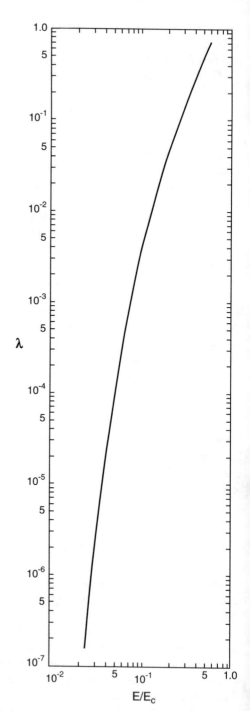

is in part because most of our understanding of basic space plasma phenomena has been developed from fluid theory and concepts. We have demonstrated that fluid description is not self-consistent and is incomplete. The physics is contained in the distribution function and obtaining self-consistent pictures of space plasma behavior will require going beyond fluid concepts and a major paradigm shift.

References

Alfvén, H.: Cosmic Plasma. D. Reidel Publishing Co., Dordrecht, Holland (1981)

Anderson, K., Milton, D.: Balloon observations of X-rays in the auroral zone, 3, high time resolution studies. J. Geophys. Res. **71**, 4617 (1964)

Bernstein, I., Tehran, S.K.: Plasma oscillations (I). Nucl. Fusion **1**, 3 (1960)

Bian, N., Kontar, E.: Parallel electric field amplification by phase mixing of Alfven waves. Astron. Astrophys. **527**, A130 (2011)

Borgogno, D., et al.: Alfvén wave filamentation and dispersive phase mixing in a high-density channel; landau fluid and hybrid simulation. Nonlinear Processes Geophys. **16**, 275 (2009)

Case, K.: Plasma oscillations. Ann. Phys. **7**, 349 (1959)

Chen, L.J., et al.: Wind observations pertaining to current disruption and ballooning instability during substorms. Geophys. Res. Lett. **30**, 1335 (2003)

Coppi, B., Laval, G., Pellat, R.: Dynamics of geomagnetic tail. Phys. Rev. Lett. **16**, 1207 (1966)

Datta, S., et al.: Analysis and modeling of microburst precipitation. J. Geophys. Res. **102**, 17325 (1997)

Dreicer, H.: Theory of runaway electrons. In: Proceedings of Tenth Annual Gaseous Electronics Conference, p. 57 (1957)

Dreicer, H.: Electron and ion runaway in fully ionized gases, I. Phys. Rev. **115**, 238 (1959)

Dreicer, H.: Electron and ion runaway in a fully ionized gas, II. Phys. Rev. **117**, 329 (1960)

Fisk, L., Gloeckler, G.: Acceleration of suprathermal tails in the solar wind. Astrophys. J. **686**, 1466 (2008)

Gjerloev, A., Hoffman, R.: Height integrated conductivity in auroral substorms, 1. Data. J. Geophys. Res. **105**, 1435 (2000)

Gloeckler, G., et al.: Sources, injection and acceleration of heliospheric ion populations. In: Mewaldt, R.A., et al. (eds.) ACE 2000 Symposium (2000)

Grad, H.: Plasmas. Phys. Today **22**, 12 (1969)

Hasegawa, A., Chen, L.: Plasma heating by Alfvén-wave phase mixing. Phys. Rev. Lett. **32**, 454 (1974)

Heyvaerts, J., Priest, E.: Coronal heating by phase-mixed shear Alfvén waves. Astron. Astrophys. **117**, 220 (1983)

Horton, W., et al.: Substorm trigger conditions. J. Geophys. Res. **104**, 22745 (1999)

Hwang, H., et al.: A case study to determine the relationship of relativistic electron events to substorm injections and ULF power. Geophys. Res. Lett. **31**, 23801 (2004)

Ichimaru, S.: Basic Principles of Plasma Physics: A statistical Approach. W. A. Benjamin Publishing Co. Inc., Reading (1973)

Krall, N., Trivelpiece, A.: Principles of Plasma Physics. McGraw Hill Book Co., New York (1973)

Lampton, M.: Daytime observations of energetic auroral-zone electrons. J. Geophys. Res. **73**, 5817 (1968)

Landau, L.: On the vibration of electronic plasma. J. Phys. **10** (1946). English Translation in JETP **16**, 574 (1964)

Laveder, D., et al.: Transverse dynamics of dispersive Alfvén waves.1. Direct numerical evidence of filamentation. Phys. Plasmas **9**, 293 (2002)

Le, G., et al.: Ionospheric response to the interplanetary magnetic field southward turning. J. Geophys. Res. **107**, SIA-2-1 (2002)

Le Contel, O., Pellat, R., Roux, A.: Self-consistent quasi-static parallel electric field associated with substorm growth phase. J. Geophys. Res. **105**, 12945 (2000)

Lee, J.J., et al.: Energy spectra of ~170–360 keV electron microbursts measured by the Korean STSAT-1. Geophys. Res. Lett. **32**, 13106 (2005)

Lee, J.J., et al.: Short-duration electron precipitation studied by test particle simulation. J. Astron. Space Sci. **32**, 317 (2015)

Lembège, B.: Selected problems in collisionless shock physics. Space Sci. Rev. **110**, 161 (2004)

Leroy, M., et al.: The structure of perpendicular bow shocks. J. Geophys. Res. **87**, 5081 (1982)

Lin, C.S., Parks, G.K.: The coupling of Alfvén and compressional waves. J. Geophys. Res. **83**, 2682 (1978)

Lin, N., et al.: Statistical study of substorm timing sequence. J. Geophys. Res. **114**, A12204 (2009)

Longmire, C.: Elementary Plasma Physics. Interscience Publishers, New York (1963)

Lorentzen, K., et al.: Precipitation of relativistic electrons by interaction with electromagnetic ion cyclotron waves. J. Geophys. Res. **105**, 5381 (2000)

Lundin, R., et al.: The contribution of the boundary layer EMF to magnetospheric substorms. In: Magnetospheric Substorms, vol. 64. American Geophysical Union, Washington (1991)

Lyons, L., et al.: Possible connection of polar cap flows to pre- and post-substorm onset PBIs and streamers. J. Geophys. Res. **116**, A12225 (2011)

Malmberg, J., Wharton, C., Gould, R., O'Neil, T.: Observation of plasma wave echoes. Phys. Fluids **11**, 1147 (1968)

Mazelle, C., et al.: Self-reformation of the quasi-perpendicular shock: CLUSTER observations. AIP Conf. Proc. **1216**, 471 (2010)

McPherron, R.L.: Substorm related changes in the geomagnetic tail: the growth phase. Planet. Space Sci. **20**, 1521 (1972)

Millan, R., Thorne, R.: Review of radiation belt relativistic electron losses. J. Atmos. Solar Terr. Phys. **69**, 362 (2007)

Milton, D., Oliven, M.: Simultaneous satellite and balloon observations of the same auroral-zone precipitation event. J. Geophys. Res. **72**, 5357 (1967)

Mouhot, C., Villani, C.: On Landau damping. Acta Math. **207**, 29 (2011)

Nakamura, R., et al.: SAMPEX observations of precipitation bursts in the outer radiation belt. J. Geophys. Res. **105**, 15875 (2000)

O'Brien, B.: High-latitude geophysical studies with satellite Injun 3: 3. Precipitation of electrons into the atmosphere. J. Geophys. Res. **69**, 13 (1964)

Ohtani, S., et al.: Large-scale electric field in the current disruption region. Geophys. Res. **107**, SMP22-1 (2002)

Oliven, M., Gurnett, D.: Microburst phenomena: 3. An association between microbursts and VLF chorus. J. Geophys. Res. **73**, 2355 (1968)

Omura, H., Zhao, Q.: Relativistic electron microbursts due to nonlinear pitch-angle scattering by EMIC triggered emissions. J. Geophys. Res. **118**, 5008 (2013)

Parks, G.K.: Spatial characteristics of auroral-zone X-ray microbursts. J. Geophys. Res. **72**, 215 (1967)

Parks, G.K.: Physics of Space Plasmas: An Introduction, 2nd edn. Westview Press, Cambridge (2004)

Parks, G.K., Winckler, J.R.: Acceleration of energetic electrons observed at the synchronous altitude during magnetospheric substorms. J. Geophys. Res. **73**, 5786 (1968)

Parks, G., Laval, G., Pellat, R.: Behavior of outer radiation zone and a new model of magnetospheric substorm. Planet. Space Sci. **20**, 1291 (1972)

Parks, G.K., et al.: Relativistic electron precipitation. Geophys. Res. Lett. **6**, 393 (1979)

Perrault, S., et al.: Current-driven electromagnetic ion cyclotron instability at substorm onset. J. Geophys. Res. **105**, 21097 (2000)

Raeder, J., et al.: Open geospace general circulation model simulation of a substorm: axial tail instability and ballooning mode preceding substorm onset. J. Geophys. Res. **115**, A00I16 (2010)

Roux, A., et al.: Auroral kilometric radiation sources: in situ and remote observations from Viking. J. Geophys. Res. **98**, 11657 (1993)

Sarkar, A.: Collisionless phase mixing of upper hybrid oscillations in a cold inhomogeneous plasma placed in an inhomogeneous magnetic field. Phys. Plasmas **20**, 052301 (2003)

Sauvaud, J.A.: Large scale responses of the magnetosphere to a southward turning of interplanetary magnetic field. J. Geophys. Res. **92**, 2365 (1987)

Schmidt, G.: Physics of High Temperature Plasmas: An Introduction. Academic, New York (1966)

Scholer, M., Matsukiyo, S.: Nonstationarity of quasi-perpendicular shocks: a comparison of full particle simulations with different ion to electron mass ratio. Ann. Geophys. **22**, 2345 (2004)

Shizgal, B.: Suprathermal particle distributions in space physics: kappa distributions and entropy. Astrophys. Space Sci. **312**, 227 (2007)

Sitnov, et al.: Distinctive features of internally driven magnetotail reconnection. Geophys. Res. Lett. **44**, 3028 (2017)

Skoug, R., et al.: Upstream and magnetosheath energetic ions with energies to ≈ 2 MeV. J. Geophys. Res. **101**, 21481 (1996)

Smith, E., et al.: Observations of interplanetary sector structure to heliographic latitudes of $16°$: pioneer 11. J. Geophys. Res. 83, 717 (1978)

Spitzer, H.: Transport phenomena in a completely ionized gas. Phys. Rev. **89**, 977 (1953)

Stasiewicz, K.: A gyroviscous model of the magnetotail current layer and the substorm mechanism. Phys. Fluids **30**, 1401 (1987)

Summers, D., Ni, B., Meredith, N.P.: Timescales for radiation belt electron acceleration and loss due to resonant wave-particle interactions: 1. Theory. J. Geophys. Res. **112**, A04206 (2007)

Tsiklauri, D., Haruki, T.: Physics of collisionless phase mixing. Phys. Plasmas **15**, 112902 (2008)

Twiss, R.: Propagation in electron-ion streams. Phys. Rev. **88**, 1392 (1952)

van Kampen, N.: On the theory of stationary waves in plasmas. Physica **21**, 949 (1955)

West, R., Parks, G.K.: ELF emissions and relativistic electron precipitation. J. Geophys. Res. **89**, 159 (1984)

Yang, Z., et al.: Full particle electromagnetic simulations of entropy generation across a collisionless shock. Astrophys. J. Lett. **793**, L11 (2014)

Additional Reading

Chapman, S., Cowling, T.: The Mathematical Theory of Nonuniform Gases. Cambridge University Press, Cambridge (1939)

Cohen, R., Spitzer, L., Routly, P.: The electrical conductivity of an ionized gas. Phys. Rev. **80**, 230 (1950)

Dory, R.A., Guest, G.E., Harris, E.G.: Unstable electrostatic plasma waves propagating perpendicular to a magnetic field. Phys. Rev. Lett. **14**, 131 (1965)

Henderson, M., et al.: Substorms during the 10–11 August 2000 Sawtooth Event. J. Geophys. Res. **111**, A06206 (2006)

Lui, A.T.Y.: Review on the characteristics of the current sheet in the earths magnetotail. In: Keiling, A., Marghitu, O., Wheatland, M. (eds.) Electric Currents in Geospace and Beyond. Wiley, Hoboken (2018)

MacDonald, W.M., Rosenbluth, M., Chuck, W.: Relaxation of a system of particles with Coulomb interactions. Phys. Rev. **107**, 350 (1957)

Rosenbluth, M., et al.: Fokker-Planck equation for an inverse-square force. Phys. Rev. **107**, 1–6 (1957)

Yoon, P., Lui, T.: Nonlinear analysis of generalized cross-field current instability. Phys. Fluids **5**, 836 (1991)

Index

© Springer Nature Switzerland AG 2018
G. K. Parks, *Characterizing Space Plasmas*, Astronomy and Astrophysics Library,
https://doi.org/10.1007/978-3-319-90041-4

Printed in the United States
By Bookmasters